AF368537

Research on the Antioxidant, Antibacterial, and Anti-Drug Properties of Plant Ingredients

Research on the Antioxidant, Antibacterial, and Anti-Drug Properties of Plant Ingredients

Editors

Călin Jianu
Georgeta Pop

MDPI • Basel • Beijing • Wuhan • Barcelona • Belgrade • Manchester • Tokyo • Cluj • Tianjin

Editors
Călin Jianu
University of Life Sciences "King
Michael I" from Timisoara
Timișoara, Romania

Georgeta Pop
University of Life Sciences "King
Michael I" from Timisoara
Timișoara, Romania

Editorial Office
MDPI
St. Alban-Anlage 66
4052 Basel, Switzerland

This is a reprint of articles from the Special Issue published online in the open access journal *Applied Sciences* (ISSN 2076-3417) (available at: https://www.mdpi.com/journal/applsci/special_issues/properties_of_plant_ingredients).

For citation purposes, cite each article independently as indicated on the article page online and as indicated below:

LastName, A.A.; LastName, B.B.; LastName, C.C. Article Title. *Journal Name* **Year**, *Volume Number*, Page Range.

ISBN 978-3-0365-8078-4 (Hbk)
ISBN 978-3-0365-8079-1 (PDF)

Contents

About the Editors

Călin Jianu

Călin Jianu became a full professor in 2015 at the Faculty of Food Engineering from the University of Life Sciences "King Michael I" from Timisoara (Romania) and has been continually involved in teaching and research activities. He studied food engineering (diploma 2000) and received his Ph.D. (2006) from the University of Agricultural Sciences and Veterinary Medicine of Banat from Timișoara (Romania). His research focuses on food additives, preservation techniques, and food safety management systems. Currently, natural product chemistry, pharmacological and biological activities of natural extracts, as well as the potential mechanisms of action underlying their effects, are his main topics of research. He has over 34 indexed publications in Web of Science (over 5 books and 6 chapters published in Romanian by two international publishing houses and 6 patents), highlighting his extensive research activity.

Georgeta Pop

Georgeta Pop is a professor at the University of Life Sciences "King Michael I" from Timisoara (Romania). She is currently head of the Department of Crop Science in the Faculty of Agriculture. She received her Ph.D. in Agronomy in 1999; since 2010, she has been a Ph.D. coordinator at the same university. Her research focuses on medicinal and aromatic plants, crop science, and the antifungal, antioxidant, and antimicrobial properties of essential oils. She is the author of 2 scientific books, 28 ICI papers, and 38 articles in scientific journals and volumes indexed in international databases. Currently, she is researching the therapeutic potential of the cannabinoid content of hemp and the acquisition of essential oils from medicinal plants of the *Asteraceae*, *Apiaceae*, and *Lamiaceae* families.

Preface to "Research on the Antioxidant, Antibacterial, and Anti-Drug Properties of Plant Ingredients"

In recent decades, the new trend of "green consumerism" has led to the investigation and application of new preservation techniques (e.g., irradiation, sonification) and increased access to natural extracts and essential oils (EOs) that can reduce or eliminate the usage of synthetic additives. Recently, the biologically active properties (e.g., antioxidant, antibacterial, antiviral, immunomodulatory effects, etc.) of volatile compounds derived from EOs isolated from various plant species have attracted interest among the scientific research community. In alignment with the efforts of academic researchers to identify and access new sources of active principles applicable to the food and pharmaceutical industries, this Special Issue collates research articles and original reviews that discuss several aspects of the biological effects of natural extracts, particularly EOs as well as the following:

- The chemistry of volatile compounds identified in natural extracts, particularly EOs;
- The antioxidant and antimicrobial properties of natural extracts and EOs and their compounds;
- An in silico evaluation of the biologically active properties of natural extracts and EOs;
- The applications of bioactive compounds from natural extracts and EOs in the food and pharmaceutical industries.

We consider this Special Issue, consisting of ten publications (eight research papers and two reviews), to be a helpful resource for academics, researchers, stakeholders, and experts who are interested in identifying, formulating, and promoting the use of new green additives in the food and pharmaceutical industries. The guest editors wish to express their appreciation and gratitude to all of the authors who contributed to this Special Issue.

Călin Jianu and Georgeta Pop
Editors

Editorial

Editorial for the Special Issue, "Research on the Antioxidant, Antibacterial, and Anti-Drug Properties of Plant Ingredients"

Călin Jianu [1,*] and Georgeta Pop [2]

[1] Faculty of Food Engineering, University of Life Sciences "King Michael I" from Timisoara, Calea Aradului 119, RO-300645 Timișoara, Romania

[2] Faculty of Agriculture, University of Life Sciences "King Michael I" from Timisoara, Calea Aradului 119, RO-300645 Timișoara, Romania; georgeta_pop@usab-tm.ro

* Correspondence: calin.jianu@gmail.com

Citation: Jianu, C.; Pop, G. Editorial for the Special Issue, "Research on the Antioxidant, Antibacterial, and Anti-Drug Properties of Plant Ingredients". *Appl. Sci.* **2023**, *13*, 6570. https://doi.org/10.3390/app13116570

Received: 16 May 2023
Revised: 19 May 2023
Accepted: 23 May 2023
Published: 29 May 2023

The increased consumer demand for safe foods has led to the development of new preservation techniques (e.g., irradiation, ultrasonic methods, and modified atmosphere packaging) and the use of natural extracts and essential oils (EOs) to reduce or eliminate the addition of synthetic additives to foodstuffs. In recent decades, the antioxidant properties of EOs and their effectiveness against pathogenic microorganisms resistant to antibiotics has been an important research topic in the food and pharmaceutical industries. It has been demonstrated that the use of essential oils protects the sensory attributes of foods, significantly improves their shelf life, and decreases food waste. Moreover, accessing them can reduce oxidative stress at the cellular level, which is characteristic of cardiovascular and neurodegenerative diseases, and in addition treat some types of cancer.

We originally produced a Special Issue of *Applied Sciences*, entitled "Research on the Antioxidant, Antibacterial, and Anti-drug Properties of Plant Ingredients", in the section "Food Science and Technology" (ISSN: 2076-3417; https://www.mdpi.com/journal/applsci, accessed on 8 June 2021). This thematic issue was dedicated to the biological properties of natural extracts and EOs, including their compounds' potential mechanisms of action. The Special Issue, now transformed into a book, comprises ten chapters with distinguished authors' significant contributions to the subject. All told, eight chapters are research papers, while the remaining two are review papers.

Chapter 1 comprises the review entitled "Antioxidant and Antimicrobial Properties of Selected Phytogenics for Sustainable Poultry Production" [1]. It regards phytogenic products as alternatives to in-feed antibiotic growth boosters in poultry production. In addition, the review delivers exhaustive information regarding the integration of phytogenic products into poultry diets in such a way as to improve feed utilization efficiency, decrease feed–food competition, and enhance the contribution of these products to eradicating hunger and food insecurity.

Chapter 2 ("Anti-Breast Cancer Activity of Essential Oil: A Systematic Review") evaluates the application of EOs compounds' applications as an alternative to breast cancer treatment [2]. In this chapter, the authors summarize terpenoids' excellent anti-breast cancer pharmacological effects. They conclude that EOs can be accessed as a primary or adjuvant breast cancer treatment due to their highly effective nature and how rarely they exhibit side effects.

Chapter 3 comprises a research paper titled "Anti-Yeast Synergistic Effects and Mode of Action of Australian Native Plant Essential Oils" [3]. In this chapter, Alderees and colleagues explore the efficacy of three Australian native EOs against weak-acid-resistant yeasts: Tasmanian pepper leaf, lemon myrtle, and anise myrtle. The tested EOs exhibited good anti-yeast and antibacterial activity. Furthermore, damage to the yeast cell membrane, ion leakage, and cell organelles was identified as being the most probable mode of action of Tasmanian pepper leaf and lemon myrtle EOs. Our findings suggest that potential applications exist for these products in the beverage and food industry as natural antimicrobials [3].

Chapter 4 addresses "Essential Oil of *Origanum vulgare* var. *aureum* L. from Western Romania: Chemical analysis, in vitro and in silico screening of its antioxidant activity" [4]. This noteworthy chapter was reserved for an *Origanum* genus member that has remained insufficiently explored, *Origanum vulgare* var. *aureum* L. The authors proceeded by assessing its EO chemical composition and antioxidant effects. The results of this investigation reveal that gamma-terpinene (22.96%) and para-cymene (14.72%) are the main compounds of the *Origanum vulgare* var. *aureum* essential oil (OEO) and that they demonstrate strong scavenging effects on the DPPH and ABTS assays and higher relative antioxidant activity in the beta-carotene/linoleic acid bleaching assay. Furthermore, in silico analysis results reveal that their main compounds may be responsible for the oil's antioxidant properties by inhibiting the ROS-producing enzymes known as lipoxygenase (LOX), and xanthin oxidase (XO).

Chapter 5 ("Chemical profile of *Ruta graveolens*, evaluation of the antioxidant and antibacterial potential of its essential oil, and molecular docking simulations") evaluate the antimicrobial and antioxidant effects of the *Ruta graveolens* L. essential oil (RGEO) [5]. The results contained in this chapter demonstrate that RGEO possesses broad-spectrum antifungal and antibacterial effects and moderate antioxidant properties. Moreover, molecular docking analysis shows that the RGEO can exert its antimicrobial activity by inhibiting the D-Alanine-d-alanine ligase and generating an in vitro antioxidant effect via cumulative XO and LOX inhibition. Therefore, the RGEO could comprise a new natural preservatives and antioxidants source with diverse food and pharmaceutical industry applications.

Chapter 6 discusses the paper "Phytochemical profile, antioxidant and wound healing potential of three artemisia species: in vitro and in ovo evaluation" [6]. This chapter proposes using ethanolic extracts of three *Artemisia* species, *Artemisia absinthium* L., *Artemisia dracunculus* L., and *Artemisia annua* L., as new efficient alternatives in injury therapy. The results demonstrates that all three *Artemisia* species can be promising low-cost, polyphenol-rich, antioxidant, and safe choices in injury treatment.

Chapter 7 is entitled "Studies Regarding the Antibacterial Effect of Plant Extracts Obtained from Epilobium parviflorum Schreb" [7]. This chapter evaluates the phytochemical contents of *E. parviflorum* Schreb. extracts (aqueous, hydroalcoholic, and ultrasonicated hydroalcoholic) and their potential antibacterial activity. The chapter records that hydroalcoholic extract have the highest content of polyphenols and flavonoids and possesses strong antibacterial properties. Additionally, the study concludes that the amount of polyphenols and flavonoids in the analyzed extracts directly influences the recorded antibacterial activity (*White Staphylococcus*, *Streptococcus mitis*, *Streptococcus sanguis*, and *Enterococcus faecalis*).

Chapter 8 concerns the paper "Evaluation of genetic damage and antigenotoxic effect of ascorbic acid in erythrocytes of *Oreochromis niloticus* and *Ambystoma mexicanum* using migration groups as a Parameter", which is marginally related to the topic of the Special Issue [8]. In this chapter, Prof. Alvarez-Moya and his research team investigate the genetic damage in erythrocytes of *Oreochromis niloticus* and *Ambystoma mexicanum* exposed to ethyl methanesulfonate and ultraviolet C radiation. The results reveal that migration groups of cell comets allow the detection of basal and induce genetic damage or damage reduction with roughly the same tail length and tail moment parameters efficiency. These findings suggest the usage of migration groups as a complementary parameter to evaluate DNA integrity in species exposed to mutagens.

Chapter 9 comprises the article "Antibacterial activity of nanoparticles of garlic (*Allium sativum*) extract against different bacteria such as *Streptococcus mutans* and *Poryphormonas gingivalis*" [9]. The increasing mortality rate of infectious diseases confirms the necessity of discovering alternative antibacterial agents to combat this threat to public health. Furthermore, garlic has been used as a medicinal plant with antibacterial properties for centuries. Therefore, ultrasonicated garlic extract is analyzed to evaluate the antibacterial activity against *Escherichia coli*, *Poryphyromonas gingivalis*, and Staphylococcus aureus sub. aureus, and *Streptococcus mutans* strains. The results demonstrate that the ultrasonicated garlic extract is a potent antibacterial agent and can help in developing novel antibiotics against bacteria that have developed resistance.

Chapter 10 is entitled "Effect of postharvest UV-C radiation on nutritional quality, oxidation and enzymatic browning of stored mature date" [10]. This chapter aims to study the influence of UV-C radiation (1, 3, and 6 kJ m^{-2}) on preservation potential after harvest of Deglet Nour dates during 5 months of storage (10 °C). The results showed that low-UV-C radiation doses contribute to the enrichment of date fruit in terms of total polyphenols, flavonoids, and tannins, increasing the antioxidant activity. The authors conclude that UV-C radiation of Deglet Nour fruits may positively impact human health by increasing the levels of certain bio-compounds, preserving their nutritional quality, and extending their shelf-life.

The chapters of this book address the undeniable importance of using EOs as new sources of bioactive compounds to eliminate the disadvantages generated by synthetic food additives and unlock new opportunities in the treatment of diseases. However, some challenges still need to be addressed, such as how to preserve their biological properties and minimize the influence on the organoleptic properties of the foods they are incorporated by, alongside testing their toxic effects or any harmful impact that may affect consumer safety and health. Nevertheless, the chapters provided by this book signify a valuable resource for researchers, decision makers, stakeholders, and professionals interested in identifying, promoting, and developing new green antioxidants and preservatives for food-related and pharmaceutical industrial applications. Moreover, following the success of this Special Issue, a new issue, entitled "Antioxidant, Antibacterial and Anti-drug Ingredients in Plants", is now open for submissions.

Author Contributions: Conceptualization, C.J. and G.P.; writing—original draft preparation, C.J. and G.P.; writing—review and editing, C.J. and G.P.; project administration, C.J. All authors have read and agreed to the published version of the manuscript.

Acknowledgments: The guest editors thank all the researchers for delivering their work to this Special Issue. We also thank all reviewers who contributed to the peer review process of submitted manuscripts by improving their quality and impact.

Conflicts of Interest: The authors declare no conflict of interest.

References

1. Mnisi, C.M.; Mlambo, V.; Gila, A.; Matabane, A.N.; Mthiyane, D.M.N.; Kumanda, C.; Manyeula, F.; Gajana, C.S. Antioxidant and Antimicrobial Properties of Selected Phytogenics for Sustainable Poultry Production. *Appl. Sci.* **2023**, *13*, 99. [CrossRef]
2. Mustapa, M.A.; Guswenrivo, I.; Zuhrotun, A.; Ikram, N.K.K.; Muchtaridi, M. Anti-Breast Cancer Activity of Essential Oil: A Systematic Review. *Appl. Sci.* **2022**, *12*, 12738. [CrossRef]
3. Alderees, F.; Mereddy, R.; Were, S.; Netzel, M.E.; Sultanbawa, Y. Anti-Yeast Synergistic Effects and Mode of Action of Australian Native Plant Essential Oils. *Appl. Sci.* **2021**, *11*, 10670. [CrossRef]
4. Jianu, C.; Lukinich-Gruia, A.T.; Rădulescu, M.; Mioc, M.; Mioc, A.; Șoica, C.; Constantin, A.T.; David, I.; Bujancă, G.; Radu, R.G. Essential Oil of *Origanum vulgare* var. *aureum* L. from Western Romania: Chemical Analysis, In Vitro and In Silico Screening of Its Antioxidant Activity. *Appl. Sci.* **2023**, *13*, 5076. [CrossRef]
5. Jianu, C.; Goleț, I.; Stoin, D.; Cocan, I.; Bujancă, G.; Mișcă, C.; Mioc, M.; Mioc, A.; Șoica, C.; Lukinich-Gruia, A.T.; et al. Chemical Profile of *Ruta graveolens*, Evaluation of the Antioxidant and Antibacterial Potential of Its Essential Oil, and Molecular Docking Simulations. *Appl. Sci.* **2021**, *11*, 11753. [CrossRef]
6. Minda, D.; Ghiulai, R.; Banciu, C.D.; Pavel, I.Z.; Danciu, C.; Racoviceanu, R.; Soica, C.; Budu, O.D.; Muntean, D.; Diaconeasa, Z.; et al. Phytochemical Profile, Antioxidant and Wound Healing Potential of Three Artemisia Species: In Vitro and In Ovo Evaluation. *Appl. Sci.* **2022**, *12*, 1359. [CrossRef]
7. Șachir, E.E.; Pușcașu, C.G.; Caraiane, A.; Raftu, G.; Badea, F.C.; Mociu, M.; Albu, C.M.; Sachelarie, L.; Hurjui, L.L.; Bartok-Nicolae, C. Studies Regarding the Antibacterial Effect of Plant Extracts Obtained from Epilobium parviflorum Schreb. *Appl. Sci.* **2022**, *12*, 2751. [CrossRef]
8. Moya, C.A.; Silva, M.R.; Ramírez, L.B.; Radillo, J.d.J.V. Evaluation of Genetic Damage and Antigenotoxic Effect of Ascorbic Acid in Erythrocytes of Orochromis niloticus and Ambystoma mexicanum Using Migration Groups as a Parameter. *Appl. Sci.* **2022**, *12*, 7507. [CrossRef]

9.   Gabriel, T.; Vestine, A.; Kim, K.D.; Kwon, S.J.; Sivanesan, I.; Chun, S.C. Antibacterial Activity of Nanoparticles of Garlic (*Allium sativum*) Extract against Different Bacteria Such as Streptococcus mutans and Poryphormonas gingivalis. *Appl. Sci.* **2022**, *12*, 3491. [CrossRef]

10. Dassamiour, S.; Boujouraf, O.; Sraoui, L.; Bensaad, M.S.; Derardja, A.E.; Alsufyani, S.J.; Sami, R.; Algarni, E.; Aljumayi, H.; Aljahani, A.H. Effect of Postharvest UV-C Radiation on Nutritional Quality, Oxidation and Enzymatic Browning of Stored Mature Date. *Appl. Sci.* **2022**, *12*, 4947. [CrossRef]

Article

Essential Oil of *Origanum vulgare* var. *aureum* L. from Western Romania: Chemical Analysis, In Vitro and In Silico Screening of Its Antioxidant Activity

Călin Jianu [1], **Alexandra Teodora Lukinich-Gruia** [2], **Matilda Rădulescu** [3,4,*], **Marius Mioc** [5], **Alexandra Mioc** [5], **Codruța Șoica** [5], **Albert Titus Constantin** [6], **Ioan David** [1], **Gabriel Bujancă** [1,*] and **Roxana Ghircău Radu** [7]

[1] Faculty of Food Engineering, University of Life Sciences "King Michael I" from Timisoara, 119 Calea Aradului, RO-300645 Timisoara, Romania; calin.jianu@gmail.com (C.J.); ioandavid@usab-tm.ro (I.D.)

[2] OncoGen Centre, County Hospital "Pius Branzeu", 156 Liviu Rebreanu Blvd., RO-300736 Timisoara, Romania; alexandra.gruia@hosptm.ro

[3] Faculty of Medicine, "Victor Babeș" University of Medicine and Pharmacy, 2nd Eftimie Murgu Square, RO-300041 Timisoara, Romania

[4] Multidisciplinary Research Center on Antimicrobial Resistance, "Victor Babeș" University of Medicine and Pharmacy, 2nd Eftimie Murgu Square, RO-300041 Timisoara, Romania

[5] Faculty of Pharmacy, "Victor Babeș" University of Medicine and Pharmacy, 2nd Eftimie Murgu Square, RO-300041 Timisoara, Romania; marius.mioc@umft.ro (M.M.); alexandra.mioc@umft.ro (A.M.); codrutasoica@umft.ro (C.Ș.)

[6] Faculty of Civil Engineering, Politehnica University Timisoara, 2nd Traian Lalescu Street, RO-300223 Timisoara, Romania; albert.constantin@upt.ro

[7] Faculty of Dental Medicine, "Vasile Goldiș" Western University of Arad, 94-96 Revolutiei Blv., RO-310025 Arad, Romania; radu.roxana@uvvg.ro

[*] Correspondence: radulescu.matilda@umft.ro (M.R.); gabrielbujanca@usab-tm.ro (G.B.)

Citation: Jianu, C.; Lukinich-Gruia, A.T.; Rădulescu, M.; Mioc, M.; Mioc, A.; Șoica, C.; Constantin, A.T.; David, I.; Bujancă, G.; Radu, R.G. Essential Oil of *Origanum vulgare* var. *aureum* L. from Western Romania: Chemical Analysis, In Vitro and In Silico Screening of Its Antioxidant Activity. *Appl. Sci.* **2023**, *13*, 5076. https://doi.org/10.3390/app13085076

Academic Editor: Monica Gallo

Received: 8 March 2023
Revised: 29 March 2023
Accepted: 13 April 2023
Published: 18 April 2023

Abstract: This investigation aims to assess the chemical composition and antioxidant properties of *Origanum vulgare* var. *aureum* L. essential oil (OEO). The oil was obtained with a 0.34% (v/w dried weight) yield and investigated by gas chromatography-mass spectrometry (GC–MS) analysis. The main compounds of the OEO were found to be gamma-terpinene (22.96%), para-cymene (14.72%), germacrene (11.64%), beta-trans-ocimene (9.81%), and cis-beta-ocimene (7.65%). Furthermore, individual antioxidant assays 1,1-diphenyl-2-picrylhydrazyl (DPPH) and [2,2′-azinobis(3-ethylbenzothiazoline-6-sulfonic acid) diammonium] (ABTS) radical scavenging activities and beta-carotene/linoleic acid bleaching were carried out. OEO demonstrated better scavenging effects on the DPPH (IC_{50} 93.12 ± 0.03 μg/mL) and ABTS (IC_{50} 27.63 ± 0.01 μg/mL) assays (significantly lower IC_{50} values; $p \leq 0.001$) than ascorbic acid (IC_{50} 127.39 ± 0.45 μg/mL). In the beta-carotene/linoleic acid bleaching assay, the OEO exhibited a higher Relative antioxidant activity (RAA %) (82.36 ± 0.14%) but lower compared with butylated hydroxyanisole (BHA) (100%), with no significant differences ($p > 0.05$) observed. According to molecular docking results, the first two main compounds of the OEO, para-cymene, and gamma-terpinene, may potentially contribute to the biological antioxidant activity of the oil by inhibiting ROS (reactive oxygen species)-producing enzymes such as lipoxygenase and xanthin oxidase. These experimental data suggest that OEO could represent a valuable new natural antioxidant source with functional properties in the food or pharmaceutical industries.

Keywords: antioxidant activity; essential oil; golden oregano; molecular docking; *Origanum vulgare* var. *aureum* L.; steam distillation

1. Introduction

Food deterioration during storage is a major environmental issue and concern of the food industry that generates over a billion tons of food waste and USD 940 billion in economic losses annually [1]. One of the leading causes of food degradation is lipid oxidation which generates off-flavors and loss of nutrients in fat-containing foods. Several techniques can be adopted to decrease auto-oxidation, such as preventing oxygen access by employing suitable packaging materials, storing food products at lower temperatures, or inactivating enzymes catalyzing oxidation [2]. However, these methods are not always applicable or economic from nutritional and technological points of view [3]. Therefore, using food antioxidants, especially synthetic ones, has become indispensable in prolonging the shelf-life of foodstuffs.

The modern food consumer does not perceive food as basic nutrition alone but also for the health benefits it can generate. Furthermore, the suspicions regarding the harmful effects on human health of synthetic antioxidants such as gallic acid esters (e.g., propyl gallate, octyl gallate, and dodecyl gallate), butylhydroxytoluene (BHT) or butylated hydroxyanisole (BHA) [4–6] have led to increased consumer demands for minimally processed foods with clean, easy-to-read labels and long shelf lives. Consequently, in the last few decades, the food industry companies were urged to access natural extracts as a source of food additives to preserve food safety without harming consumers' health. Through their extracts, especially essential oils (EO), aromatic and medicinal plants are a valuable source of biologically active compounds [7–9]. Moreover, recent investigations have revealed that EOs possess promising biological properties, such as antioxidant, antimicrobial, and antiviral effects [8,10–12], recommending them as an alternative to synthetic food additives.

The *Origanum* (Lamiaceae) genus is an annual, perennial, and shrubby herb extensively found in the Mediterranean region [13]. *Origanum* is classified into three groups, ten sections and thirty-eight species (six subspecies and seventeen hybrids) [14]. Two commercially important varieties, *Origanum vulgare* subsp. *hirtum* (Greek oregano) and *Origanum onites* (Turkish oregano) are also among the most marketed and used spice plants [14–16]. The flavoring properties of oregano are mainly associated with its aromatic substances, especially its EO [14,17]. Carvacrol is the main volatile oil compound in oregano herbs, responsible for the characteristic "oregano" scent. Other compounds that dominate the EOs of genus members are thymol, para-cymene, and gamma-terpinene. Several investigations reported that oregano oil has a powerful antimicrobial effect against bacteria, yeast, and fungi. Furthermore, EOs extracted from *Origanum* sp. are known to possess cytotoxic activities against cancer cells, to protect cells from reactive oxygen species (ROS) degradation, and to generate an antioxidant and anti-inflammatory biological effect via key enzymes (lipoxygenase) inhibition [18,19]. Also, oregano oil possesses antioxidant properties effective in slowing color loss and lipid oxidation in fatty foods and scavenging free radicals [20–22].

However, some genus members remain insufficiently studied, such as *Origanum vulgare* var. *aureum* L., for which scientific literature provides insufficient data about its EO antioxidant properties. Furthermore, the probable mechanism of action against lipid oxidation of *Origanum vulgare* var. *aureum* essential oil (OEO) has not been investigated previously. Considering the background, we aimed to investigate the chemical composition and evaluate the antioxidant properties of OEO. Moreover, we aimed to analyze in silico the antioxidant biological effects based on the inhibitory capacity of OEO components against key enzymes involved in producing reactive oxygen species utilizing a molecular docking-based approach.

2. Materials and Methods

2.1. Raw Material and Chemicals

O. vulgare var. *aureum* L. stems, leaves, and flowers were harvested at full blossom from the village of Ludești de Jos, Hunedoara County (Romania) (45°05′ N 22°19′ E) in July 2021. A voucher specimen (VSNH.BUASTM-93/2) was taken in the Herbarium of the Faculty of Agronomy, University of Life Sciences "King Michael I" from Timișoara. Anhydrous sodium sulphate (Na_2SO_4), potassium persulfate ($K_2S_2O_8$), hexane, methanol, chloroform, ethanol, and ascorbic acid were obtained from Sigma-Aldrich (Germany). In addition, 1,1-diphenyl-2-picrylhydrazyl radical (DPPH), β-carotene, 2,2′-azinobis (3-ethylbenzothiazoline-6-sulfonic acid) diammonium salt (ABTS), C_8-C_{20} alkane standard mixture and butylated hydroxyanisole (BHA) were obtained from Merck Company (Darmstadt, Germany).

2.2. Extraction of OEO

The collected specimens were air dried in the dark at laboratory temperature (end of 2021/beginning of 2022) and cut into ca. 1.5 cm long parts before the distillation. OEO was obtained by steam distillation operating a modified Craveiro-type apparatus described previously by Jianu et al. [23]. After extraction, the OEO was collected, dried over anhydrous Na_2SO_4, and kept in amber-sealed vials ($-18\ ^\circ C$) for future analysis (yielding 0.34% v/w).

2.3. Gas Chromatography-Mass Spectrometry Analysis

This assay was performed on a gas chromatograph (HP6890 Gas-Chromatograph) coupled with a mass spectrometer (HP5973 Mass Spectrometer). The oil sample was diluted 1:1000 in hexane before 1 µL was injected in a splitless mode. The sample passed through a capillary Br-5MS column with a 5% Phenyl-arylene-95% Dimethylpolysiloxane phase; the column was 30 m long, with a 0.25 mm internal diameter, 0.25 µm film thickness (Bruker, Billerica, MA, USA), with a helium flow rate of 1 mL/min. The column was heated at a program starting from 50 °C to 300 °C at 6 °C/min, where the final hold was 5 min. The MS Quad temperature was set at 150 °C, and the ionization energy was 70 eV. After 3 min, solvent delay started scanning compounds between 50 to 550 amu. The resulting peaks represented compounds found in the oil sample and were identified by comparing their mass spectra with the ones from the NIST2.0 library (USA National Institute of Science and Technology software, Version NIST 2.0 library). The percentage area of each compound was calculated by dividing its area by the total area and multiplying by 100. Furthermore, their retention indices (RIs) were calculated using the linear equation obtained from a calibration curve of a standard alkane C_8-C_{20} and then compared to previously published values in the literature to confirm the identity of the identified components [24].

2.4. Antioxidant Activity by 1,1-Diphenyl-2-Picrylhydrazyl (DPPH) Radical Scavenging Assay

A DPPH radical scavenging assay was conducted using the stable radical DPPH [25]. A methanolic solution of DPPH radical at different concentrations, from 1.5 mg/mL to 2.93 µg/mL, was prepared. A ratio of 1:10 DPPH/sample (v/v) was plated in triplicates on 96 well plates and incubated in the dark at room temperature. After a 30 min reaction period, absorbances were registered at 515 nm on a spectrophotometer Tecan i-control, 1.10.4.0 infinite 200Pro. BHA and ascorbic acid were used as positive controls. The following equation computed the DPPH scavenging activity: % DPPH scavenging activity = (A_{blank} − A_{OEO})·$100/A_{blank}$, where: A_{blank} and A_{OEO} are the absorbances of the control and the tested oil. IC_{50} (µg/L) value, defined as the concentration of oil required to scavenge the formation of 50% of free radicals, was calculated using the BioDataFit 1.02 program (Chang Broscience Inc, Castro Valley, CA, USA) and expressed as means ± standard deviation (SD) of three independent experiments.

2.5. Antioxidant Activity by [2,2′-Azinobis(3-Ethylbenzothiazoline-6-Sulfonic Acid) Diammonium] (ABTS) Radical Scavenging Assay

ABTS radical scavenging activity was conducted using the method previously reported by Rădulescul et al. (2021) [26] with a slight modification. Briefly, ABTS (7.0 mM) and $K_2S_2O_8$ (2.5 mM) were added to an amber-colored bottle for the preparation of ABTS cation ($ABTS^+$) and kept in the dark for 14 h at 21 °C. First, the $ABTS^+$ solution was diluted in ethanol (approximately 1:80) to an absorbance of 0.700 ± 0.035 at 734 nm [27]. Subsequently, 100 µL of the OEO in methanol, with various concentrations (from 1.5 mg/mL to 9.3 µg/mL), were mixed with 1 mL of $ABTS^+$ solution. The absorbances were measured at 734 nm after 30 min (at 21 °C in the dark). BHA and ascorbic acid served as positive controls. Finally, the ABTS scavenging activity was analyzed by IC_{50} (µg/L) value as the inhibitory concentration of the OEO required to scavenge the formation of 50% of ABTS, expressed as means $\pm$ standard deviation (SD) of three independent experiments.

2.6. Beta-Carotene/Linoleic Acid Bleaching Assay

The β-Carotene/Linoleic acid bleaching test was assessed employing the method described by Rădulescu et al. (2021) [26]. First, a stock solution of β-carotene was obtained by dissolving 0.5 mg β-carotene in 1 mL of $CHCl_3$, 25 µL linoleic acid, and 200 mg Tween 40. A vacuum rotary evaporator removed chloroform under a vacuum (at 40 °C for 5 min), and then 100 mL of distilled water was added to the residue to form a transparent yellowish emulsion. Next, 350 µL of oil in methanol (2 mg/mL) was stirred exhaustively with 2.5 mL of β-carotene stock solution and incubated for 48 h at 21 °C. The same methodology was repeated for the synthetic antioxidant (BHA) serving as positive control and a blank sample (350 µL of methanol). After this incubation period, the absorbance of the samples was read at 490 nm in three independent experiments. The following equation computed the Relative Antioxidant Activity (RAA %): RAA = A_{OEO}/A_{BHA}, where A_{OEO} and A_{BHA} are the absorption of OEO and the absorbance of the BHA, respectively.

2.7. In Silico Prediction of Bioactivity and Molecular Docking Studies

Molecular docking simulations were performed using a method previously described [23]. In brief, docking targets were optimized using the 3D crystallographic structure of proteins available from the RCSB Protein Data Bank [28] (Table 1). The 38 OEO components were downloaded from the Pubchem repository [29] as SDF files and were later converted to PDBQT files using Autodocktools [30]. The PyRx v0.8 virtual screening software (The Scripps Research Institute, La Jolla, CA, USA) and Vina's embedded scoring function [31] were used for molecular docking. Each input molecular structure's target protein was docked with the default number of conformers (eight per each ligand structure). The structure of each protein native ligand (NL) (Table 1) was retrieved from their respective PDB file and converted to the PDBQT format as described above. In order to validate our protocol, the calculated root means square deviation (RMSD) between predicted and actual native ligand docking conformation for every case could not exceed a 2 Å threshold. The coordinates and size of the docking grid box were selected to match the active binding domain (Table 1) perfectly. The software generated ΔG binding energy values (kcal/mol) as docking scores for each docked molecule. Protein-ligand binding interactions were also investigated using Accelrys Discovery Studio 4.1. (Dassault Systems BIOVIA, San Diego, CA, USA).

Table 1. Docking parameters used for the in silico evaluation of the 38 OEO components.

Protein	PDB ID	Native Ligand	Grid Box Center Coordinates and Size (Å)
Lipoxygenase	1N8Q	Protocatechuic acid	center_x = 20.6778 center_y = 2.2697 center_z = 19.5423 size_x = 8.5358 size_y = 9.1096 size_z = 9.9453

Table 1. *Cont.*

Protein	PDB ID	Native Ligand	Grid Box Center Coordinates and Size (Å)
CYP2C9	1OG5	S-Warfarin	center_x = −20.4200 center_y = 85.2723 center_z = 38.2181 size_x = 9.6524 size_y = 10.6625 size_z = 11.5535
NADPH-oxidase	2CDU	Adenosine-5′-diphosphate	center_x = 18.4683 center_y = −5.1659 center_z = −0.0901 size_x = 12.3286 size_y = 15.3831 size_z = 15.1287
Xanthine oxidase	3NRZ	Hypoxanthine	center_x = 37.6450 center_y = 19.2898 center_z = 17.5578 size_x = 11.6388 size_y = 9.8519 size_z = 10.1251
Myeloperoxidase	5QJ2	7-{[3-(1-methyl-1H-pyrazol-3-yl) phenyl]methoxy}-1H-[1,2,3]triazolo[4,5-b]pyridin-5-amine	center_x = −20.1951 center_y = 11.9649 center_z = 32.9278 size_x = 9.7712 size_y = 12.5445 size_z = 11.9903

2.8. Statistical Analysis

Conventional statistical methods were employed to compute the mean ± standard deviation (SD) of three independent experiments carried out separately for the antioxidant activity assays. In addition, a post hoc test (Tukey) was applied to compare the significant differences between the mean values obtained from the antioxidant activity measurements, with a significance of 0.05 ($p < 0.05$). Analysis was performed using the IBM SPSS 25.0 package (SPSS Inc., Chicago, IL, USA).

3. Results and Discussion

3.1. OEO Yield and Chemical Composition

A greenish-yellow oil with a pungent smell was obtained from dry *O. vulgare* var. *aureum* L. (stems, leaves, and flowers), with a 0.34% (*v/w*) yield. Our results agree with those mentioned in the literature, which record yields of obtaining EOs from different oregano species as lower than 1% [32].

Table 2 summarizes the chemical compositions of the EO isolated from *O. vulgare* var. *aureum* L. by steam distillation. The GC-MS analysis identified 38 compounds in the analyzed sample (Figure 1), with gamma-terpinene (22.96%), para-cymene (14.72%), germacrene (11.64%), beta-trans-ocimene (9.81%), and cis-beta-ocimene (7.65%) as main compounds of the oil (Table 2). In addition, beta-linalool (3.61%), isothymol methyl ether (3.34%), beta-phellandrene (2.44%), beta-caryophyllene (2.35%), alpha-terpinene (1.84%), beta-cadinene (1.78%), alpha-farnesene (1.55%), (-)-spathulenol (1.35%), beta-myrcene (1.31%) and alpha-himalachene (1.29%) were identified in smaller quantities.

Table 2. Chemical composition of OEO as determined by GC-MS analysis.

No.	Compound Name	MW (g/mol)	RI$_{exp.}$	Area %
1	Alpha-thujene	136.23	912	0.83
2	Alpha-pinene	136.23	919	0.35
3	Camphene	136.23	934	0.16
4	Beta-phellandrene	136.23	955	2.44
5	Beta-pinene	136.23	960	0.26
6	Beta-myrcene	136.23	970	1.31
7	Alpha-phellandrene	136.23	987	0.15
8	3-Carene	136.23	990	0.05
9	Alpha-terpinene	136.23	998	1.84
10	Para-cymene	134.21	1007	14.72
11	Beta-terpinyl acetate	196.29	1011	0.35
12	Beta-trans-ocimene	136.23	1018	9.81
13	Cis-beta-ocimene	136.23	1030	7.65
14	Gamma-terpinene	136.23	1044	22.96
15	Beta-linalool	154.25	1087	3.61
16	Isothymol methyl ether	164.24	1243	3.34
17	Dihydroedulan	194.31	1299	1.01
18	Carvacrol	150.22	1309	0.62
19	Alpha-cubebene	204.35	1363	0.06
20	Alpha-copaene	204.35	1394	0.27
21	Beta-bourbonene	204.35	1402	0.97
22	Beta-elemene	204.35	1408	0.71
23	Beta-caryophyllene	204.35	1439	2.35
24	Beta-cubebene	204.35	1450	0.45
25	Alpha-caryophyllene	204.35	1476	0.46
26	Alloaromadendrene	204.35	1481	0.75
27	Germacrene D	204.35	1503	11.64
28	Gamma-elemene	204.35	1517	0.98
29	Alpha-muurolene	204.35	1520	0.52
30	Alpha-farnesene	204.35	1525	1.55

Table 2. *Cont.*

No.	Compound Name	MW (g/mol)	RI$_{exp.}$	Area %
31	Alpha-himalachene	222.37	1529	1.29
32	Gamma-cadinene	206.37	1534	0.51
33	Beta-cadinene	206.37	1540	1.78
34	(-)-Spathulenol	220.35	1595	1.35
35	Caryophyllene oxide	220.35	1600	0.35
36	Alpha-cadinol	222.37	1669	0.49
37	Chamazulene	184.28	1735	0.14
38	Isoaromadendrene epoxide	220.35	1806	0.51
			Total identified (%):	**98.59%**

The retention experimental (RI$_{exp.}$) was determined on a Br-5MS column using a homologous series of n-alkanes (C$_8$-C$_{20}$); MW—molecular weight.

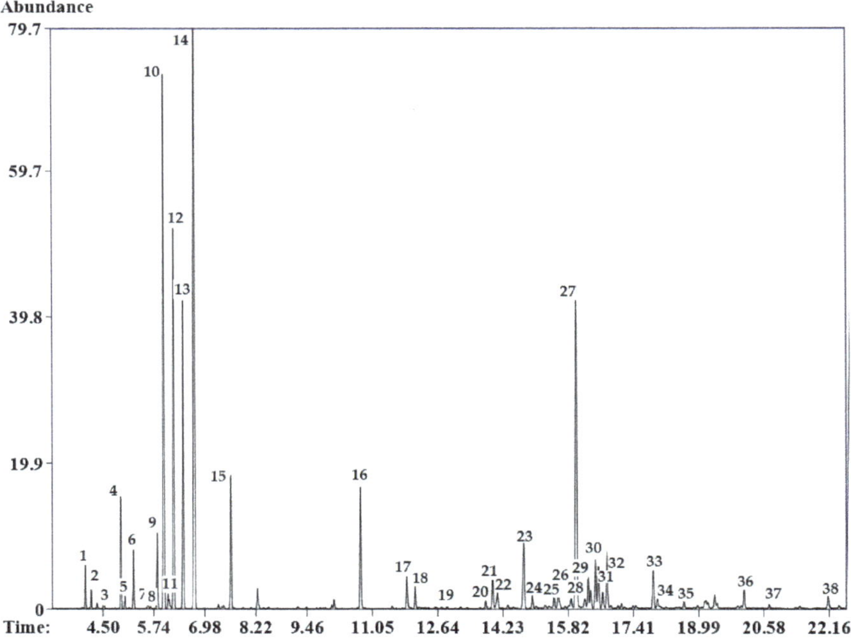

Figure 1. Gas chromatogram of OEO from western Romania.

Popa et al. [33] reported linalool (26.54%), para-cymene (20.81%), and gamma-terpinene (13.73%) as the major compounds for the oil isolated from Romanian *O. aureum*. Furthermore, high contents of linalool (25.5%), para-cymene (20.7%), and gamma-terpinene (15.66%) were also documented for the golden oregano oil by Moisa et al. [34]. In contrast, the oil obtained from Polish *O. aureum* contained mainly cis-sabinene hydrate (18.7%), germacrene D (16.5%), and thymol (12.2%), respectively [35]. The variability of oil content and the chemical composition of the oregano herbs EOs is strongly dependent on the environment and local conditions of the plants, climatic conditions, and geographical distribution of the plant collections, as previously documented [35–38].

3.2. Assessment of Antioxidant Activity

It is well-known that free radicals initiate the process of lipid peroxidation and the propagation of the chain of radical structures. Therefore, several methods have been developed to determine compounds' antioxidant properties [39–41]. For example, 2,2-diphenyl-1-picrylhydrazyl (DPPH) and 2,2′-azino-bis(3-ethylbenzthiazoline-6-sulfonic acid) radical (ABTS) assays are two spectrophotometer techniques widely employed for assessing the antioxidant properties of natural extracts. Briefly, the DPPH method is based on reducing the purple DPPH radical to 1,1-diphenyl-2-picryl-hydrazine (DPPH-H). In contrast, the ABTS method implicates the reduction or radical scavenging of a blue/green ABTS radical to a colorless sulfonic acid [42]. Butylated hydroxyanisole (BHA) and ascorbic acid were used as the positive control. The IC_{50} values were computed employing BioDataFit 1.02 program (Chang Broscience Inc, Castro Valley, CA, USA). In our study, OEO reduced the stable free radical DPPH with an IC_{50} value of 93.12 ± 0.03 µg/mL (Table 3). Tukey's test shows that the OEO's DPPH values are significantly lower than ascorbic acid ($p < 0.01$). Previously, Popa et al. documented low antioxidant capacity for Romanian *O. vulgare* var. *aureum* volatile oil, with a $58.18 \pm 0.07\%$ inhibition [33]. Comparable effects were reported by Moisa et al. for oils isolated from *O. vulgare* var. *aureum* leaves and flowers with $63.1 \pm 0.7\%$ and $66.4 \pm 2.23\%$ inhibitions [34]. Better results were documented for oils extracted from *O. vulgare* var. *aureum* stems and whole plant, respectively, with inhibitions ranging between $80.3 \pm 0.01\%$ to $88.6 \pm 0.1\%$ [34]. However, because of the various modes of result expression used in different studies, directly comparing our findings with those reported in the literature is difficult. Nevertheless, the DPPH radical scavenging ability registered for the studied OEO is in accord with those previously described for *O. vulgare* (46.66–97.61 µg/mL), *Poliomintha longiflora* (83.70 ± 4.12 µg/mL), *O. onites* L. (116.74–132.93 µg/mL) and *O. syriacum* (91.45 ± 2.30 µg/mL) EOs [43–46].

Table 3. The antioxidant activity of essential oil extracted from *O. vulgare* var. *aureum*.

Samples Tested	Parameters		
	DPPH, IC_{50} (µg/mL)	ABTS, IC_{50} (µg/L)	β-Carotene/Linoleic Acid (% Inhibition Rate)
OEO	93.12 ± 0.03	27.63 ± 0.01	82.36 ± 0.14
BHA	10.11 ± 0.01	8.71 ± 0.01	100
Ascorbic acid	127.39 ± 0.45	35.89 ± 0.05	n.d.

n.d.—not determined.

In the ABTS assay, OEO exhibited a good ability to scavenge the ABTS radical (IC_{50} 27.63 ± 0.01 µg/mL) (Table 3), significantly more potent ($p < 0.01$) than standard ascorbic acid (IC_{50} 35.89 ± 0.05 µg/mL) according to the Tukey's test. In contrast, BHA revealed significantly ($p < 0.001$) higher antioxidant capability (IC_{50} 8.71 ± 0.01 µg/mL) compared to OEO and ascorbic acid. Nevertheless, our results reveal that OEO was more effective in scavenging ABTS radicals than other oregano species: *O. tyttanthum*, *O. compactum*, *O. vulgare*, and *L. graveolens* [47–50].

Previously published research investigating the DPPH and reactive oxygen species scavenging capability of different extracts discovered that these extracts have similar scavenging performance in both cases [51,52]. This is to be expected, given that it is dependent on the extract components' capacity to reduce free radicals. Based on this, OEO could be a good source of antioxidants that can also quench endogenous reactive oxygen species.

The beta-carotene/linoleic acid bleaching assay is based on the discoloration of beta-carotene determined by its reaction with the radicals resulting from the oxidation of linoleic acid in an emulsion. The presence of an antioxidant can delay the rate of beta-carotene discoloration. Table 3 displays the inhibition of beta-carotene bleaching by the OEO and the positive control used (BHA). In the case of the beta-carotene-linoleic acid bleaching assay, OEO was able to inhibit the linoleic acid oxidation (82.36 ± 0.14%) effectively but lower than BHA (100%) ($p > 0.05$). No previous investigations were documented concerning OEO activity in *the* beta-carotene/linoleic acid system to allow us to make direct comparisons.

3.3. In Silico Prediction of the Mechanism by Molecular Docking Analysis

In recent years, computer simulation methods such as molecular docking and molecular dynamics simulation have been widely used to study structure-function relationships and interaction mechanisms [53,54].

The OEO's potential protein-targeted antioxidant effect was evaluated using molecular docking. The obtained docking scores for the 38 OEO components are depicted in Table 4.

Table 4. Docking scores for compounds **1–38** (binding energy, kcal/mol); compounds with better docking scores than the target native ligand score is highlighted.

Target PDB ID	1N8Q	1OG5	2CDU	3NRZ	5QJ2
Docked OEO Component ID	**Binding Free Energy ΔG (kcal/mol)**				
Native ligand	−5.7	−9.8	−9.3	−6.7	−8.5
1	−6.4	−5.7	−5.5	−5	−6.3
2	−5.3	−5.6	−5.1	−0.9	−5.6
3	−4	−5.6	−5.2	0.1	−5.7
4	−5.7	−6.2	−5.8	−6.5	−6
5	−5	−5.6	−5.2	−0.9	−5.8
6	−5.5	−5.4	−5	−5.8	−5.5
7	−5.7	−6.2	−5.6	−6.6	−6.3
8	−6.4	−5.7	−5.5	−2.8	−6.1
9	−5.9	−6.1	−5.6	−6.7	−6.4
10	−5.9	−6.2	−5.7	−7	−6.5
11	−3.1	−6.6	−5.8	−2.7	−6.5
12	−5.2	−5.5	−5.2	−6.2	−5.7
13	−4.9	−5.5	−5	−5.8	−5.6
14	−5.7	−6.2	−5.6	−6.8	−6.5
15	−4.9	−5.3	−5.2	−5	−5.8
16	−4.6	−6.3	−6.1	−6.9	−6.4
17	−2.2	−7.1	−6.3	1.4	−7.1
18	−6	−6.2	−6	−7.1	−7
19	−1.7	−7.3	−6.6	2.7	−7.7
20	−2.4	−7.5	−6.3	2.1	−7.3
21	−3.9	−7	−7	−1.5	−7.7
22	−2.3	−7.1	−6.4	−1.2	−6.9
23	−2.9	−7.4	−6.9	1.6	−7
24	−1.7	−7.5	−6.5	2.9	−7.7
25	−1.2	−7.1	−5.9	3.4	−6.8
26	−1.2	−7.4	−6.5	5.5	−7.3
27	−2.2	−7.4	−6.5	−1.4	−7.6
28	−1.1	−7	−5.8	−0.1	−7.2
29	−2.6	−7.2	−6.3	6.2	−7.3
30	−4.1	−6.8	−6.2	−7.3	−6.5
31	0.2	−7.3	−6.5	9.9	−7.3
32	−3.5	−7.2	−7.2	−1.6	−7.3
33	−3	−7.5	−7.4	−1.3	−7.6
34	0.5	−7.1	−6.5	5.9	−6.9
35	−0.9	−7.4	−6.5	5.1	−6.5
36	−2.3	−7.1	−6.9	−0.3	−6.9
37	−4.9	−7.9	−7.5	−6.1	−7.8
38	−0.2	−7.2	−7.1	6.8	−7

Background color explicitly presents the results obtained.

Our in silico-based method aimed to identify protein targets that can be inhibited by multiple components of our OEO or, at the very least, by the EO's major constituents. In order to achieve this goal, all 38 OEO components were investigated in silico for their inhibitory potential against proteins involved in producing reactive oxygen species (ROS) [55]. The proteins selected for the docking-based virtual screening of the OEO components were Lipoxygenase, CYP2C9, NADPH-oxidase, Xanthine oxidase, and Myeloperoxidase. Docking scores show that of the five investigated proteins, two cases stand out (lipoxygenase—1N8Q and xanthine oxidase—3NRZ), where several components recorded the same or better docking scores than the respective protein's NL score. However, in potentially assessing a biological effect of a multicomponent mixture such as the present OEO, one must consider the quantity of each molecule in that mixture.

Figuring out correlations between docking scores is challenging since every protein has distinct binding site features, and native ligands yield different docking scores, resulting in varying control values for each score set. To compensate for these drawbacks, we determined each docking score as a percentage of its corresponding native ligand's score (considered 100%). These percentages were illustrated as a radar chart, with the scores for every compound being indicated by plot lines while the protein targets formed the chart's corners. Compounds with a calculated percentage of 10% or below were all assigned a 10% value because positive ΔG values would give negative percentages, and the centre of the chart would become crowded and unintelligible. When a chart plot line stretches toward a specific protein target corner, this would indicate that the oil components scored close to or higher than that protein's NL. The final result should reveal the most likely targeted protein by the EO's constituents.

Given that a particular compound may have excellent docking scores, but only a small quantity of that molecule is contained in the volatile oil, we first plotted the components that exceed 1% of the oil content (Figure 2A). Aside from a few compounds, the same trend can be seen in this figure, where most constituents show higher affinity towards the same two proteins (lipoxygenase—1N8Q and xanthine oxidase—3NRZ) where compounds registered higher docking scores than the native ligands used as positive controls in this setting. If further compounds are removed from the same chart so that only major components that exceed a 5% weight content of the OEO remain, the trend, as mentioned earlier, becomes more visible. Aside from compound 27 (Germacrene D), which performed poorly against all five proteins, compounds 10 (para-cymene) and 14 (gamma-terpinene) stand out as the most abundant OEO components that also performed better than the native ligands when docked in both lipoxygenase (1N8Q) and xanthine oxidase (3NRZ). It was previously demonstrated that para-cymene has antinociceptive and anti-inflammatory activity associated with lipoxygenase inhibition [56]. Another study investigating the antioxidant activity of thyme oils found that the EO's lipoxygenase inhibition was primarily due to thymol, para-cymene, and linalool [57]. Similarly, gamma-terpinene was previously reported as an in vitro lipoxygenase inhibitor [58]. Given their highly similar structures, except for para-cymene having an aromatic ring and gamma-terpinene having a cyclohexadiene ring, both compounds exhibit a similar binding pattern within the lipoxygenase binding domain near the iron site, where both compounds interact through multiple hydrophobic interactions with the surrounding amino acids (Figure 3).

No relevant literature is available regarding the direct inhibition activity of para-cymene against xanthine oxidase. However, in the case of gamma-terpinene, a study investigating the antioxidant and xanthine oxidase inhibitory activity of various sunflower essential oils discovered the monoterpene, as mentioned above, as one of the EO's major constituents. According to a Pearson correlation analysis, gamma-terpinene was one of the constituents associated with antioxidant and anti-xanthine oxidase activity [59]. Similar to the previous case, both compounds exhibit nearly the same conformation in the active domain of xanthine oxidase, interacting with the same amino acids via hydrophobic interactions. The binding patterns shown by both compounds are depicted in Figure 4. Given that six compounds scored higher than the co-crystallized ligand (the endogenous hypoxanthine) in the case of xanthine oxidase, five of which are major components of OEO, accounting for approximately half of the OEO weight content, we can easily predict that the oil could have a significant inhibitory activity towards xanthine oxidase.

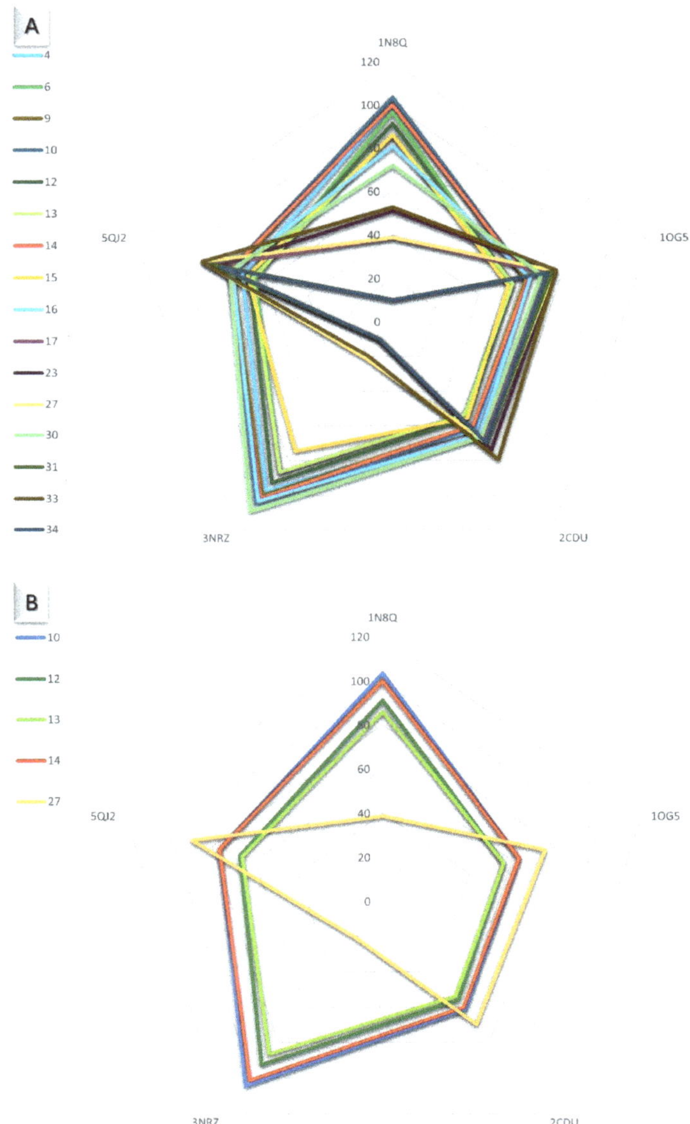

Figure 2. Graphical representation of docking scores related to antioxidant protein targets, corresponding to the components above 1% of the OEO content (**A**) and OEO constituents above 5% of the OEO content (**B**) (representing above 65% of the volatile oil); lines are generated for each compound based on the docking scores calculated for each target protein as a percentage of the NL's docking score used as the positive control (100%); the lines and are plotted as series in a radar chart, where the proteins represent the corners of the chart.

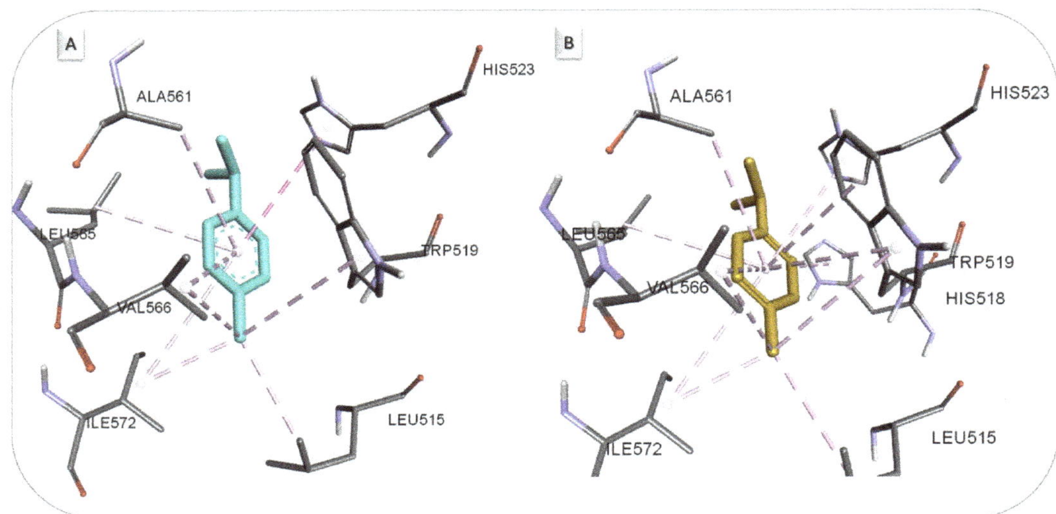

Figure 3. Lipoxygenase (1N8Q) structure in complex with docked compound (**A**) para-cymene (cyan) and (**B**) gamma-terpinene (gold) with interacting amino acids (grey sticks) via hydrophobic interactions (pink-dotted lines). Interatomic length between interacting atoms varied between 3.53–5.63 Å.

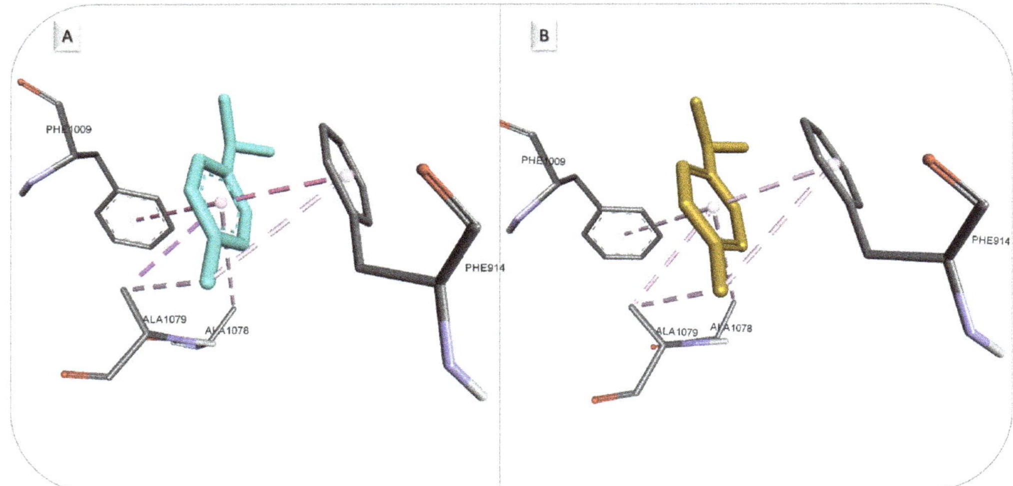

Figure 4. Xanthine oxidase (3NRZ) structure in complex with docked compound (**A**) para-cymene (cyan) and (**B**) gamma-terpinene (gold) with interacting amino acids (grey sticks) via hydrophobic interactions (pink/purple-dotted lines). Interatomic length between interacting atoms varied between 3.61–5.21 Å.

4. Conclusions

The present investigation demonstrated that OEO revealed better antioxidant properties (significantly lower IC50 values; $p \leq 0.001$) than ascorbic acid in vitro assays, such as DPPH and ABTS radical scavenging. Furthermore, according to molecular docking results, the first two major components of the volatile oil, para-cymene and gamma-terpinene, may contribute to the EO's biological antioxidant activity by inhibiting ROS-producing enzymes such as lipoxygenase and xanthin oxidase. Due to the effective antioxidant activity recorded by OEO, it can represent a new natural source of antioxidants with potential applications in the food or pharmaceutical industry. However, further investigations are needed to elucidate the mechanism of action against oxidation reactions and to establish the safety of usage doses.

Author Contributions: Conceptualization and methodology, C.J., C.Ș. and M.M.; investigation, G.B., A.T.L.-G., R.G.R., I.D., A.M., M.M. and M.R.; statistical analysis, A.T.C.; writing—original draft preparation, C.J., G.B., R.G.R. and C.Ș.; writing—review and editing, C.J., M.R. and M.M. All authors have read and agreed to the published version of the manuscript.

Funding: This research received no external funding.

Institutional Review Board Statement: Not applicable.

Informed Consent Statement: Not applicable.

Data Availability Statement: The data published in this research are available on request from the first author and corresponding authors.

Acknowledgments: This paper is published from the own funds of the University of Life Sciences "King Michael I" from Timisoara.

Conflicts of Interest: The authors declare no conflict of interest.

References

1. FAO. Members Call for Action against Food Loss and Waste in the World's Hungriest Region—Asia and the Pacific. 2021. Available online: http://www.fao.org/asiapacific/news/detail-events/en/c/1403168/ (accessed on 11 December 2022).
2. Dorman, H.; Peltoketo, A.; Hiltunen, R.; Tikkanen, M. Characterisation of the antioxidant properties of de-odourised aqueous extracts from selected Lamiaceae herbs. *Food Chem.* **2003**, *83*, 255–262. [CrossRef]
3. Viuda-Martos, M.; Ruiz Navajas, Y.; Sánchez Zapata, E.; Fernández-López, J.; Pérez-Álvarez, J.A. Antioxidant activity of essential oils of five spice plants widely used in a Mediterranean diet. *Flavour Fragr. J.* **2010**, *25*, 13–19. [CrossRef]
4. Park, S.; Lee, J.-Y.; Lim, W.; You, S.; Song, G. Butylated hydroxyanisole exerts neurotoxic effects by promoting cytosolic calcium accumulation and endoplasmic reticulum stress in astrocytes. *J. Agric. Food Chem.* **2019**, *67*, 9618–9629. [CrossRef] [PubMed]
5. Liang, X.; Zhao, Y.; Liu, W.; Li, Z.; Souders, C.L., II; Martyniuk, C.J. Butylated hydroxytoluene induces hyperactivity and alters dopamine-related gene expression in larval zebrafish (*Danio rerio*). *Environ. Pollut.* **2020**, *257*, 113624. [CrossRef] [PubMed]
6. Ham, J.; Lim, W.; Park, S.; Bae, H.; You, S.; Song, G. Synthetic phenolic antioxidant propyl gallate induces male infertility through disruption of calcium homeostasis and mitochondrial function. *Environ. Pollut.* **2019**, *248*, 845–856. [CrossRef] [PubMed]
7. Ashok, B.; Hariram, N.; Siengchin, S.; Rajulu, A.V. Modification of tamarind fruit shell powder with in situ generated copper nanoparticles by single step hydrothermal method. *J. Bioresour. Bioprod.* **2020**, *5*, 180–185. [CrossRef]
8. Tagnaout, I.; Zerkani, H.; Bencheikh, N.; Amalich, S.; Bouhrim, M.; Mothana, R.A.; Alhuzani, M.R.; Bouharroud, R.; Hano, C.; Zair, T. Chemical Composition, Antioxidants, Antibacterial, and Insecticidal Activities of *Origanum elongatum* (Bonnet) Emberger & Maire Aerial Part Essential Oil from Morocco. *Antibiotics* **2023**, *12*, 174. [CrossRef] [PubMed]
9. Yuan, Y.; Sun, J.; Song, Y.; Raka, R.N.; Xiang, J.; Wu, H.; Xiao, J.; Jin, J.; Hui, X. Antibacterial activity of oregano essential oils against Streptococcus mutans in vitro and analysis of active components. *BMC Complement. Med. Ther.* **2023**, *23*, 61. [CrossRef] [PubMed]
10. Mancianti, F.; Ebani, V.V. Biological activity of essential oils. *Molecules* **2020**, *25*, 678. [CrossRef]
11. Saeed, K.; Pasha, I.; Jahangir Chughtai, M.F.; Ali, Z.; Bukhari, H.; Zuhair, M. Application of essential oils in food industry: Challenges and innovation. *J. Essent. Oil Res.* **2022**, *34*, 97–110. [CrossRef]
12. Kačániová, M.; Galovičová, L.; Ivanišová, E.; Vukovic, N.L.; Štefániková, J.; Valková, V.; Borotová, P.; Žiarovská, J.; Terentjeva, M.; Felšöciová, S. Antioxidant, antimicrobial and antibiofilm activity of coriander (*Coriandrum sativum* L.) essential oil for its application in foods. *Foods* **2020**, *9*, 282. [CrossRef]
13. Pezzani, R.; Vitalini, S.; Iriti, M. Bioactivities of *Origanum vulgare* L.: An update. *Phytochem. Rev.* **2017**, *16*, 1253–1268. [CrossRef]
14. Skoula, M.; Harborne, J.B. The taxonomy and chemistry of Origanum. In *Oregano: The genera Origanum and Lippia*; CRC Press: Boca Raton, FL, USA, 2002; Volume 67.
15. Shayista, C.; Zahoor, A.K.; Phalestine, S. Medicinal importance of genus Origanum: A review. *J. Pharmacogn. Phytother.* **2013**, *5*, 170–177.
16. Verma, R.S.; Padalia, R.C.; Chauhan, A. Analysis of the hydrosol aroma of Indian oregano. *Med. Aromat. Plants* **2012**, *1*, 112. [CrossRef]

17. Yan, F.; Azizi, A.; Janke, S.; Schwarz, M.; Zeller, S.; Honermeier, B. Antioxidant capacity variation in the oregano (*Origanum vulgare* L.) collection of the German National Genebank. *Ind. Crops Prod.* **2016**, *92*, 19–25. [CrossRef]
18. Karioti, A.; Milošević-Ifantis, T.; Pachopos, N.; Niryiannaki, N.; Hadjipavlou-Litina, D.; Skaltsa, H. Antioxidant, anti-inflammatory potential and chemical constituents of Origanum dubium Boiss., growing wild in Cyprus. *J. Enzym. Inhib. Med. Chem.* **2015**, *30*, 38–43. [CrossRef] [PubMed]
19. Coccimiglio, J.; Alipour, M.; Jiang, Z.H.; Gottardo, C.; Suntres, Z. Antioxidant, Antibacterial, and Cytotoxic Activities of the Ethanolic *Origanum vulgare* Extract and Its Major Constituents. *Oxidative Med. Cell. Longev.* **2016**, *2016*, 1404505. [CrossRef]
20. Castilho, P.C.; Savluchinske-Feio, S.; Weinhold, T.S.; Gouveia, S.C. Evaluation of the antimicrobial and antioxidant activities of essential oils, extracts and their main components from oregano from Madeira Island, Portugal. *Food Control* **2012**, *23*, 552–558. [CrossRef]
21. Şahin, F.; Güllüce, M.; Daferera, D.; Sökmen, A.; Sökmen, M.; Polissiou, M.; Agar, G.; Özer, H. Biological activities of the essential oils and methanol extract of *Origanum vulgare* ssp. vulgare in the Eastern Anatolia region of Turkey. *Food Control* **2004**, *15*, 549–557. [CrossRef]
22. Botsoglou, N.; Florou-Paneri, P.; Christaki, E.; Fletouris, D.; Spais, A. Effect of dietary oregano essential oil on performance of chickens and on iron-induced lipid oxidation of breast, thigh and abdominal fat tissues. *Br. Poult. Sci.* **2002**, *43*, 223–230. [CrossRef]
23. Jianu, C.; Rusu, L.-C.; Muntean, I.; Cocan, I.; Lukinich-Gruia, A.T.; Goleţ, I.; Horhat, D.; Mioc, M.; Mioc, A.; Şoica, C.; et al. In Vitro and In Silico Evaluation of the Antimicrobial and Antioxidant Potential of *Thymus pulegioides* Essential Oil. *Antioxidants* **2022**, *11*, 2472. [CrossRef]
24. Adams, R.P. *Identification of Essential Oil Components by Gas Chromatography/Mass Spectrometry*; Allured Publishing Corporation: Carol Stream, IL, USA, 2007; Volume 456.
25. Jianu, C.; Mihail, R.; Muntean, S.G.; Pop, G.; Daliborca, C.V.; Horhat, F.G.; Nitu, R. Composition and antioxidant capacity of essential oils obtained from *Thymus vulgaris*, *Thymus pannonicus* and *Satureja montana* grown in Western Romania. *Rev. Chim.* **2015**, *66*, 2157–2160.
26. Rădulescu, M.; Jianu, C.; Lukinich-Gruia, A.T.; Mioc, M.; Mioc, A.; Şoica, C.; Stana, L.G. Chemical composition, in vitro and in silico antioxidant potential of Melissa officinalis subsp. officinalis essential oil. *Antioxidants* **2021**, *10*, 1081. [CrossRef]
27. Xin, X.; Fan, R.; Gong, Y.; Yuan, F.; Gao, Y. On-line HPLC-ABTS•+ evaluation and HPLC-MS n identification of bioactive compounds in hot pepper peel residues. *Eur. Food Res. Technol.* **2014**, *238*, 837–844. [CrossRef]
28. Berman, H.M.; Westbrook, J.; Feng, Z.; Gilliland, G.; Bhat, T.N.; Weissig, H.; Shindyalov, I.N.; Bourne, P.E. The Protein Data Bank. *Nucleic Acids Res.* **2000**, *28*, 235–242. [CrossRef]
29. Kim, S.; Chen, J.; Cheng, T.; Gindulyte, A.; He, J.; He, S.; Li, Q.; Shoemaker, B.A.; Thiessen, P.A.; Yu, B.; et al. PubChem 2023 update. *Nucleic Acids Res.* **2022**, *51*, D1373–D1380. [CrossRef] [PubMed]
30. Morris, G.M.; Huey, R.; Lindstrom, W.; Sanner, M.F.; Belew, R.K.; Goodsell, D.S.; Olson, A.J. AutoDock4 and AutoDockTools4: Automated docking with selective receptor flexibility. *J. Comput. Chem.* **2009**, *30*, 2785–2791. [CrossRef]
31. Trott, O.; Olson, A.J. AutoDock Vina: Improving the speed and accuracy of docking with a new scoring function, efficient optimization, and multithreading. *J. Comput. Chem.* **2010**, *31*, 455–461. [CrossRef]
32. Burdock, G.A. *Fenaroli's Handbook of Flavor Ingredients*; CRC Press: Boca Raton, FL, USA, 2016.
33. Popa, C.L.; Lupitu, A.; Mot, M.D.; Copolovici, L.; Moisa, C.; Copolovici, D.M. Chemical and biochemical characterization of essential oils and their corresponding hydrolats from six species of the Lamiaceae family. *Plants* **2021**, *10*, 2489. [CrossRef]
34. Moisa, C.; Copolovici, L.; Pop, G.; Lupitu, A.; Ciutina, V.; Copolovici, D. Essential oil composition, total phenolic content, and antioxidant activity-Determined from leaves, flowers and stems of *Origanum vulgare* L. Var. *Aureum*. *Sciendo* **2018**, *1*, 555–561. [CrossRef]
35. Baj, T.; Sieniawska, E.; Ludwiczuk, A.; Widelski, J.; Kiełtyka-Dadasiewicz, A.; Skalicka-Woźniak, K.; Głowniak, K. Thin-layer chromatography—Fingerprint, antioxidant activity, and gas chromatography—Mass spectrometry profiling of several *Origanum* L. species. *JPC-J. Planar Chromatogr.-Mod. TLC* **2017**, *30*, 386–391. [CrossRef]
36. Vinciguerra, V.; Rojas, F.; Tedesco, V.; Giusiano, G.; Angiolella, L. Chemical characterization and antifungal activity of *Origanum vulgare*, Thymus vulgaris essential oils and carvacrol against Malassezia furfur. *Nat. Prod. Res.* **2019**, *33*, 3273–3277. [CrossRef] [PubMed]
37. Verma, R.S.; Padalia, R.C.; Chauhan, A. Volatile constituents of *Origanum vulgare* L., 'thymol' chemotype: Variability in North India during plant ontogeny. *Nat. Prod. Res.* **2012**, *26*, 1358–1362. [CrossRef] [PubMed]
38. Laothaweerungsawat, N.; Sirithunyalug, J.; Chaiyana, W. Chemical compositions and anti-skin-ageing activities of *Origanum vulgare* L. essential oil from tropical and mediterranean region. *Molecules* **2020**, *25*, 1101. [CrossRef]
39. Antolovich, M.; Prenzler, P.D.; Patsalides, E.; McDonald, S.; Robards, K. Methods for testing antioxidant activity. *Analyst* **2002**, *127*, 183–198. [CrossRef]
40. Niki, E. Assessment of antioxidant capacity in vitro and in vivo. *Free. Radic. Biol. Med.* **2010**, *49*, 503–515. [CrossRef]
41. Foti, M.C. Use and Abuse of the DPPH• Radical. *J. Agric. Food Chem.* **2015**, *63*, 8765–8776. [CrossRef]
42. Gulcin, İ. Antioxidants and antioxidant methods: An updated overview. *Arch. Toxicol.* **2020**, *94*, 651–715. [CrossRef]

43. Gruľová, D.; Caputo, L.; Elshafie, H.S.; Baranová, B.; De Martino, L.; Sedlák, V.; Gogaľová, Z.; Poráčová, J.; Camele, I.; De Feo, V. Thymol chemotype *Origanum vulgare* L. essential oil as a potential selective bio-based herbicide on monocot plant species. *Molecules* **2020**, *25*, 595. [CrossRef]
44. Cid-Pérez, T.S.; Ávila-Sosa, R.; Ochoa-Velasco, C.E.; Rivera-Chavira, B.E.; Nevárez-Moorillón, G.V. Antioxidant and antimicrobial activity of Mexican oregano (Poliomintha longiflora) essential oil, hydrosol and extracts from waste solid residues. *Plants* **2019**, *8*, 22. [CrossRef]
45. Ozkan, G.; Baydar, H.; Erbas, S. The influence of harvest time on essential oil composition, phenolic constituents and antioxidant properties of Turkish oregano (*Origanum onites* L.). *J. Sci. Food Agric.* **2010**, *90*, 205–209. [CrossRef]
46. Shehadeh, M.; Jaradat, N.; Al-Masri, M.; Zaid, A.N.; Hussein, F.; Khasati, A.; Suaifan, G.; Darwish, R. Rapid, cost-effective and organic solvent-free production of biologically active essential oil from Mediterranean wild Origanum syriacum. *Saudi Pharm. J.* **2019**, *27*, 612–618. [CrossRef] [PubMed]
47. Morshedloo, M.R.; Mumivand, H.; Craker, L.E.; Maggi, F. Chemical composition and antioxidant activity of essential oils in *Origanum vulgare* subsp. gracile at different phenological stages and plant parts. *J. Food Process. Preserv.* **2018**, *42*, e13516. [CrossRef]
48. Sharopov, F.; Braun, M.S.; Gulmurodov, I.; Khalifaev, D.; Isupov, S.; Wink, M. Antimicrobial, antioxidant, and anti-inflammatory activities of essential oils of selected aromatic plants from Tajikistan. *Foods* **2015**, *4*, 645–653. [CrossRef]
49. Babili, F.E.; Bouajila, J.; Souchard, J.P.; Bertrand, C.; Bellvert, F.; Fouraste, I.; Moulis, C.; Valentin, A. Oregano: Chemical analysis and evaluation of its antimalarial, antioxidant, and cytotoxic activities. *J. Food Sci.* **2011**, *76*, C512–C518. [CrossRef] [PubMed]
50. Alvarez, M.V.; Ortega-Ramirez, L.A.; Silva-Espinoza, B.A.; Gonzalez-Aguilar, G.A.; Ayala-Zavala, J.F. Antimicrobial, antioxidant, and sensorial impacts of oregano and rosemary essential oils over broccoli florets. *J. Food Process. Preserv.* **2019**, *43*, e13889. [CrossRef]
51. Hazra, B.; Sarkar, R.; Biswas, S.; Mandal, N. Comparative study of the antioxidant and reactive oxygen species scavenging properties in the extracts of the fruits of *Terminalia chebula*, *Terminalia belerica* and *Emblica officinalis*. *BMC Complement. Altern. Med.* **2010**, *10*, 20. [CrossRef] [PubMed]
52. Wettasinghe, M.; Shahidi, F. Scavenging of reactive-oxygen species and DPPH free radicals by extracts of borage and evening primrose meals. *Food Chem.* **2000**, *70*, 17–26. [CrossRef]
53. Ghannay, S.; Aouadi, K.; Kadri, A.; Snoussi, M. GC-MS Profiling, Vibriocidal, Antioxidant, Antibiofilm, and Anti-Quorum Sensing Properties of *Carum carvi* L. Essential Oil: In Vitro and In Silico Approaches. *Plants* **2022**, *11*, 1072. [CrossRef]
54. Wu, X.; Hu, Q.; Liang, X.; Fang, S. Fabrication of colloidal stable gliadin-casein nanoparticles for the encapsulation of natamycin: Molecular interactions and antifungal application on cherry tomato. *Food Chem.* **2022**, *391*, 133288. [CrossRef]
55. Costa, J.d.S.; Ramos, R.d.S.; Costa, K.d.S.L.; Brasil, D.d.S.B.; Silva, C.H.T.d.P.d.; Ferreira, E.F.B.; Borges, R.d.S.; Campos, J.M.; Macêdo, W.J.d.C.; Santos, C.B.R.d. An In Silico Study of the Antioxidant Ability for Two Caffeine Analogs Using Molecular Docking and Quantum Chemical Methods. *Molecules* **2018**, *23*, 2801. [CrossRef] [PubMed]
56. de Souza Siqueira Quintans, J.; Menezes, P.P.; Santos, M.R.V.; Bonjardim, L.R.; Almeida, J.R.G.S.; Gelain, D.P.; Araújo, A.A.d.S.; Quintans-Júnior, L.J. Improvement of p-cymene antinociceptive and anti-inflammatory effects by inclusion in β-cyclodextrin. *Phytomedicine* **2013**, *20*, 436–440. [CrossRef] [PubMed]
57. Cutillas, A.-B.; Carrasco, A.; Martinez-Gutierrez, R.; Tomas, V.; Tudela, J. Thyme essential oils from Spain: Aromatic profile ascertained by GC–MS, and their antioxidant, anti-lipoxygenase and antimicrobial activities. *J. Food Drug Anal.* **2018**, *26*, 529–544. [CrossRef] [PubMed]
58. Baylac, S.; Racine, P. Inhibition of 5-lipoxygenase by essential oils and other natural fragrant extracts. *Int. J. Aromather.* **2003**, *13*, 138–142. [CrossRef]
59. Huang, C.-Y.; Chang, Y.-Y.; Chang, S.-T.; Chang, H.-T. Xanthine Oxidase Inhibitory Activity and Chemical Composition of Pistacia chinensis Leaf Essential Oil. *Pharmaceutics* **2022**, *14*, 1982. [CrossRef]

applied sciences

Article

Evaluation of Genetic Damage and Antigenotoxic Effect of Ascorbic Acid in Erythrocytes of *Orochromis niloticus* and *Ambystoma mexicanum* Using Migration Groups as a Parameter

Carlos Alvarez Moya [1,*], Mónica Reynoso Silva [1,*], Lucia Barrientos Ramírez [2] and José de Jesús Vargas Radillo [2]

[1]   Environmental Mutagenesis Laboratory, Cellular and Molecular Department, University of Guadalajara, Guadalajara 45200, JAL, Mexico

[2]   Departments of Wood, Pulp and Paper Cucei, University of Guadalajara, Guadalajara 45200, JAL, Mexico; lucia.barrientos@academicos.udg.mx (L.B.R.); vradillo@gmail.com (J.d.J.V.R.)

*   Correspondence: calvarez@cucba.udg.mx (C.A.M.); monica.reynoso@cucba.udg.mx (M.R.S.); Tel.: +52-377771121 (C.A.M.); +52-3337771121 (M.R.S.)

Abstract: The comet assay system is an efficient method used to assess DNA damage and repair; however, it currently provides the average result and, unfortunately, the heterogeneity of DNA damage loses relevance. To take advantage of this heterogeneity, migration groups (MGs) of cell comets can be formed. In this study, genetic damage was quantified in erythrocytes of *Oreochromis niloticus* and *Ambystoma mexicanum* exposed to ethyl methanesulfonate (ethyl methanesulfonate (EMS) 2.5, 5, and 10 mM over two hours) and ultraviolet C radiation (UV-C) for 5, 10, and 15 min using the tail length, tail moment, and migration group parameters. Additionally, blood cells were exposed to UV-C radiation for 5 min and treated post-treatment at 5, 10, and 15 mM ascorbic acid (AA) for two hours. With the MG parameter, it was possible to observe variations in the magnitude of genetic damage. Our data indicate that MGs help to detect basal and induced genetic damage or damage reduction with approximately the same efficiency of the tail length and tail moment parameters. MGs can be a complementary parameter used to assess DNA integrity in species exposed to mutagens.

Keywords: comet assay; migration groups; bioassays; genetic damage; ascorbic acid

Citation: Moya, C.A.; Silva, M.R.; Ramírez, L.B.; Radillo, J.d.J.V. Evaluation of Genetic Damage and Antigenotoxic Effect of Ascorbic Acid in Erythrocytes of *Orochromis niloticus* and *Ambystoma mexicanum* Using Migration Groups as a Parameter. *Appl. Sci.* **2022**, *12*, 7507. https://doi.org/10.3390/app12157507

Academic Editors: Ionel Calin Jianu and Georgeta Pop

Received: 24 June 2022
Accepted: 23 July 2022
Published: 26 July 2022

Publisher's Note: MDPI stays neutral with regard to jurisdictional claims in published maps and institutional affiliations.

1. Introduction

The comet assay system can evaluate DNA damage and repair [1–4] and, due to its great sensitivity, versatility, and rapidity [5,6], has become a valuable tool for understanding the mechanism of action of chemical, physical, or biological agents [7]. It can detect and quantify genetic damage in individual cells, allowing information to be obtained on the heterogeneity of DNA damage or repair within a subpopulation of cells [8–11]. The heterogeneity of genetic damage has been defined as the variance of DNA damage [12] and was first demonstrated by Ostling and Johanson [13] who observed that the variable cellular penetration of an anticancer substance caused the appearance of undamaged cells as well as cells with extensive DNA break. These studies indicated that the heterogeneity of genetic damage could reflect the variation of DNA lesions from one cell to another in the same population [14]. The heterogeneity in DNA migration allowed the observation of responses of small subpopulations to various treatments [12,15] as well as its relationship with cell size, differences in sensitivity to genotoxic agents, or differences in the type of manifested genetic damage [2,10,15–17].

Two classifications were reported in 1995 that used DNA damage heterogeneity to form cell groups with different degrees of damage by visual scoring [18,19]. Visual scoring implies the categorization of comets according to the intensity of the tail and among its advantages were its simplicity, speed, low cost, and the quantification of genetic damage without the use of sophisticated image analysis programs [20]. This proved to be comparable with computerized analysis programs [21]. However, despite these research advances, it is currently considered that the full potential of the comet test to detect heterogeneity in response to genetic damage has not been exploited, since only the average result is used and heterogeneity loses relevance [10,22]. Reynoso-Silva et al. [22] proposed a new parameter to detect genetic damage, called MGs, which take advantage of the heterogeneity of genetic damage and do not use only the average migration of the tail. It is based on the formation of groups of comets with a similar tail length and allows the use of visual scoring analysis and computerized programs. It was also was observed that the number of MGs is directly proportional to the genetic damage as well as the concentration of the genotoxic agent. MGs have not been deeply studied and could be related to the different sensitivities of cells in response to DNA damage and/or repair due to a specific agent [22] so it is necessary to investigate the advantages that this new parameter can have. Here, the comet assay system was used to evaluate the genetic damage induced by EMS and UV-C, as well as the post-treatment protective effects of AA in UV-C-exposed blood cells of *Oreochromis niloticus* and *Ambystoma mexicanum* using the tail length, the tail moment, and the MG parameter.

2. Materials and Methods

2.1. Chemical and Physical Agents

EMS (CAS 66-27-3) y AA (CAS 50-81-7) were obtained from Sigma Chemical Co. (Guadalajara, Jalisco, México). Dimethyl sulfoxide (DMSO, CAS 67-68-5) and disodium salt EDTA (CAS 60-00-4) were obtained from J.T. Baker (Ciudad de México, México). UV-C radiation was generated from a 254 nm wavelength lamp with a force of 10 volts, which was placed at a distance of 70 cm from the slides containing the blood cells (BCs).

2.2. Obtaining and Preparing the Sample

2.2.1. *Ambystoma Mexicanum* Blood Cells

Eight specimens of *Ambystoma mexicanum* were acquired from an Environmental Management Unit and were kept in captivity following the recommendations of the Basic Manual for the captive care of the Xochimilco de González y Zamora axolotl [23]. The axolotls were acclimatized in 50 L fish tanks with tap water under dark conditions with the following physicochemical conditions: salinity—0, temperature—$15 \pm 2\,^\circ$C, pH—7.3 ± 0.2, and dissolved oxygen—8.1 ± 0.5 mg/L. They fed red worm and brine shrimp on alternate days. From each specimen, 100 µL of the whole blood was collected through a cut of approximately 5 mm of the third right branchial arch [24] and placed in a test tube containing 4 mL of phosphate solution (PBS) (160 mM NaCl, 8 mM Na_2HPO_4, 4 mM NaH_2PO_4, and 50 mM EDTA, pH 7) and immediately centrifuged at 3000 rpm for 5 min. The supernatant was removed and the pellet was suspended in 1 mL of PBS and immediately at $4\,^\circ$C until use. Positive and negative controls were used as recommended [25].

2.2.2. *Oreochromis niloticus* Blood Cells

The fish obtained from the National Genomic Bank were acclimatized in aquariums of 5000 L of tap water following the recirculation of air, under a natural photoperiod with the following physicochemical conditions: salinity—0, temperature—$20 \pm 1\,^\circ$C, pH—7.3 ± 0.2, and dissolved oxygen—8.1 ± 0.5 mg/L. Fish were fed fish roe every other day. Blood was extracted by gill puncture from samples 15–23 cm in length. Then, 100 µL of blood sample was obtained from each individual and treated as mentioned in axolotls.

Use in animals was in accordance with the Guide for the Care and Use of Laboratory Animals (National Research Council 1996). The specimens were sacrificed following the recommendations of Close et al. [26] and NOM-062-ZOO-1999 [27]. All procedures in both species were approved by the Institutional Ethics Committee.

2.3. Blood Cell Treatment

2.3.1. EMS and UV-C-Induced Genetic Damage

Next, 50 μL of each cell suspension, previously obtained, was mixed in individual tubes *v/v* with EMS-PBS solution, taking care of the final concentrations of EMS 2.5, 5, and 10 mM and that of PBS, for 2 h at 4 °C. At the end of the time, two centrifugations were carried out at 3000 rpm for 10 min with PBS to completely eliminate the genotoxic residues. Finally, the precipitate was re-suspended in 0.5 mL of PBS and stored at 4 °C until it was used in the comet test.

To evaluate the genetic damage induced by UV-C, the slides were prepared with agar containing the BC placed at a distance of 70 cm from the UV-C lamp, exposing them separately at 1, 3, and 5 min. After the exhibition, the comet test was carried out.

2.3.2. Post-Treatment Antigenotoxic Effect of AA on Blood Cells Exposed to UV-C

To evaluate the post-treatment antigenotoxic effect of AA, the slides were prepared with agar containing the BC, as indicated above, and exposed to UV-C for 5 min, before being subsequently immersed in different concentrations of AA (5, 10 and 15 mM) for two hours. At the end of the time, three five-minute washes with distilled water were carried out to remove AA residues.

2.4. Alkaline Comet Test

The comet assay was carried out using the method of Speit and Hartmann [28]. Slides were covered with a 1% normal melting point (NMP) agarose, allowing the solidification and removal of a completely clean surface. Then, a 0.6% low melting point (LMP) agarose layer was placed on the slide. Once solidified, another layer of agarose was added (5 μL of the previously obtained cell suspension and 95 μL of the 0.5% LMP agarose). Finally, a third layer of 0.5% LMP agarose was added to cover the second layer [29]. Afterwards, the slides were immersed in lysis solution (2.5 mM NaCl, 10 mM Na_2 EDTA, 10 mM Tris-HCl, 1% lauryl sarcosinate, 1% Triton X-100, and 10% DMSO, pH 10) for 24 h at 4 °C. Subsequently, they were placed in a horizontal electrophoresis system with electrophoresis buffer (NaOH 300 mM and Na_2 EDTA 1 mM) for 45 min and electrophoresis was performed for 15 min (1.0 V/cm with an amperage of ~300 mA (Labconco, model 4333280). After electrophoresis, all slides were gently washed with distilled water to remove the alkaline solution, and all were immersed in neutralization buffer (0.4 M Tris nase, pH 7.5) for 5 min. The gels were stained with ethidium bromide (100 μL to 20 μg/mL) for 3 min and then washed three times with distilled water to remove possible bromide residues.

The slides were analyzed using a fluorescence microscope (Zeiss, model Axioskop 40) with an excitation filter of 515–560 nm. Comets were observed at 10X and their tail length was determined using the comet II test software (Zeeiz Sinoptic Mikro SA De CV México, 2012) that automatically indicates the start and end of the comet's length [30]. Approximately 100 comet cells were analyzed per sample (2 slides per experimental treatment).

To take advantage of the heterogeneity of genetic damage observed in the comet test, the grouping method proposed by Reynoso-Silva et al. was used [22]. Additionally, Microsoft Excel 2019 software was utilized to form MGs with comet cells.

2.5. Statistical Analysis

The Shapiro–Wilk and Levene tests were performed to analyze the normality of the data and its homoscedasticity. Analysis of variance based on Kruskal–Wallis ranks was performed. Dunn's posterior tests were then performed to identify which treatment made the difference. All statistical analyses were performed using the statistical software Sigma Plot 12.0 [31] and Stat-graphics of Nau [32]. $p \leq 0.05$ indicated significance.

3. Results

Basal and induced genetic damage by EMS and UV-C in erythrocytes of *Oreochromis niloticus* and *Ambystoma mexicanum* are presented in Figure 1. In *Oreochromis niloticus*, the tail moment (Figure 1A) and the tail length (Figure 1C) showed significant genetic damage ($p < 0.05$) induced by EMS and UV-C compared to the negative control. EMS showed dose-proportional genetic damage in both parameters, although the 10 mM concentration did not increase as expected, which could be attributed to excessive DNA destruction. In the UV-C case, the tail length was visually proportional to the exposure time, although the difference between UV-3 and UV-5 was not significant. Both tail moment and migration groups were also proportional to UV-1 and UV-3; however, UV-5 did not show significant differences with UV-3. The behavior can be considered uniform in the three parameters since these are not 100% compatible and lead to slight statistical variations. Similar to what was observed with tail moment and tail length, the MG (Figure 1E) shows significant genetic damage ($p < 0.05$) induced by EMS and UV-C with respect to the negative control. It was observed that the concentration of EMS 10 mM decreases the number of MGs; however, the difference is not significant with respect to EMS 2.5 and 5 mM. In the case of UV-C, differences were observed in the number of MGs with an increasing dose.

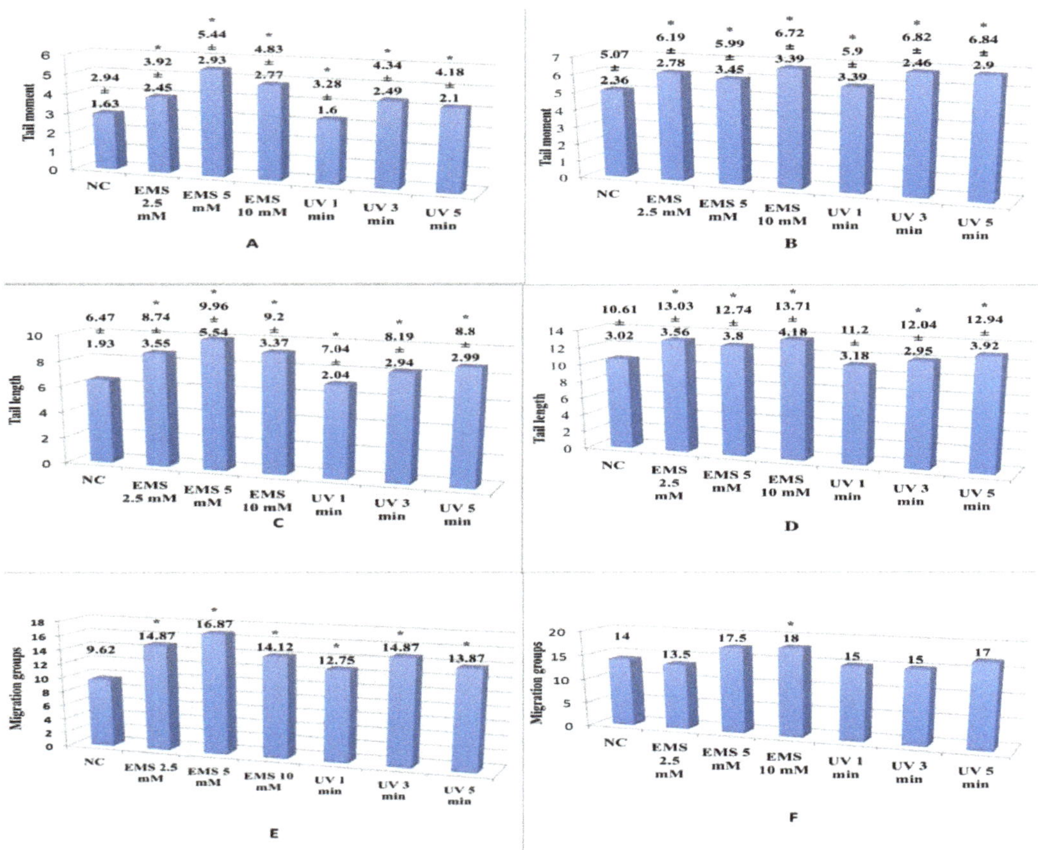

Figure 1. Induced genotoxic effect by different times or concentrations of UV-C and EMS evaluated with the parameters tail moment, tail length, and MGs in comet cells of *Oreochromis niloticus* (**A,C,E**) and *Ambystoma mexicanum* (**B,D,F**). * Significant differences ($p < 0.05$) regarding the negative control (NC).

In *Ambystoma mexicanum*, the tail moment (Figure 1B) and tail length also indicated genetic damage. Both EMS and UV-C induce significant genetic damage with respect to the negative control ($p < 0.05$). EMS presented a behavior similar to that reported in *Oreochromis niloticus* only in the tail moment parameter. No dose-response relationship is observed in the tail length. Regarding UV-C, the tail moment and the tail length were proportional to the exposure time. MGs did not reveal EMS or UV-C-induced genetic damage (Figure 1F). Only the 10 mM EMS treatment presented significant differences compared to the negative control ($p < 0.05$).

The genetic damage induced by EMS and UV-C and the protective effect of ascorbic acid after the exposure of *Oreochromis niloticus* erythrocytes to these agents is presented in Table 1. The genotoxic treatment with UV-C in *Oreochromis niloticus* erythrocytes increases significantly ($p < 0.05$) and in direct proportion to the exposure time, the average length of the tail, and the number of MGs with respect to the negative control. No significant differences were observed with respect to the negative control (NC) in the migration group containing the highest number of comet cells. The EMS showed the same behavior except for the 10 mM EMS concentration where the number of MGs (6) and the average tail length (9.2 μm) were obtained. A significant decrease ($p < 0.05$) compared to NC was also shown in the number of cells in the migration group that contained the highest number of comet cells. The antigenotoxic treatment with different concentrations of AA after exposing the cells to 5 min of UV-C showed significant differences in the number of MGs only in UV-C 5 min + 5 mM and in the case of tail length in UV-C 5 min + 5 mM and 15 mM AA. The tail length detected an increase in genetic damage compared to positive control (PC), but not a decrease, while MGs detected a significant decrease at least for UV-C 5 min + AA 5 mM. The number of MGs containing the highest number of comet cells increases significantly ($p < 0.05$) compared to the PC.

Table 1. Genotoxic and antigenotoxic treatments in *Oreochromis niloticus* erythrocytes. The genetic damage was observed using the number of MGs, migration groups with the highest comet cell number, and the mean tail length observed in each treatment.

	Genotoxic Treatment			Ascorbic Acid Antigenotoxic Post-Treatment		
	Number of Migration Groups	Migration Group Containing the Highest Comet Cells Number	Mean Tail Length in μm	Number of Migration Groups	Migration Group Containing the Highest Comet Cells Number	Mean Tail Length in μm
NC	9.6	155	6.47	9.62 *	155	6.47
EMS 2.5 mM	14.87	108 *	8.7 *			
EMS 5 mM	16.87	118 *	9.96 *			
EMS 10 mM	14.12	106	9.2 *			
UV-C 1 min	12.75	182	7.04 *			
UV-C 3 min	14.87	130	8.19 *			
UV-C 5 min	13.87	153	8.8 *			
PC (UV-C 5 min)	13.87			10.87	153	7.04
UV-C, 5 min + AA 5 mM				9.6 *	178 *	7.43 *
UV-C, 5 min + AA 10 mM				10.25	251 *	7.33
UV-C, 5 min + AA 15 mM				13.87	260 *	8.28 *

* Significant differences ($p < 0.05$) regarding the negative control.

The antigenotoxic effects of AA on blood cells *Oreochromis niloticus* and *Ambystoma mexicanum* exposed to UV-C are shown in Figure 2. In *Oreochromis niloticus*, both the tail moment (Figure 2A) and tail length (Figure 2C) show that the AA treatments at 5, 10, and

15 mM are significantly different from the negative control ($p < 0.05$), but a reduction in genetic damage with respect to the PC is observed in the parameter tail length. In the MG (3E), UV-C treatment 5 min + AA 10 mM caused a statistically significant reduction in genetic damage compared to the positive control ($p < 0.05$), even at the basal level, since no statistical difference was observed with respect to the negative control ($p < 0.05$). In *Ambystoma mexicanum*, the tail moment and tail length (Figure 2B,D) showed significant differences in the different treatments with respect to the negative control ($p < 0.05$), but only the post-treatment AA 15 mM reduced the genetic damage compared to the positive control ($p < 0.05$), although it did not reach the level of basal genetic damage. In the MG, no significant differences were observed between the treatments.

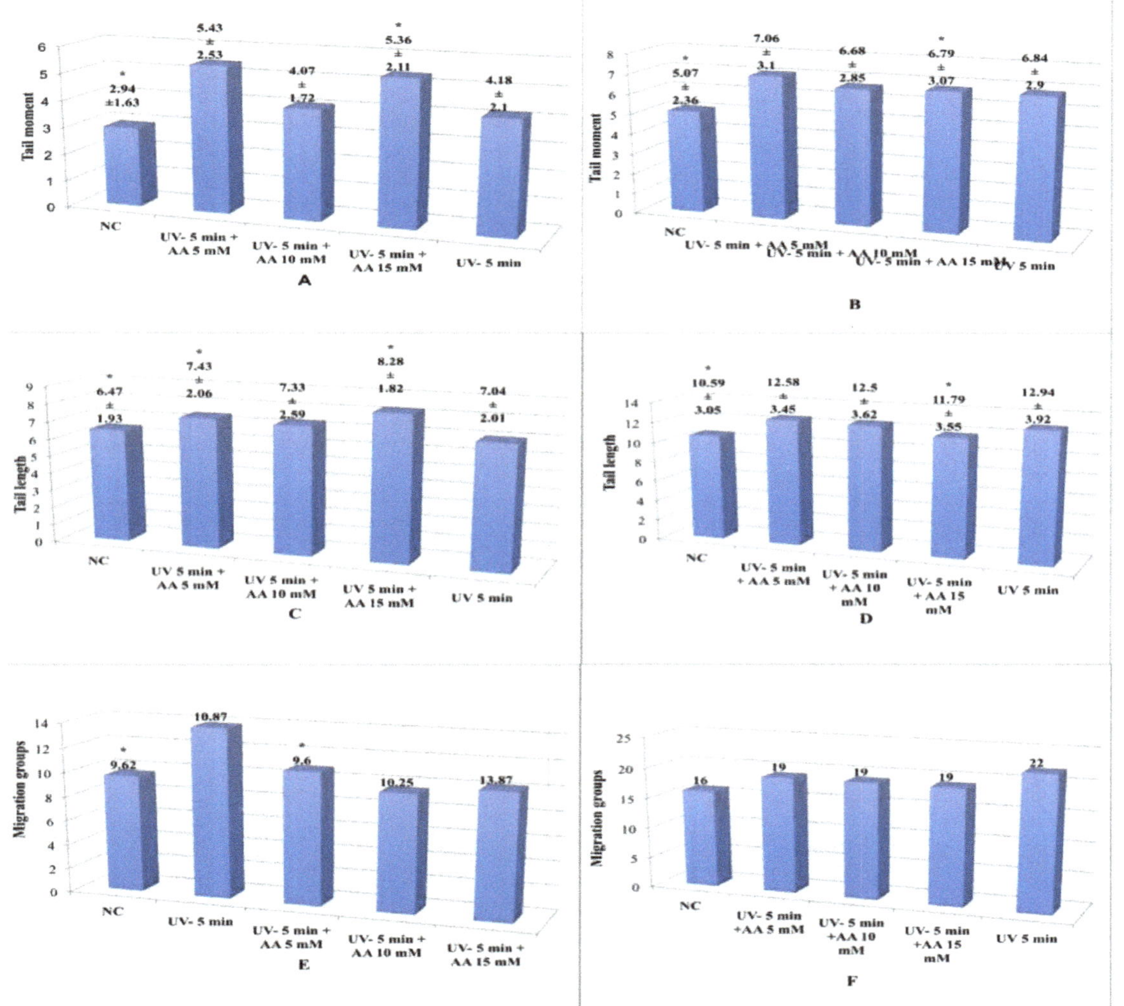

Figure 2. Genetic damage observed using the parameters: tail moment, tail length, and MGs in comet cells of *Oreochromis niloticus* (**A,C,E**) and Ambystoma mexicanum (**B,D,F**) exposed post-treatment to UV-C 5 min (PC) and UV-C + AA. * Significant differences ($p < 0.05$) regarding the negative control.

The genetic damage induced by EMS and UV-C and the protective effect of ascorbic acid after exposure of *Ambystoma mexicanum* erythrocytes to these agents is presented in Table 2. UV-C genotoxic treatment in *Ambystoma mexicanum* erythrocytes significantly increased ($p < 0.05$) the average length of the tail, but not the MG with respect to the NC. No significant differences with respect to NC were observed in the MG containing the highest number of comet cells. The EMS showed practically the same behavior, but in the 10 mM EMS, a significant decrease was observed ($p < 0.05$) with respect to the NC in the number of cells of the migration group that contained the highest number of comets.

Table 2. Genotoxic and antigenotoxic treatments in *Ambystoma mexicanumn* erythrocytes. The genetic damage was observed using the number of migration groups, the migration groups with the highest comet cell number, and mean tail length observed in each treatment.

	Genotoxic Treatment			Ascorbic Acid Antigenotoxic Post-Treatment		
	Number of Migration Groups	Migration Group Containing the Highest Comet Cells Number	Mean Tail Length in μm	Number of Migration Groups	Migration Group Containing the Highest Comet Cells Number	Mean Tail Length in μm
NC	13	123	10.61	16	123	10.59 *
EMS 2.5 mM	14	90	13.03 *			
EMS 5 mM	11	105	12.74 *			
EMS 10 mM	14	88 *	13.71 *			
UV-C 1 min	11	132	11.2			
UV-C 3 min	11	127	12.04 *			
UV-C 5 min	11	95	12.94 *			
PC (UV-C 5 min)				22	95	12.94
UV-C, 5 min + AA 5 mM				19 *	115	12.58
UV-C, 5 min + AA 10 mM				19 *	113	12.5
UV-C, 5 min + AA 15 mM				19 *	119 *	11.79 *

* Significant differences ($p < 0.05$) regarding the negative control.

The antigenotoxic treatment with different concentrations of AA after exposing the cells to 5 min of UV-C showed a significant decrease ($p < 0.05$) with respect to PC in the number of MGs in all concentrations. The tail length decreased genetic damage compared to PC only for 5 min UV-C + 15 mM AA. The number of MGs containing the highest number of comet cells significantly increased ($p < 0.05$) compared to PC.

4. Discussion

The comet test is a popular tool used to detect genetic damage [1] and there are several parameters used in this test such as tail length and tail moment [33]. Although the comet test is a well-established test in genetic research [3], its full potential to quantify heterogeneity in response to genotoxic agents has not been exploited [10]; for this reason, Reynoso-Silva et al. [22] proposed a new parameter to detect genetic damage, called the MG. In an effort to assess the ability of the MG as a parameter to detect genetic damage, *Oreochromis niloticus* and *Ambystoma mexicanum* erythrocytes were exposed in vitro to the genotoxic agents EMS and UV-C; additionally, the antigenotoxic effect of AA was assessed. The use of EMS and UV-C radiation as genotoxic models was previously reported in different test systems [34–39] and our data with tail moment and tail length confirm these reports.

The evaluation of MGs, using the tail length, as a product of the heterogeneity in the length of the comets, allows characteristic genetic damage to be evaluated in response to specific genotoxic agents [22]. For example, UV-C generates MGs with comet cells between 130 and 182, while EMS does it between 106 and 118.

Treatment with both genotoxic agents in *Oreochromis niloticus* and *Ambystoma mexicanum* erythrocytes showed that MG is as efficient parameter as tail length can be used to assess genetic damage [22]. The decrease in tail length observed in *Oreochromis niloticus* erythrocytes with 10 mM EMS treatment can be attributed to excess DNA damage, leading to the underestimation of tail length [40] (Alvarez et al., 2001) and a decrease in the number of MGs. This excess damage and the significant decrease in the number of cells in the MG containing the highest number of comet cells in *Ambystoma mexicanum* suggest that although the average genetic damage increases, the number of cells with high damage decreases. This behavior has not been previously reported. The formation of MGs makes it possible to identify cells with different levels of genetic damage and compare them in the same sample, so that they do not lose relevance as a consequence of the average effect [22].

The large amount of MGs observed in *Oreochromis niloticus* erythrocytes, after applying different treatments, is due to the fact that not all cell nuclei suffer the same damage [41]. The MG helps to identify groups with the highest number of comets and the number of cells per group [22], which could be the subpopulations most affected by each treatment and could be related to the specific damage that provokes each agent. In the erythrocytes of *Ambystoma mexicanum*, we did not observe this behavior with MGs; despite the genetic damage observed, the erythrocytes did not show great heterogeneity of genetic damage, which could be due to the fact that this species has a large number of mechanisms to preserve genome stability [42]; however, we currently have little information about it. This is of great importance [10] because a genotoxic agent might damage only a small fraction of the cells in a tissue, but it is those cells that are most likely to cause a tumor.

AA has been reported as an antigenotoxic agent [43,44]; therefore, it was used in this study and the MG was able to detect a decrease in genetic damage. In *Oreochromis niloticus*, a reduction in genetic damage was observed through the tail length and MG parameters, as has been reported for other bioassays or comet test parameters. Cases where no reduction is observed could be related to poor DNA repair in fish compared to mammals [45–48]. In *Ambystoma mexicanum* erythrocytes, it was not possible to evaluate differences between treatments and MGs also showed the ability to detect UV-C-induced genetic damage and post-treatment using AA in erythrocytes from both organisms. Furthermore, tail length could not detect a decrease in tail magnitude with respect to PC. These results agree with what was reported by Reynoso-Silva et al. [22]. The MG containing the highest number of comet cells significantly increased ($p \leq 0.05$) the number of cells compared to the PC. This may be due to the specific damage that is manifested by the action of high-concentration AA [49]. AA antigenotoxicity is notable at concentrations as low as 5 mM, as observed using MGs, but AA concentrations equal to or greater than 10 mM do not show an antigenotoxic effect.

Despite the average effect of the comet test, the number of MGs observed through the tail length parameter is directly proportional to the genetic damage, which coincides with that reported by Reynoso-Silva et al. [22], indicating that the frequency of comets per group is related to the damage and depends specifically on the genotoxic agent and not only on the concentration.

A limitation of the comet assay is that the cellular response to a genotoxic treatment not only depends on the average amount of DNA damage in the cells; it also depends on the response of a small population of cells [10]. This work shows that MGs allow the subpopulations most affected by agents that damage the genetic material to be observed, which increases the sensitivity of the comet test and complements the existing parameters.

5. Conclusions

Although there are established methods to detect DNA damage, it is necessary to make modifications to the procedures in order to increase their sensitivity and versatility. To use the MG parameter, it was possible to observe induction and a reduction in genetic damage in *Oreochromis niloticus* erythrocytes, but the same effectiveness was not observed in *Ambystoma mexicanum*, which suggests that differences in the sensitivity to genotoxic agents in the organisms studied or the presence of subpopulation cellular highly affected each treatment. MGs can be used as a complementary parameter to assess DNA integrity in species exposed to mutagens.

Author Contributions: Conceptualization, methodology, validation, formal analysis, investigation, resources, and writing—original draft preparation, C.A.M. and M.R.S.; visualization, project administration, funding acquisition, and investigation, L.B.R.; methodology and formal analysis, J.d.J.V.R. All authors have read and agreed to the published version of the manuscript.

Funding: This research was funded by UNIVERSIDAD DE GUADALAJARA, México (grant number BCyM/258/2021), and DBCyM (P3e) funded the APC.

Institutional Review Board Statement: The study was conducted and approved by the Institutional Review Board of DBCyM (DBCyM/200/2020; 11 February 2020).

Informed Consent Statement: Not applicable.

Data Availability Statement: The data presented in this study are available on request from the corresponding author.

Acknowledgments: The authors are grateful for the collaboration of CONACYT, who provided financial support to the students involved in this study.

Conflicts of Interest: The authors declare no conflict of interest.

References

1. Uno, Y.; Kojima, H.; Omori, T.; Corvi, R.; Honma, M.; Schechtman, L.; Hayashi, M. JaCVAM-organized international validation study of the in vivo rodent alkaline comet assay for detection of genotoxic carcinogens: II. Summary of definitive validation study results. *Mutat. Res. Genet. Toxicol. Environ. Mutagenesis* **2015**, *786–788*, 45–76. [CrossRef] [PubMed]
2. Koppen, G.; Azqueta, A.; Pourrut, B.; Brunborg, G.; Collins, A.R.; Langie, S.A.S. The next three decades of the comet assay: A report of the 11th International Comet Assay Workshop. *Mutagenesis* **2017**, *32*, 397–408. [CrossRef] [PubMed]
3. Kirkland, D.; Uno, Y.; Luijten, M.; Beevers, C.; van Benthem, J.; Burlinson, B.K.; Lovell, D.P. In vivo genotoxicity testing strategies: Report from the 7th International workshop on genotoxicity testing (IWGT). *Mutat. Res. Genet. Toxicol. Environ. Mutagenesis* **2019**, *847*, 403035. [CrossRef]
4. Glei, M.; Schneider, T.; Schlörmann, W. Comet assay: An essential tool in toxicological research. *Arch. Toxicol.* **2016**, *90*, 2315–2336. [CrossRef] [PubMed]
5. Martus, H.J.; Froetschl, R.; Gollapudi, B.; Honma, M.; Marchetti, F.; Pfuhler, S.; Schoeny, R.; Uno, Y.; Yauk, C.; Kirkland, D.J. Summary of major conclusions from the 7th International Workshop on Genotoxicity Testing (IWGT), Tokyo, Japan. *Mutat. Res. Genet. Toxicol. Environ. Mutagenesis* **2020**, *852*, 503134. [CrossRef] [PubMed]
6. Rodríguez-Rey, A.; Noris-García, E.; Fundora-Torres, M.T. Principios y relevancia del ensayo cometa. *Rev. Cuba. Investig. Bioméd.* **2016**, *35*, 184–194.
7. Singh, N.P. The comet assay: Reflections on its development, evolution and applications. *Mutat. Res. Rev. Mutat.* **2016**, *767*, 23–30. [CrossRef]
8. Hughes, C.M.; Lewis, S.E.; Mckelvey-Martin, V.J.; Thompson, W. Reproducibility of human sperm DNA measurements using the alkaline single cell gel electrophoresis assay. *Mutat. Res. Fundam. Mol. Mech. Mutagenesis* **1997**, *374*, 261–268. [CrossRef]
9. Olive, P.L. The Comet Assay: An Overview of Techniques. In Situ Detection of DNA Damage. *Methods Mol. Biol.* **2002**, *203*, 179–194. [CrossRef]
10. Olive, P.L.; Durand, R.E. Heterogeneity in DNA damage using the comet assay. *Cytometry Part A* **2005**, *66*, 1–8. [CrossRef]
11. Nahon, P.; Allaire, M.; Nault, J.C.; Paradis, V. Characterizing the mechanism behind the progression of NAFLD to hepatocellular carcinoma. *Hepatic Oncol.* **2020**, *7*, HEP36. [CrossRef]
12. Seidel, C.; Lautenschläger, C.; Dunst, J.; Müller, A.C. Factors influencing heterogeneity of radiation-induced DNA-damage measured by the alkaline comet assay. *Radiat. Oncol.* **2012**, *7*, 61. [CrossRef]
13. Östling, O.; Johanson, K. Bleomycin, in Contrast to Gamma Irradiation, Induces Extreme Variation of DNA Strand Breakage from Cell to Cell. International Journal of Radiation Biology and Related Studies in Physics. *Chem. Med.* **1987**, *52*, 683–691. [CrossRef]
14. Tronov, V.A.; Grin'ko, E.V.; Afanas'ev, G.G.; Filippovich, I.V. Study of DNA damage and heterogeneity of cells by a gel microelectrophoresis method. *Biofizika* **1994**, *39*, 810–819.

15. Olive, P.L.; Banáth, J.P.; Durand, R.E. Detection of subpopulations resistant to DNA-damaging agents in spheroids and murine tumours. *Mutat. Res. Fundam. Mol. Mech. Mutagenesis* **1997**, *375*, 157–165. [CrossRef]

16. Koppen, G.; Toncelli, L.; Triest, L.; Verschaeve, L. The comet assay: A tool to study alteration of DNA integrity in developing plant leaves. *Mech. Ageing Dev.* **1999**, *110*, 13–24. [CrossRef]

17. Alvarez-Moya, C.; Reynoso-Silva, M.; Canales-Aguirre, A.A.; Chavez-Chavez, J.O.; Castañeda-Vázquez, H.; Feria-Velasco, A.I. Heterogeneity of genetic damage in cervical nuclei and lymphocytes in women with different levels of dysplasia and cancer-associated risk factors. *Biomed. Res. Int.* **2015**, *2015*, 293408. [CrossRef]

18. Kobayashi, H. A comparison between manual microscopic analysis and computerized image analysis in the single cell gel electrophoresis. *MMS Commun.* **1995**, *3*, 103–115. Available online: https://scholar.google.com/scholar?hl=es&as_sdt=0%2C5&q=Kobayashi%2C+H.+%281995%29.+A+comparison+between+manual+microscopic+analysis+and+computerized+image+analysis+in+the+single+cell+gel+electrophoresis.+MMS+Commun.%2C+3%2C+103-115.&btnG= (accessed on 1 August 2021).

19. Collins, A.R.; Ai-guo, M.; Duthie, S.J. The kinetics of repair of oxidative DNA damage (strand breaks and oxidised pyrimidines) in human cells. *Mutat. Res. DNA Repair.* **1995**, *336*, 69–77. [CrossRef]

20. Collins, A.; Dušinská, M.; Franklin, M.; Somorovská, M.; Petrovská, H.; Duthie, S.; Fillion, L.; Panayiotidis, M.; Rašlová, K.; Vaughan, N. Comet assay in human biomonitoring studies: Reliability, validation, and applications. *Environ. Mol. Mutagen.* **1997**, *30*, 139–146. [CrossRef]

21. Bolognesi, C.; Cirillo, S.; Chipman, J.K. Comet assay in ecogenotoxicology: Applications in *Mytilus* sp. *Mutat. Res. Genet. Toxicol. Environ. Mutagenesis* **2019**, *842*, 50–59. [CrossRef]

22. Reynoso-Silva, M.; Álvarez-Moya, C.; Ramírez-Velasco, R.; Sámano-León, A.G.; Arvizu-Hernández, E.; Castañeda-Vásquez, H.; Ruíz-Lopez, M.A. Migration Groups: A Poorly Explored Point of View for Genetic Damage Assessment Using Comet Assay in Human Lymphocytes. *Appl. Sci.* **2021**, *11*, 4094. [CrossRef]

23. Gonzáles, H.M.; Zamora, E.S. *Manual Básico Para el Cuidado en Cautiverio del Axolote de Xochimilco (Ambystoma mexicanum)*; Editorial; Universidad Nacional Autónoma de México Instituto de Biología: Ciudad de Méxoco, México, 2014; pp. 18–25. Available online: http://www.ibiologia.unam.mx/barra/publicaciones/manual_axolotes.pdf (accessed on 1 January 2020).

24. Barriga-Vallejo, C.; Aguilera, C.; Cruz, J.; Banda-Leal, J.; Lazcano, D.; Mendoza, R. Ecotoxicological Biomarkers in Multiple Tissues of the Neotenic Ambystoma spp. for a Non-lethal Monitoring of Contaminant Exposure in Wildlife and Captive Populations. *Water Air Soil Pollut.* **2017**, *228*, 1–11. [CrossRef]

25. Møller, P. The comet assay: Ready for 30 more years. *Mutagenesis* **2018**, *33*, 1–7. [CrossRef] [PubMed]

26. Close, B.; Banister, K.; Baumans, V.; Bernoth, E.M.; Bromage, N.; Bunyan, J.; Erhardt, W.; Flecknell, P.; Gregory, N.; Hackbarth, H.; et al. Recommendations for euthanasia of experimental animals: Part 2. *Lab. Anim. UK* **1997**, *31*, 1–32. [CrossRef] [PubMed]

27. NOM-062-ZOO-1999 Norma Oficial Mexicana. Especificaciones Técnicas para la Producción, Cuidado y Uso de los Animales de Laboratorio. México, D.F.: Secretaría de Agricultura, Ganadería, Desarrollo Rural, Pesca y Alimentación. 1999. Available online: https://www.fmvz.unam.mx/fmvz/principal/archivos/062ZOO.PDF (accessed on 11 February 2021).

28. Speit, G.; Hartmann, A. The comet assay (single-cell gel test). In *DNA Repair Protocols*; Humana Press: Totowa, NJ, USA, 1999; pp. 203–212. [CrossRef]

29. Singh, N.P.; McCoy, M.T.; Tice, R.R.; Schneider, E.L. A simple technique for quantitation of low levels of DNA damage in individual cells. *Exp. Cell Res.* **1988**, *175*, 184–191. [CrossRef]

30. Ma, T.H.; Cabrera, G.L.; Cebulska-Wasilewska, A.; Chen, R.; Loarca, F.; Vandenberg, A.L.; Salamone, M.F. Tradescantia stamen hair mutation bioassay. *Mutat. Res. Fundam. Mol. Mech. Mutagenesis* **1994**, *310*, 211–220. [CrossRef]

31. Parcela, S. Sigmaplot (12.0) [Software Data Analysis]. Software. Systal. 2011. Available online: http://www.sigmaplot.co.uk/products/sigmaplot/produpdates/prod-updates18.php (accessed on 1 March 2021).

32. Nau, R. Statgraphics (Versión 5) [Overview & tutorial guide]. Fuqua School of Business. Duke University. 2005. Available online: https://statgraphics.net/ (accessed on 1 April 2021).

33. Kumaravel, T.; Jha, A.N. Reliable Comet assay measurements for detecting DNA damage induced by ionising radiation and chemicals. *Mutat. Res. Genet. Toxicol. Environ. Mutagenesis* **2006**, *605*, 7–16. [CrossRef]

34. Kodym, A.; Afza, R. Physical and Chemical Mutagenesis. In *Plant Functional Genomics*; Humana Press: Totowa, NJ, USA, 2003; pp. 189–202. [CrossRef]

35. Wyatt, M.D.; Pittman, D.L. Methylating Agents and DNA Repair Responses: Methylated Bases and Sources of Strand Breaks. *Chem. Res. Toxicol.* **2006**, *19*, 1580–1594. [CrossRef]

36. Gocke, E.; Bürgin, H.; Müller, L.; Pfister, T. Literature review on the genotoxicity, reproductive toxicity, and carcinogenicity of ethyl methanesulfonate. *Toxicol. Lett.* **2009**, *190*, 254–265. [CrossRef]

37. Buonanno, M.; Stanislauskas, M.; Ponnaiya, B.; Bigelow, A.W.; Randers-Pehrson, G.; Xu, Y.; Shuryak, I.; Smilenov, L.; Owens, D.M.; Brenner, D.J. 207-nm UV Light—A Promising Tool for Safe Low-Cost Reduction of Surgical Site Infections. II: In-Vivo Safety Studies. *PLoS ONE* **2016**, *11*, e0138418. [CrossRef]

38. Byrns, G.; Barham, B.; Yang, L.; Webster, K.; Rutherford, G.; Steiner, G.; Petras, D.; Scannell, M. The uses and limitations of a hand-held germicidal ultraviolet wand for surface disinfection. *J. Occup. Environ. Hyg.* **2017**, *14*, 749–757. [CrossRef]

39. Narita, K.; Asano, K.; Morimoto, Y.; Igarashi, T.; Hamblin, M.R.; Dai, T.; Nakane, A. Disinfection and healing effects of 222-nm UVC light on methicillin-resistant Staphylococcus aureus infection in mouse wounds. *J. Photochem. Photobiol. B Biol.* **2018**, *178*, 10–18. [CrossRef]

40. Alvarez-Moya, C.; Santerre-Lucas, A.; Zúñiga-González, G.; Torres-Bugarín, O.; Padilla-Camberos, E.; Feria-Velasco, A. Evaluation of genotoxic activity of maleic hydrazide, ethyl methane sulfonate, and N-nitroso diethylamine in Tradescantia. *Salud Pública México* **2001**, *43*, 6–12. Available online: http://www.scielo.org.mx/scielo.php?script=sci_arttext&pid=S0036-36342001000600007 (accessed on 30 March 2022). [CrossRef]

41. Olive, P.L.; Banáth, J.P. The comet assay: A method to measure DNA damage in individual cells. *Nat. Protoc.* **2006**, *1*, 23–29. [CrossRef]

42. García-Lepe, U.O.; Cruz-Ramírez, A.; Bermúdez-Cruz, R.M. DNA repair during regeneration in *Ambystoma mexicanum*. *Dev. Dynam.* **2021**, *250*, 788–799. [CrossRef]

43. Yen, G.C.; Duh, P.D.; Tsai, H.L. Antioxidant and pro-oxidant properties of ascorbic acid and gallic acid. *Food Chem.* **2002**, *79*, 307–313. [CrossRef]

44. Maeda, J.; Allum, A.J.; Mussallem, J.T.; Froning, C.E.; Haskins, A.H.; Buckner, M.A.; Miller, C.D.; Kato, T.A. Ascorbic Acid 2-Glucoside Pretreatment Protects Cells from Ionizing Radiation, UVC, and Short Wavelength of UVB. *Genes* **2020**, *11*, 238. [CrossRef]

45. Miller, M.R.; Blair, J.B.; Hinton, D.E. DNA repair synthesis in isolated rainbow trout liver cells. *Carcinogenesis* **1989**, *10*, 995–1001. [CrossRef]

46. Bailey, G.S.; Williams, D.E.; Hendricks, J.D. Fish Models for Environmental Carcinogenesis: The Rainbow Trout. *Environ. Health Perspect.* **1996**, *104*, 5–21. [CrossRef]

47. Willett, K.L.; Lienesch, L.A.; Di Giulio, R.T. No detectable DNA excision repair in UV-exposed hepatocytes from two catfish species. *Comp. Biochem. Physiol. Part C Toxicol. Pharm.* **2001**, *128*, 349–358. [CrossRef]

48. Kienzler, A.; Bony, S.; Devaux, A. DNA repair activity in fish and interest in ecotoxicology: A review. *Aquat. Toxicol.* **2013**, *134–135*, 47–56. [CrossRef]

49. Nefić, H. The genotoxicity of vitamin C in vitro. *Bosnian. J. Basic Med. Sci.* **2008**, *8*, 141–146. [CrossRef]

applied sciences

MDPI

Article

Effect of Postharvest UV-C Radiation on Nutritional Quality, Oxidation and Enzymatic Browning of Stored Mature Date

Saliha Dassamiour [1,2], Ourida Boujouraf [2], Linda Sraoui [2], Mohamed Sabri Bensaad [1,3], Ala eddine Derardja [4], Sultan J. Alsufyani [5], Rokayya Sami [6,*], Eman Algarni [6], Huda Aljumayi [6] and Amani H. Aljahani [7]

[1] Laboratory of Biotechnology of Bioactive Molecules and Cellular Physiopathology (LBMBPC), Faculty of Natural and Life Sciences, University Batna 2, Fesdis, Batna 05078, Algeria; s.dassamiour@univ-batna2.dz (S.D.); m.bensaad@univ-batna2.dz (M.S.B.)

[2] Department of Microbiology and Biochemistry, Faculty of Natural and Life Sciences, University Batna 2, 53, Route de Constantine, Fesdis, Batna 05078, Algeria; katrykorina@gmail.com (O.B.); onzukaeikich87@yahoo.com (L.S.)

[3] Laboratory of Cellular and Molecular Physio-Toxicology-Pathology and Biomolecules (LPTPCMB), Department of Biology of Organisms, Faculty of Natural and Life Sciences, University Batna 2, Fesdis, Batna 05078, Algeria

[4] Laboratoire Bioqual, Institut National de l'Alimentation, la Nutrition et des Technologies Agro-Alimentaires (INATAA), Université Frères Mentouri Constantine 1, Route de Ain El-Bey, Constantine 25000, Algeria; alaeddine.derardja@umc.edu.dz

[5] Department of Physics, College of Sciences, Taif University, P.O. Box 11099, Taif 21944, Saudi Arabia; s.alsufyani@tu.edu.sa

[6] Department of Food Science and Nutrition, College of Sciences, Taif University, P.O. Box 11099, Taif 21944, Saudi Arabia; eman1400@tu.edu.sa (E.A.); huda.a@tu.edu.sa (H.A.)

[7] Department of Physical Sport Science, College of Education, Princess Nourah bint Abdulrahman University, P.O. Box 84428, Riyadh 11671, Saudi Arabia; ahaljahani@pnu.edu.sa

* Correspondence: rokayya.d@tu.edu.sa

Citation: Dassamiour, S.; Boujouraf, O.; Sraoui, L.; Bensaad, M.S.; Derardja, A.e.; Alsufyani, S.J.; Sami, R.; Algarni, E.; Aljumayi, H.; Aljahani, A.H. Effect of Postharvest UV-C Radiation on Nutritional Quality, Oxidation and Enzymatic Browning of Stored Mature Date. *Appl. Sci.* **2022**, *12*, 4947. https://doi.org/10.3390/app12104947

Academic Editors: Georgeta Pop and Ionel Calin Jianu

Received: 27 April 2022
Accepted: 10 May 2022
Published: 13 May 2022

Publisher's Note: MDPI stays neutral with regard to jurisdictional claims in published maps and institutional affiliations.

Abstract: The effect of three doses of UV-C radiation (1, 3 and 6 kJ m^{-2}) on conservation potential after harvest of the Deglet-Nour date for five months of storage at 10 °C was studied. Contents of water, total sugar, carotenoids, proteins, total polyphenols, flavonoids and condensed tannins, as well as browning index, enzyme activities of polyphenoloxidase and peroxidase and antioxidant capacity of samples were monitored during storage using standard methods. Doses 1 and 6 kJ m^{-2} significantly slowed the water loss of samples until the second month of storage, with 17.68% and 16.02% of loss compared to control (31.45%). In the second month of storage, a significant increase in carotenoids was also observed for doses 1 and 6 kJ m^{-2}, with values of 4.17 and 4.02 mg kg^{-1} versus the control (3.45 mg kg^{-1}), which resulted in deceleration in carotenoid degradation. A gradual decrease in total sugar content was noted for all samples; it was slower within irradiated ones at the second month, where the slowing down of sugar consumption was significantly favored in the samples irradiated at 1 and 6 kJ m^{-2}, which was marked by decreases of 4.98% and 4.57% versus 8.96% in the control. Protein content of irradiated samples (3 and 6 kJ m^{-2}) increased at the third month, giving 1.70 and 2.41 g kg^{-1} compared to 1.29 g kg^{-1} for the control. An important decrease in enzymatic activity of polyphenoloxidase was detected, in addition to a fluctuation in peroxidase during storage. The browning index was lower in the irradiated sample until the fourth month of storage, where the result was more significant. An increase in the content of condensed tannins was detected, especially during the two first months, and while the significant increase in the content of flavonoids was read at the last month, it was detected from the first month for polyphenols. This was more significant for the highest dose, were the content reached 0.537 g kg^{-1} versus 0.288 g kg^{-1} in control at the first month. A dose-dependent increase in antiradical activity was noted during the last months of storage, while the increase in iron-reducing power was detected at the first month. UV-C delayed installation of Deglet-Nour browning and enriched it with antioxidants.

Keywords: *Phoenix dactylifera* L.; ultraviolet radiation; phenolics; polyphenoloxidase; darkening

1. Introduction

Phoenix dactylifera L. is a vital plant for the desert region in Algeria, where it plays a very important ecological and socio-economic role for the population living in this region. Indeed, the fruit is the subject of important commercial activity, in particular the famous variety 'Deglet-Nour', which is a worldwide commercial variety of choice because it is flavorful, mellow and has an exquisite taste. Deglet-Nour derives its vernacular name from the expression finger of light, it is the most popular cultivar in all the palm groves of southeastern Algeria, and it reaches maturity in October–November.

During its maturation, it goes through four essential stages called Kimri, Khalal, Rutab and Tamar, where the fruit gradually loses its water and intensifies in color. In the final Tamar stage, Deglet-Nour undergoes a decrease in astringency, and it is light red in color with yellowish tints. Polyphenols are micronutrients, particularly abundant in fruit, grains and vegetables [1]. Their involvement in biological activities of dates was confirmed [2,3]. Degradation of fruit phenolics alters their nutritional quality, especially since these molecules characterize the sensorial properties of fruits and vegetables [4]. Their preservation within fruit and vegetables is a great necessity, since their oxidation by polyphenoloxidase (PPO) and peroxidase (POD) actions leads to installation of fruit browning [5], which drops their market value. The first experiments of food irradiation were carried out to avoid germination of fruits and vegetables. Moreover, food irradiation can control insect infestations and delay ripening of fresh fruit and vegetables [6]. Otherwise, it should be noted that UV-C radiation has not been tested to slow browning and degradation of related molecules, phenolics, in fruit and vegetables.

The main objective of this work was to study the effect of UV-C radiation on the quality of the Deglet-Nour date by monitoring changes in some of its organoleptic and pharmacological properties during five months of storage after irradiation. The color alteration was targeted by study of fruit browning, in addition to other quality criteria such as changes in antioxidant activity and content of major components.

2. Materials and Methods

2.1. Chemical, Reagents and Materials

All solvents and reagents used in the present study were purchased from Sigma Aldrich, Steinheim, Germany. This includes: acetone, methanol, β-carotene, phenol, sulfuric acid, glucose, polyvinylpyrrolidone, Bradford reagent, bovine serum albumin (BSA), 4-methyl-catechol, guaiacol, sodium fluoride (NaF), BHT, aluminum trichloride (AlCl$_3$), quercetin, vanillin, catechin, gallic acid, Folin–Ciocalteu reagent, 2,2-diphenyl 1-picrylhydrazyl (DPPH), potassium ferricyanide (K$_3$[Fe(CN)$_6$]), sodium carbonate (Na$_2$CO$_3$), KH$_2$PO$_4$, K$_2$HPO$_4$, trichloroacetic acid (TCA), ferric chloride (FeCl$_3$).

The spectrophotometer used in quantification of molecules and monitoring of enzymatic and antioxidant activities was anUV/VIS-7220G, Shenzhen ThreeNH Technology, Shenzhen, China; the centrifuge was an EBA 200 model, Hettich, Montévrain, France.

2.2. Plant Material and Experimental Design

A semi-soft date fruit, 'Deglet-Nour', was used in this study. Samples were harvested at full maturity, at Tamar stage of ripening in November 2017. They were from a young 10-year-old orchard in the southeast of Algeria. Dates were sorted to keep only the fruits with homogeneity of color and ripeness degree. Fruits were divided into five batches, three of them each received a different dose of UV-C ray (1 kJ m^{-2}, 3 kJ m^{-2} and 6 kJ m^{-2}); the fourth was used as control sample while the fifth was used to determine all experiment parameters before irradiation by D$_0$ analysis. The UV-C apparatus, used to irradiate samples, consisted of two lamps emitting quasi-monochromatic UV radiation at 254 nm and having a normal output power of 60 W. All the samples were placed in perforated low density polyethylene bags and stored at 10 °C until the time of analysis. Sample storage was up to five months. In addition to D$_0$ analysis, other analyses were carried out each month to ensure the monitoring of evolution of contents of water, carotenoids, total sugars, proteins, total polyphenols, flavonoids and condensed tannins, as well as browning index, enzyme activities of polyphenoloxidase and peroxidase, in addition to the antiradical activity and reducing power.

2.3. Visual Analysis and Determination of Water Content

Monitoring of changes in the external color of the fruit was carried out by taking pictures just after irradiation and at the end of storage. The water content was determined by drying a known weight of the sample in an isothermal oven (Red LINE, Paris, France) at 80 °C $\pm$ 2 °C and at atmospheric pressure until obtaining a constant mass sample. The water content is equal to the loss of fruit mass in the measurement conditions.

2.4. Determination of Carotenoid Content

The extraction and quantification of carotenoids were carried out by macerating 1 g of pulp of each sample in 2 mL of acetone containing 0.05 mL of BHT (0.1%) for 24 h. the mixture was filtered, and the operation was repeated until a clear coloration of the extract was observed. Absorbance of the filtrate was determined at 450 nm against a blank (acetone). The concentrations of carotenoids were estimated by reference to the standard curve of β-carotene and the results were expressed in milligrams per kilogram of fruit dry weight (mg kg^{-1}).

2.5. Determination of Total Sugar Content

The approach of Dubois et al. [7] was followed to determine total sugar using phenol and concentrated sulfuric acid. An amount of 0.2 g of date pulp for each sample was macerated in 3 mL of distilled water for 24 h, and the extract was centrifuged. An intake of 50 µL of the supernatant was diluted by addition of 10 mL of distilled water. An amount of 1 mL from solution of each sample, 0.05 mL of the phenol solution (80%) and 2 mL of concentrated sulfuric acid were homogenized and incubated at 30 °C for 15 min, and the reaction was stopped by a stream of cold water. Absorbance was measured at 490 nm.

The concentrations of total sugars in date extracts were calculated by reference to a calibration curve obtained using glucose; the results were expressed in grams glucose equivalent per kilogram dry weight.

2.6. Determination of Total Protein Content, PPO and POD Activities

Bradford method [8] was used to estimate total protein content in Deglet-Nour samples. Preparation of protein extracts was performed using the Lichanporn and Techavuthiporn [9] approach; 0.4 g of date paste was macerated in 4 mL of 0.05 M phosphate buffer pH 6.8 containing 0.08 g of polyvinylpyrrolidone. After 24 h, the mixture was centrifuged at 8000× g for 10 min. An amount of 250 µL of Bradford reagent was added to 1 mL of the supernatant, stirred and incubated for 30 min in the dark at ambient temperature and absorbance was measured at 595 nm. Protein concentrations were calculated by reference to the calibration curve obtained using bovine serum albumin (BSA) as standard. The results were expressed in grams of BSA equivalent per kilogram of dry weight (g kg^{-1}).

Both enzymes' activity was assayed using the protein extracts. PPO activity was assessed by measuring oxidation of 4-methyl-catechol as substrate at 410 nm, as described by Dassamiour et al. [4]. The increase in the absorbance at 410 nm at 30 °C was automatically recorded for 5 min. POD activity was determined using guaiacol as substrate, as previously reported by Daas Amiour and Hambaba [5]. One unit of enzyme activities was defined as the amount of the enzyme which caused a change of 0.01 in absorbance/min. Results were expressed in enzymatic unit per gram of protein (U g^{-1}).

2.7. Determination of Browning Index and Contents of Flavonoids and Condensed Tannins

The browning index is a good indicator of color intensification of samples. It was determined by homogenizing 1 g of Deglet-Nour pulp in 2 mL of methanol containing NaF (4 mM). The homogenate was macerated for 30 min then filtered and centrifuged at

$10,000 \times g$ for 15 min. The supernatant was used directly to measure absorbance at 430 nm, as a browning index per one gram of dry weight.

The extract was also used in flavonoid and tannin quantification; the first was carried out using the $AlCl_3$ method as reported by Daas Amiour et al. [3]. Flavonoid concentration was calculated by reference to the calibration curve obtained using quercetin. The results were expressed in gram quercetin equivalent per kilogram of fruit pulp's dry weight $(g\ kg^{-1})$. The vanillin assay was carried out to determine condensed tannin content, which was calculated using a calibration curve of catechin.

2.8. Determination of Total Polyphenol Content

Date paste (0.5 g) of different irradiation doses was macerated in a mixture of solvents (30 mL H_2O/30 mL methanol/40 mL acetone). After 24 h of maceration, the extract was centrifuged at $10,000 \times g$ for 10 min. Total phenolics content was assayed using Folin–Ciocalteu reagent, following the method described by Daas Amiour and Hambaba [5]. Absorbance was read at 760 nm and total phenolic concentrations were expressed in gram gallic acid equivalents per kilogram of dry weight $(g\ kg^{-1})$.

2.9. Evaluation of Antioxidant Activity

Antioxidant activity was conducted on the extracts prepared for the determination of total phenolic compounds, using two tests.

2.9.1. Anti-Free-Radical Activity (DPPH Test)

Antiradical capacity was evaluated by the approach of Masuda et al. [10] using the extracts prepared for determination of phenol content. A volume of 2.45 mL of each extract was introduced into a test tube, to which 50 µL of 5 mM DPPH methanol solution was added and after 30 min of incubation at room temperature, the optical density was determined at 517 nm, against methanol as blank. The anti-free-radical activity was estimated in percentage using the following formula:

$$\text{DPPH scavenging activity}(\%) = \left[\frac{\left(\text{Abs}_{\text{control}} - \text{Abs}_{\text{sample}} \right)}{\text{Abs}_{\text{control}}} \right] \times 100 \tag{1}$$

2.9.2. Ferric-Reducing Power Test

Oyaizu's [11] method was used for this approach. A volume of 200 µL of different dose extracts was mixed with 200 µL of 0.2 M phosphate buffer solution (pH 6.6) and 200 µL of solution of $K_3Fe(CN)_6$ (1%). The whole was incubated at 50 °C for 30 min then 200 µL of 10% TCA were added to stop the reaction; preparations were centrifuged at $3750 \times g$ for 10 min. An amount of 400 µL of supernatant was combined with 400 µL of distilled water and 90 µL of aqueous solution of $FeCl_3$ (0.1%) and after 1 h of incubation, the absorbance was read at 700 nm. Note that a stronger absorbance indicates increased reducing power.

2.10. Statistical Analysis

Each measurement was performed in triplicate for each sample starting from the extraction step. Data were analyzed by Graph pad prism using the One-way ANOVA test followed by Tukey's post hoc test. Values are considered significant at $p < 0.01$. Correlation tests were performed using Excel.

3. Results and Discussion

3.1. Changes in External Color and in Carotenoid Content

The visual appearance of all samples, just after irradiation and at the end of every month of storage, is given in Figure 1 as photos taken in those times. Indeed, it appears clearly that browning was less pronounced on irradiated samples. This result can be explained by the inactivation, by ultraviolet radiation, of the polyphenoloxidase located in the surface tissue of the treated fruits. The intensification of the brown coloration in the irradiated samples is not visually distinguishable between the three doses in this case, but the results which follow have made it possible to demonstrate the difference.

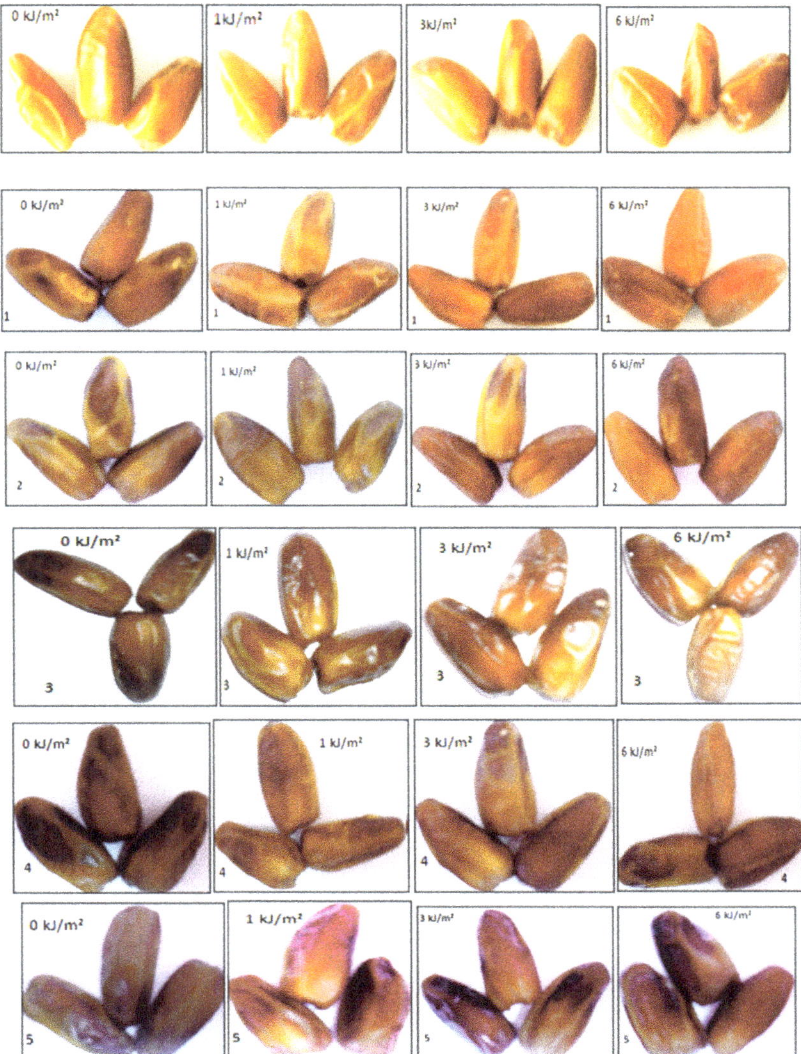

Figure 1. Changes in color appearance of date, control and UV-C treated samples, during 5 months of storage at 10 °C (the number at the left corner of each photo indicates the storage month no.).

Changes in carotenoid levels revealed a moderate elevation in treated samples during the first two months of storage versus a slow decline in all samples during the remainder of the period (Table 1), indicating progressive degradation of these pigments. In the second month the increase was significant; it was observed for doses 1 and 6 kJ m^{-2} with the values of 4.17 and 4.02 mg kg^{-1} versus the control (3.45 mg kg^{-1}). Similar results were observed in fresh-cut carrot [12]. Carotenoids are subject, indeed, to isomerization and oxidation during processing and storage of foods because they are highly unsaturated. Degradation of carotenoids increases with length and severity of the processing conditions, temperature, storage time, transmission of light and exposition to O_2 [13]. Moreover, the increase in their level may be due to their biosynthesis in response to stress caused by UV-C. It has been confirmed in several studies that the level of carotenoids increases following exposure to abiotic stress, such as drought, light, etc. The most important carotenoids identified in this cultivar of date are ß-carotene and lutein [14].

Table 1. Changes in carotenoid contents of Deglet-Nour irradiated samples and control during five months of storage at 10 °C.

Storage Period (Months)	Irradiation Doses (kJ m^{-2})	Carotenoid Content (mg kg^{-1})
0	-	3.46 ± 0.33
1	0	3.41 ± 0.22
	1	3.49 ± 0.20
	3	3.47 ± 0.10
	6	3.62 ± 0.07
2	0	3.45 ± 0.12 [a]
	1	4.17 ± 0.07 [b]
	3	3.51 ± 0.21 [a]
	6	4.02 ± 0.06 [b]
3	0	3.35 ± 0.16
	1	3.58 ± 0.02
	3	3.48 ± 0.25
	6	3.77 ± 0.17
4	0	3.33 ± 0.25
	1	3.44 ± 0.25
	3	3.00 ± 0.17
	6	3.56 ± 0.13
5	0	3.20 ± 0.17
	1	3.06 ± 0.03
	3	2.86 ± 0.06
	6	2.97 ± 0.08

Values are expressed as mean $\pm$ SEM (n = 3); [a,b]: different letters indicate a significant difference between the four samples at each storage period ($p < 0.01$), while the absence of the letters indicates the absence of difference.

However, Freitas et al. [15] found that UV-C radiation significantly increased the content of carotenoids in pineapple core, which is agreement with the obtained result.

3.2. Changes in Moisture, Total Sugar and Protein Contents

In order to note the irradiation effect on water retention capacity, the variation in humidity rate within the fruit was measured, as it is a quality criterion of dates. The changes in water content of Deglet-Nour samples were obtained during storage period (Figure 2).

Water content decreased in all treated samples and the control. Expressed in percentage, it ranged from 25.13 in the first month to 12.42% at the end of the storage period. Some authors have explained the decrease in moisture by evaporation of date water following the destruction of the cellulose walls and the pecto-cellulosic substances [16], probably during postharvest handling. The decrease in moisture content is less pronounced in the case of irradiated samples when referring to control, especially for doses 1 and 6 kJ m^{-2} during the first two months of storage, which can be explained by water retention and reduction in fruit sweating after irradiation. After one month of storage, the decreases were 2.52%, 10% and 5.82%, corresponding to doses of 1, 3 and 6 kJ m^{-2}, respectively, versus 11.28% in the control. Doses 1 and 6 kJ·m^{-2} significantly slowed water loss of samples until the second month of storage, with 17.68% and 16.02% of loss compared to control (31.45%). It is certain that dehydrated Deglet-Nour loses its market value, as it normally has a soft to semi-soft consistency, and the slowing down of the water loss by this treatment participates in preserving the fruit quality. Radiation's effect on water content found in this study is in perfect agreement with results obtained by Jemni et al. [17], who noted that water content was more important in the irradiated samples with a decrease rate between 174 and 130 g kg^{-1} of fresh weight for the same date variety.

The evolution of sugar content obtained for control and irradiated samples of the Deglet-Nour date, during five months of storage, is shown in Figure 2. The decrease in total sugars was slight but more pronounced in the case of the control until the second month of storage, where the difference appears significant. The diminution in total sugar content can be explained by the consumption of sugars as respiratory substrates [4] to ensure the energy supply necessary for the metabolism of the date fruit during postharvest storage.

Until the second month, the slowing down of sugar consumption was significantly favored in the samples irradiated at 1 and 6 kJ m^{-2}, which was marked by decreases of 4.98% and 4.57% versus 8.96% in the control.

The obtained results of total sugar contents are in agreement with those of Lemoine and collaborators [18] on broccoli and those of Pan et al. [19] on strawberry. In fact, they found that the total sugar content decreased slightly immediately after treatment with 4.1 kJ m^{-2} of UV-C, and also during storage when referring to control group.

During the first month of storage, the difference between total sugar content of irradiated samples and that of control does not appear to be significant, probably because water loss was greater in control and thus respiration lower, since an increase in fruit breathing rates has been observed with higher moisture content [4,20]. Indeed, it was reported that Deglet-Nour fruit with 27% moisture had a respiration rate five times higher than that of fruit at 20–22% of moisture [21]. Slight decreases in Deglet-Nour sugar content do not alter its nutritional value, as it contains large amounts at the full mature stage Tamar, most of which are glucose and fructose.

It is also interesting to report that the increase in total sugar was observed in UV-C-treated sweet oranges [22], and that supports our obtained results of irradiated samples when compared to control.

Proteins belong to one of the most important classes of molecules present in all living organisms. They provide the essential functions of the cell. Variation in total protein content in control and irradiated samples of the Deglet-Nour date during storage are shown in Figure 2.

Analysis of the results shows that there is an increase in the total protein levels during storage compared to the control. This increase in protein content is more marked in the case of the irradiated sample at 6 kJ m^{-2} in the third month of storage. Indeed, samples of 3 and 6 kJ·m^{-2} gave 1.70 and 2.41 g·kg^{-1} versus 1.29 g kg^{-1} of the control. The increase may be due to the activation of biosynthesis of enzymes involved in synthesis of defense molecules against stress caused by UV-C. Most likely, the increase is concomitant with the slowing down of protein degradation, in particular by inhibition of the action of proteases, which is associated with the ripening and senescence of fruit, as has been reported in many works [23]. Therefore, it appears that the degree of slowing of protein decomposition is higher in the irradiated samples than in the control.

Other hypotheses cited by Pickering and Davies [24] indicate that protein synthesis or protection against degradation can be linked to maintaining the activity of antioxidant enzymes. Furthermore, Costa et al. and Lemoine et al. [18,25] confirmed that the protein content of broccoli is increased after irradiation at 8 kJ m^{-2} for 21 d of storage at 4 °C but at a rate of about 5.05 g kg^{-1} of fresh weight, which is greater than our result (2.40 g kg^{-1}), recalling that the date generally contains small amounts of proteins.

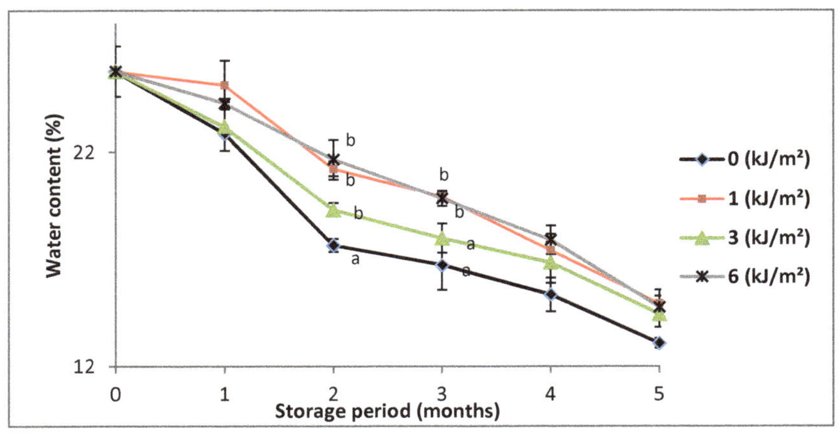

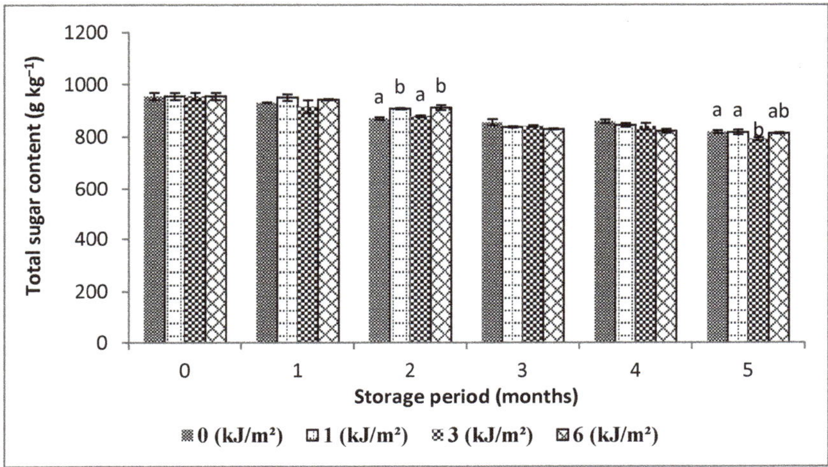

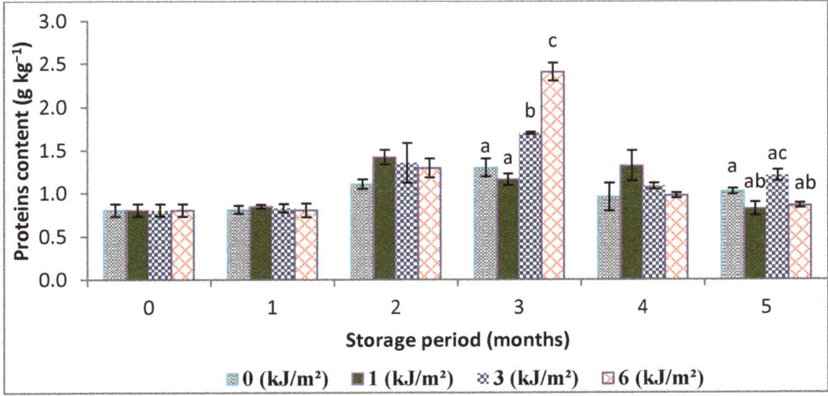

Figure 2. Variations in water, total sugar and total protein contents of Deglet-Nour samples, during five months of storage at 10 °C. Vertical bars represent standard error ($n = 3$), different letters indicate a significant difference between the four samples at each storage period ($p < 0.01$), while the absence of the letters indicates the absence of difference.

Another study conducted by Xie et al. [26] on strawberry reported that the group of fruit which was treated with UV-C was marked by a significant decrease in the levels of simple sugars, namely glucose, fructose and sucrose, but also in the levels of organic acids, especially those of citric and ascorbic acids, while a slight decrease was observed for in malic acid. The researchers also noted that the growth, physicochemical quality, firmness and moisture of UV-C-treated samples were preserved when compared to control. The concentrations of nine types of essential amino acids were preserved in coconut water samples which were irradiated by UV-C [27].

3.3. Enzymatic Activity of Polyphenoloxidase, Peroxidase and Browning Index

The presence of PPO and POD enzymes in the Deglet-Nour date has been already confirmed via the demonstration and the determination of their catalytic activity within this fruit [4,5]. A significant decrease in PPO activity of 6 kJ m^{-2} irradiated samples was noted throughout the whole storage period (Figure 3); it was affected by UV-C treatment. It may be that the expression of PPO gene was inhibited; indeed, the decrease in PPO activity in certain tissues of *Agaricus bisporus* has been correlated with the inhibition of expression of the PPO gene in these tissues [28]. The actual results are in agreement with those of Mohammadi et al. [29]. By applying doses of 0.25, 0.5 and 0.75 kJ m^{-2} to strawberry fruit, these authors observed a decrease in PPO activity, and they obtained the lowest activity after 5 d of storage at 10 °C using 0.75 kJ m^{-2}. Indeed, irradiation has been used to control the adverse browning phenomenon by inhibition of PPO activity [30].

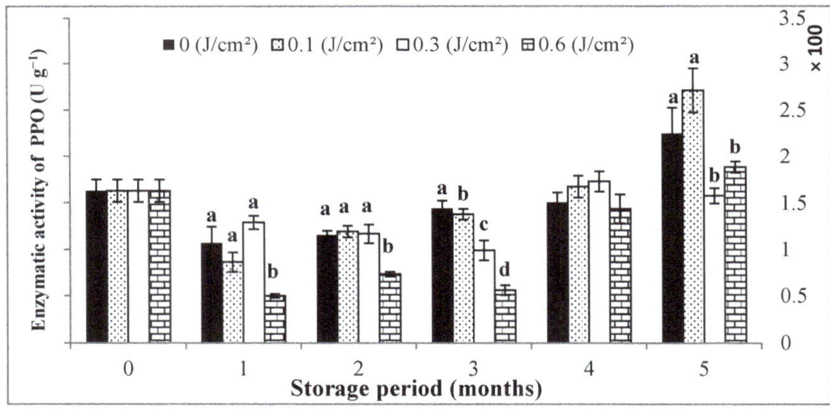

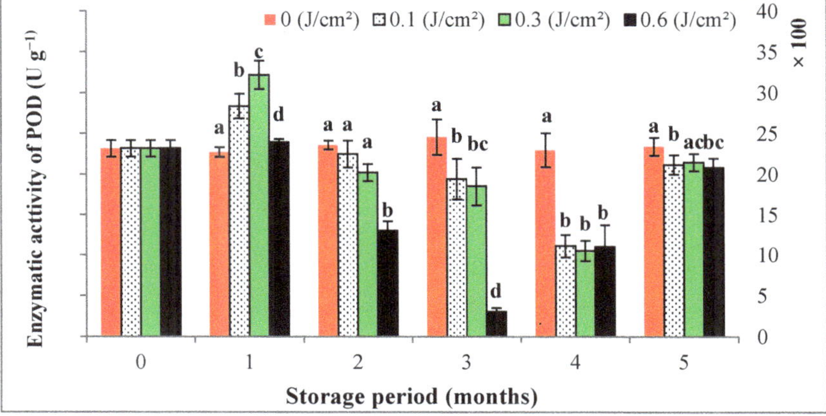

Figure 3. *Cont.*

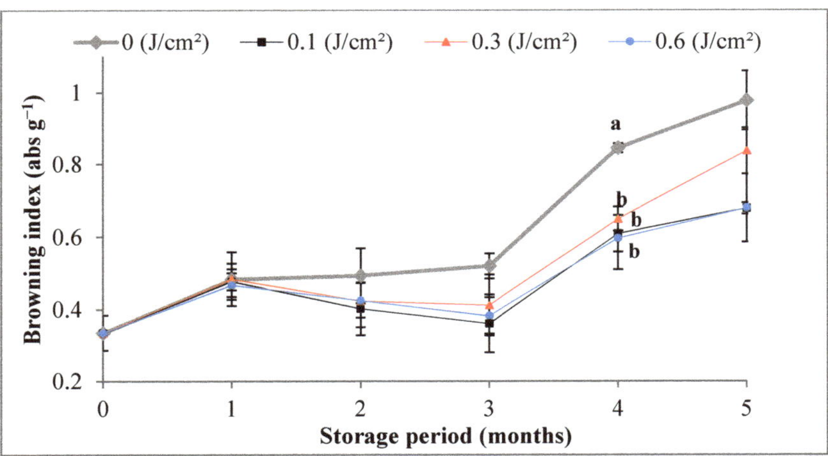

Figure 3. Variations in PPO and POD activities and in the browning index values of Deglet-Nour samples during five months of storage at 10 °C. Vertical bars represent standard error ($n = 3$), different letters indicate a significant difference between the four samples at each storage period ($p < 0.01$), while the absence of the letters indicates the absence of difference.

The obtained results showed, for kJ m^{-2}, a decrease in POD activity versus control from the second month until the end of storage (Figure 3). However, 1 kJ m^{-2} and 3 kJ m^{-2} induced a slowdown of POD activity for the last three months of storage. POD activity increased in irradiated samples during the first month, as generally peroxidase activity is stimulated by the readily available phenolic substrates, wounding and physiological stress, which were induced in this case by radiation and manipulation of samples.

No statistical significance was observed between the browning index values of treated and control dates during the first three months of storage, although the indexes of irradiated samples appeared to be lower (Figure 3). It was only in the fourth month that this difference was more significant in favor of a considerable reduction in browning indexes of irradiated samples, irrespective of the dose, in comparison with the control. Noting that, the analysis of results showed increased values of the browning index of all samples during storage, compared with the results of D$_0$ as initial analysis. However, this increase was more pronounced in the case of control samples, proving that UV-C slowed browning of Deglet–Nour dates, which resulted in a longer shelf-life during storage. Indeed, ascorbic acid is a highly effective inhibitor of enzymatic browning, especially since generally, its rate increases within stored Deglet-Nour dates following its exposure to abiotic stress such as storage under modified atmospheres [4]. The result is consistent with that of Gonzalez–Aguilar et al. [31], who found that browning was significantly reduced during 21 d of storage at 5 °C of peaches (*Prunus persica* cv. Jefferson) following exposure to 3, 5 and 10 min to UV-C radiation.

Another study using comparable doses of UV-C reported the effectiveness of the low dose (0.5 kJ m^{-2}) to preserve grapefruit from decay development, while the untreated control sample was marked by severe rind browning and necrotic peel [30]. It is also interesting to underline that the combination of UV-A and UV-C rays could exert a noticeable inactivation effect on PPO activity in apple juice, and the reduction in this enzyme activity was estimated at 32.58% [32].

A study conducted on coconut water revealed the effectiveness of UV-C irradiation to reduce the activity of PPO and POD enzymes in the exposed samples by 94 and 93%, respectively [27].

The exposition of potato slices to a combination composed of ascorbic acid, calcium chloride and UV-C significantly reduced the activities of PPO and POD [33]. A negligible browning process was reported in fresh-cut lotus 5–10 min following it exposition to UV-C radiation, with a remarkable inactivation of the previously cited enzymes [34].

3.4. Total Polyphenol Content (TPC), Flavonoids and Condensed Tannins

During the storage period, total polyphenol content followed an increasing evolution for all irradiated and control samples compared to the initial analysis (D_0) (Table 2). Total polyphenols in irradiated samples increased more than the control, regardless of dose, during the first month of storage. This increase is highly significant for the three doses but more considerable for the highest one. Indeed, this content reached 0.537 g kg^{-1} at the first month versus 0.288 g kg^{-1} in the control. This is similar to the result obtained by Winter and Rostàs [35], who observed that TPC content in soybean samples rapidly increased as a result of UV-C exposure. UV-C also induced the accumulation of phenolic compounds in tomatoes [36,37] and in grapes [38]. The increase in our case was followed by a decrease compared to the first month, but the levels remained higher than that of control for the remainder of the storage period. In comparison, Wu et al. [39] found a diminution of phenolic compounds after UV-C treatment of mushroom which led to its browning by, most likely, oxidation of these compounds. Perkins-Veazie et al. [40] reported that no change was observed in the phenolic content of blueberry samples irradiated at 2 and 4 kJ m^{-2}. The nature of phenols, the constitution and the physiology of each plant, in addition to storage conditions, may be the source of this diversity of results.

Table 2. Changes in contents of total phenols (TP), flavonoids and condensed tannins of Deglet-Nour irradiated samples and control during five months of storage at 10 °C.

Storage Period (Months)	Irradiation Doses (kJ m^{-2})	TP Content (g kg^{-1})	Flavonoids Content (g kg^{-1})	Tannins Content (g kg^{-1})
0	-	0.274 ± 0.014	0.034 ± 0.003	0.109 ± 0.010
1	0	0.288 ± 0.027 [a]	0.035 ± 0.003	0.131 ± 0.003 [a]
	1	0.469 ± 0.007 [b]	0.038 ± 0.004	0.299 ± 0.003 [b]
	3	0.453 ± 0.009 [b]	0.046 ± 0.003	0.370 ± 0.002 [c]
	6	0.537 ± 0.014 [c]	0.039 ± 0.002	0.283 ± 0.020 [b]
2	0	0.274 ± 0.021 [a]	0.027 ± 0.003	0.134 ± 0.019 [a]
	1	0.306 ± 0.014 [a]	0.030 ± 0.004	0.251 ± 0.011 [b]
	3	0.339 ± 0.018 [b]	0.032 ± 0.003	0.272 ± 0.011 [c]
	6	0.345 ± 0.018 [b]	0.035 ± 0.001	0.274 ± 0.015 [c]
3	0	0.308 ± 0.025 [a]	0.027 ± 0.003	0.141 ± 0.015 [a]
	1	0.317 ± 0.013 [a]	0.027 ± 0.004	0.149 ± 0.016 [a]
	3	0.388 ± 0.012 [b]	0.034 ± 0.005	0.236 ± 0.005 [b]
	6	0.400 ± 0.006 [b]	0.038 ± 0.003	0.238 ± 0.003 [b]
4	0	0.329 ± 0.010 [a]	0.030 ± 0.002 [a]	0.147 ± 0.023 [a]
	1	0.366 ± 0.020 [a]	0.050 ± 0.004 [b]	0.144 ± 0.013 [a]
	3	0.364 ± 0.030 [a]	0.045 ± 0.003 [b]	0.162 ± 0.010 [b]
	6	0.437 ± 0.012 [b]	0.046 ± 0.002 [b]	0.164 ± 0.006 [b]
5	0	0.337 ± 0.013 [a]	0.032 ± 0.005 [a]	0.139 ± 0.008 [a]
	1	0.411 ± 0.020 [b]	0.045 ± 0.003 [b]	0.147 ± 0.014 [a]
	3	0.409 ± 0.012 [b]	0.041 ± 0.005 [b]	0.164 ± 0.007 [b]
	6	0.417 ± 0.036 [b]	0.050 ± 0.006 [b]	0.135 ± 0.019 [a]

Values are expressed as mean ± SEM (n = 3), different letters indicate a significant difference between the four samples at each storage period ($p < 0.01$) while the absence of the letters indicates the absence of difference.

The results illustrated in Table 2 show a significant increase, regardless of dose, in flavonoid contents of irradiated samples in the last two months of storage compared to values detected in the first three months, where there is no significant difference in flavonoid concentration between control and treated samples. This result is somewhat similar to those obtained by some researchers. According to Gonzalez-Aguilar et al. [41], an accumulation of phenols and total flavonoids is noted during storage, following the application of 2.46 and 4.93 kJ m^{-2} doses on *Mangifera indica*. In addition, exposure to a dose of 4.1 kJ m^{-2} induced an overproduction of total phenols in *Fragaria × ananassa* Duch. Cv. *Toyonoka* [42]. Flavonoids could play a protective role against pathogens and environmental stress, due, for example, to drought or ultraviolet radiation, as has been noted in several works. Generally, the increase in phenol levels following UV-C exposure is associated with stimulation of the expression of the phenylalanine ammonia lyase gene; this was also confirmed by Sheng and collaborators [38] when studying the effect of UV-C treatment on table grape in addition to other genes involved in the phenylpropanoid pathway. However, the application of increasing UV-C doses of 3.2, 9.6 and 19.2 kJ m^{-2} on *Lycopersicom esculentum* cv. Durita increased total phenol levels during storage [43]. In contrast, Costa et al. [23] observed a decrease in total phenol and flavonoid content in *Brassica oleracea* L. var. Italica, cv. Cicco following exposure to UV-C doses of 4, 7, 10 and 14 kJ m^{-2}.

Contents of condensed tannins increased, varying considerably between the three doses of irradiation versus control, where the increase reached a maximum value of the order of 0.37 g kg^{-1} in the case of the 0.3 J cm^{-2} dose (Table 2). This augmentation is justified by a response to stress caused by UV-C light as a self-protecting function of the fruit. Several studies confirmed the biosynthesis of tannins by plants when exposed to abiotic stresses such as low cadmium stress [44] and arid conditions [45]. Compared to irradiated samples at the first two months of storage, a decrease in the content of condensed tannins was noticed towards the end of storage; it reached a minimum value of the order of 0.135 g kg^{-1} at the dose 0.6 J cm^{-2}. However, the decrease relative to the control was not established. The decrease in tannin levels is probably due to the transformation of these compounds from the soluble form to the insoluble form, which leads to their precipitation, and consequently to the decrease in astringency, which is an index of advanced ripening in date fruit. This study confirms that tannins are capable of complexing proteins and other macromolecules such as nucleic acids, which can be involved in slowing down the enzymatic activity of POD and PPO during storage of samples. Indeed, the slowdown in these activities clearly appears from the third month; this coincides with the decrease in the content of tannins due to their combination with PPO and POD, which has limited their extractability and thus altered the PPO and POD function.

It is also interesting to note that the exposure of acerola to slight UV-C radiation during postharvest storage significantly decreased the rhythm of degradation of vitamin C and phenolic compounds but also preserved the nutraceutical quality of the fruit, unlike the control not exposed to UV-C [46]. These researchers suggested a possible retention of these elements in this fruit via a possible alteration of the metabolism process of these compounds, but also suggested an elevation in mitochondrial activity and an upregulation of the antioxidant system in order to correctly fight free radicals, especially ROS, a known agent for fruit rot [47].

Another study conducted by Park and Kim [48] reported that the exposition of peeled garlic cloves to low UV-C radiation considerably increased polyphenol and flavonoid content, especially apigenin and quercetin, considered as important antioxidant agents. This may explain the slight deterioration of this group marked by a significant decrease in the microbial population, while the control sample surprisingly developed a yellow color, indicating a possible acceleration in the maturation process [49]. A comparable elevation in total phenolic and flavonoid concentrations was also observed in lettuce exposed to UV-C [50].

3.5. Anti-Free-Radical Activity and Ferric-Reducing Power

The general reading of the results of the scavenging effect analysis of DPPH radical (Figure 4) showed an increased rate of DPPH scavenging from the fourth month in favor of samples irradiated at doses 0.3 and 0.6 J cm^{-2}. At the end of the first three months of storage, no significant distinction could be drawn between the antiradical effect of irradiated samples and that of the control, despite the apparent increase in the case of treated samples.

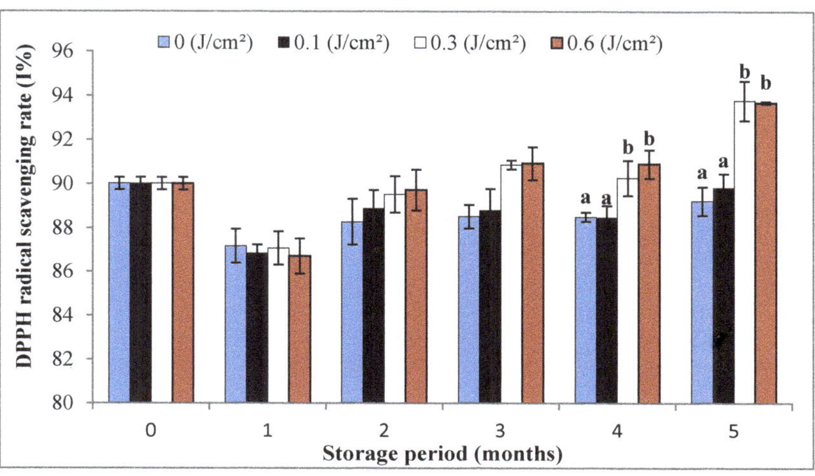

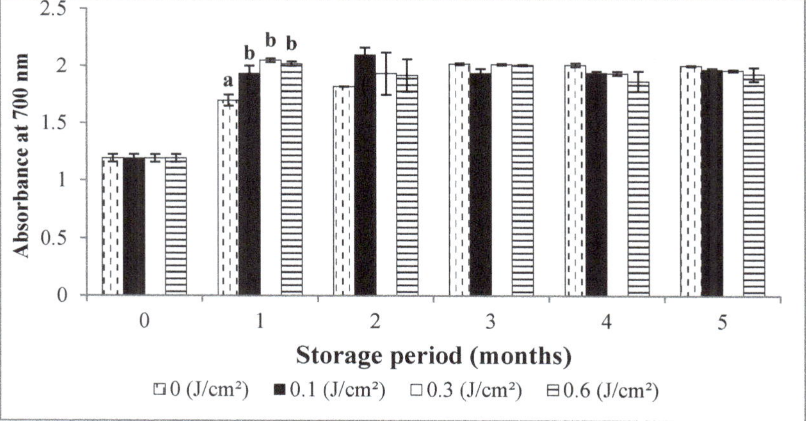

Figure 4. Changes in DPPH scavenging activity and iron-(III)-reducing power of the of Deglet-Nour samples during five months of storage at 10 °C. Vertical bars represent standard error ($n = 3$); a, b: different letters indicate a significant difference between the four samples at each storage period ($p < 0.01$), while the absence of the letters indicates the absence of difference.

The increase in antioxidant activity during the storage of dates is probably due to biosynthesis and accumulation of phenolic compounds that play a very important role in free radical scavenging. The combination of these results with those of the total polyphenol (PPT) assays revealed a very significant linear correlation between the anti-free-radical capacity of the extracts and their PPT content ($R^2 = 0.80$) from the third month of storage, which means that the antioxidant activity is most likely due to the phenol composition of Deglet-Nour extracts. Most of these compounds have antioxidant activities, due to the presence of many hydroxyls in their structures, which can react with free radicals.

The results of Kim et al. [43] on Citrus pomaces, as well as those of Lemoine et al. and Martínez-Hernández et al. [51,52] on broccoli and Vicente et al. [53] on sweet pepper, revealed that the increase in antioxidant capacity could be linked to the elevation in total phenols within UV irradiated samples, which is similar to ours. The rates of ferric reduction using Deglet-Nour samples, expressed as absorbance in 700 nm, are mentioned in Figure 4.

The analysis of the results makes it possible to observe a significant increase in absorbance of irradiated samples after one month of storage relative to the control. An increase in absorbance means an increase in the reducing power of tested extracts. The increase in iron (III) reductive activity in all samples compared to the initial (D_0) analysis would be due to the concentration of antioxidant molecules in the extracts as a result of water loss that was much higher in the control samples. Linear correlation between ferric-reducing power and phenolic contents was medium ($R^2 = 0.5$), thus involving part of these compounds. Generally, the reducing power of date extracts is due to the presence of substances having an ideal structure for free radical scavenging: possessing free hydroxyl groups and acting as hydrogen donors. Similar results have been observed by Gonzalez-Aguilar et al. [41] on mango, Perkins-Veazie et al. [40] on blueberries and Sheng et al. [38] on table grape, where the increase in PPT content led to increased antioxidant activity.

In addition to phenolics, involvement of selenium can be established, as date fruit is a good source of this element, 2.4–4 mg kg^{-1} [2]. Indeed, Molan et al. [54] found that the increase in the level of selenium in green tea leads to an increase in its antioxidant activities FRAP and DPPH. Otherwise, implication of carotenoids in these reductive activities cannot be considered, especially since the correlation between carotenoid content and antioxidant activity rate is low ($R^2 = 0.13; 0.01$).

Moreover, the exposition of lettuce to UV-C radiation can significantly decrease DPPH scavenging activity, but also helps this plant to adapt and increase its tolerance to salinity stress [50].

In general, and from research studies that realized fruits and vegetables using UV-C radiation, results are encouraging, since quality was altered but improved in in most cases. Indeed, astudy conducted by Hosseini et al. [55] showed the effectiveness of low UV-C to preserve the physicochemical and sensory parameter of pistachio. Furthermore, low doses are sufficient for some fruits and vegetables, and indeed these researchers noted that the dose of 2.1 kJ m^{-2} was more active than 4.5 kJ m^{-2}, and pistachio were in this case lighter, redder and less yellow than the highest tested dose.

Another study demonstrated that the exposition of a freshly extracted tomato juice to UV-C could be a good alternative to preserve the industrial quality of this preparation by slightly increasing some physicochemical properties such as total soluble solid, water activity, pH, color, titratable acidity and clarity [56]. The color parameter of orange juice was also preserved after UV-C exposition, and the researchers noted a non-negligible antimicrobial effect against *Saccharomyces cerevisiae* [57] which is very promising for future industrial application. Almost the same physicochemical results were observed for white grape juice, and this time the UV-C radiation was active against the *Escherichia coli* K-12 strain [58].

4. Conclusions

UV-C radiation low doses contributed to enrichment of date fruit with total polyphenols, flavonoids and tannins, which led to the increase in the antioxidant activity. The enrichment in component contents of irradiated samples reached a maximum of 0.537 g kg^{-1} versus the control (0.288 g kg^{-1}) for PPT, and a maximum of 0.370 g kg^{-1} versus the control (0.131 g kg^{-1}) of tannins during the first month of storage. On the other hand, a maximum of flavonoids of 0.050 g kg^{-1} versus the control (0.035 g kg^{-1}) was obtained after 5 and 6 months of storage. This physical technique has delayed the enzymatic browning within this fruit. Indeed, the dose (0.6 J cm^{-2}) resulted in the minimum browning index (0.59 abs g^{-1}) after four months of storage, which was comparable to the other doses. UV-C radiation of Deglet-Nour fruit may have a positive impact on human health by increasing the levels of certain bio-compounds in addition to preserving its nutritional quality and extending its shelf-life. No distinguishable difference between the effects of the three doses can be confirmed, thus there is a need to test other doses and other parameters, to be able to pick out the most adequate for the preservation of nutritional quality and organoleptic properties of Deglet-Nour fruit.

Author Contributions: Data curation, E.A.; Formal analysis, S.D. and A.H.A.; Funding acquisition, S.J.A.; Investigation, A.e.D.; Methodology, O.B., L.S. and M.S.B.; Supervision, R.S.; Visualization, H.A.; Writing—review & editing, H.A. All authors have read and agreed to the published version of the manuscript.

Funding: This research received no external funding.

Institutional Review Board Statement: Not applicable.

Informed Consent Statement: Not applicable.

Data Availability Statement: Available upon request from the corresponding author.

Acknowledgments: The author expresses thanks to the Algerian Ministry of Higher Education. Princess Nourah bint Abdulrahman University Researchers Supporting Project Number (PNURSP2022R249), Princess Nourah bint Abdulrahman University, Riyadh, Saudi Arabia.

Conflicts of Interest: The authors declare no conflict of interest.

References

1. El Gharras, H. Polyphenols: Food sources, properties and applications—A review. *Int. J. Food Sci.* **2009**, *44*, 2512–2518. [CrossRef]
2. Baliga, M.S.; Baliga, B.R.V.; Kandathil, S.M.; Bhat, H.P.; Vayalil, P.K. A review of the chemistry and pharmacology of the date fruits (*Phoenix dactylifera* L.). *Food Res. Int.* **2011**, *44*, 1812–1822. [CrossRef]
3. Daas Amiour, S.; Alloui-Lombarkia, O.; Bouhdjila, F.; Ayachi, A.; Hambaba, L. Étude de l'implication des composés phénoliques des extraits de trois variétés de datte dans son activité antibactérienne. *Phytothérapie* **2014**, *12*, 135–142. [CrossRef]
4. Dassamiour, S.; Vidal, V.; Laurent, S.; Sallanon, H.; Charles, F. Effect of gaseous pretreatment on enzymatic browning of mature date after cold storage. *Fruits* **2018**, *73*, 243–251. [CrossRef]
5. Daas Amiour, S.; Hambaba, L. Effect of pH, temperature and some chemicals on polyphenoloxidase and peroxidase activities in harvested Deglet Nour and Ghars dates. *Postharvest Biol. Technol.* **2016**, *111*, 77–82. [CrossRef]
6. Xuetong, F.; Sommers, C.H. Irradiation of Fresh and Fresh-Cut Fruits and Vegetables: Quality and Shelf Life. Ch. 15. In *Food Irradiation Research and Technology*, 2nd ed.; Blackwell Publishing and the Institute of Food Technologists: Hoboken, NJ, USA, 2012; pp. 271–293. [CrossRef]
7. Dubois, M.; Gilles, K.A.; Hamilton, J.K.; Rebers, P.T.; Smith, F. Colorimetric method for determination of sugars and related substances. *Anal. Chem.* **1956**, *28*, 350–356. [CrossRef]
8. Bradford, M.M. A rapid and sensitive method for the quantitation of microgram quantities of protein utilizing the principle of protein-dye binding. *Anal. Biochem.* **1976**, *72*, 248–254. [CrossRef]
9. Lichanporn, I.; Techavuthiporn, C. The effects of nitric oxide and nitrous oxide on enzymatic browning in longkong (*Aglaia dookkoo* Griff.). *Postharvest Biol. Technol.* **2013**, *86*, 62–65. [CrossRef]
10. Masuda, T.; Yonemori, S.; Oyama, Y.; Takeda, Y.; Tanaka, T.; Andoh, T.; Nakata, M. Evaluation of the antioxidant activity of environmental plants: Activity of the leaf extracts from seashore plants. *J. Agric. Food Chem.* **1999**, *47*, 1749–1754. [CrossRef]
11. Oyaizu, M. Antioxidative activities of browning products of glucosamine fractionated by organic solvent and thin-layer chromatography. *Nippon Shokuhin Kogyo Gakkaishi J. Jpn. Soc. Food Sci.* **1988**, *35*, 771–775. [CrossRef]
12. Alegria, C.; Pinheiro, J.; Duthoit, M.; Gonçalves, E.M.; Moldão-Martins, M.; Abreu, M. Fresh-cut carrot (cv. Nantes) quality as affected by abiotic stress (heat shock and UV-C irradiation) pre-treatments. *LWT Food Sci. Technol.* **2012**, *48*, 197–203. [CrossRef]
13. Rodriguez-Amaya, D.B. Changes in carotenoids during processing and storage of foods. *Arch. Latinoam. Nutr.* **1999**, *49*, 38S–47S. [PubMed]
14. Boudries, H.; Kefalas, P.; Hornero-Méndez, D. Carotenoid composition of Algerian date varieties (*Phoenix dactylifera*) at different edible maturation stages. *Food Chem.* **2007**, *101*, 1372–1377. [CrossRef]
15. Freitas, A.; Moldão-Martins, M.; Costa, H.S.; Albuquerque, T.G.; Valente, A.; Sanches-Silva, A. Effect of UV-C radiation on bioactive compounds of pineapple (*Ananas comosus* L. Merr.) by-products. *J. Sci. Food Agric.* **2015**, *95*, 44–52. [CrossRef]
16. Mutlak, H.H.; Mann, J. Darkening of dates: Control by microwave heating. *Date Palm J.* **1984**, *3*, 303–316.
17. Jemni, M.; Gómez, P.A.; Souza, M.; Chaira, N.; Ferchichi, A.; Otón, M.; Artés, F. Combined effect of UV-C, ozone and electrolyzed water for keeping overall quality of date palm. *LWT Food Sci. Technol.* **2014**, *59*, 649–655. [CrossRef]
18. Lemoine, M.L.; Civello, P.M.; Martínez, G.A.; Chaves, A.R. Influence of postharvest UV-C treatment on refrigerated storage of minimally processed broccoli (*Brassica oleracea* var. Italica). *J. Sci. Food Agric.* **2007**, *87*, 1132–1139. [CrossRef]
19. Pan, J.; Vicente, A.R.; Martínez, G.A.; Chaves, A.R.; Civello, P.M. Combined use of UV-C irradiation and heat treatment to improve postharvest life of strawberry fruit. *J. Sci. Food Agric.* **2004**, *84*, 1831–1838. [CrossRef]
20. Yahia, E.M.; Kader, A.A. Date (*Phoenix dactylifera* L.). In *Postharvest Biology and Technology of Tropical and Subtropical Fruits*, 1st ed.; Woodhead Publishing: Sawston, UK, 2011. [CrossRef]

21. Rygg, G.L. Date development, handling and packing in the United States. In *Agricultural Handbook No. 482*, 1st ed.; Agricultural Research Service, US Department of Agriculture Date Development: Washington, DC, USA, 1975.

22. Hu, L.; Yang, C.; Zhang, L.; Feng, J.; Xi, W. Effect of light-emitting diodes and ultraviolet irradiation on the soluble sugar, organic acid, and carotenoid content of postharvest sweet oranges (*Citrus sinensis* (L.) Osbeck). *Molecules* **2019**, *24*, 3440. [CrossRef]

23. Barka, E.A.; Kalantari, S.; Makhlouf, J.; Arul, J. Impact of UV-C irradiation on the cell wall-degrading enzymes during ripening of tomato (*Lycopersicon esculentum* L.) fruit. *J. Agric. Food Chem.* **2000**, *48*, 667–671. [CrossRef]

24. Pickering, A.M.; Davies, K.J. Degradation of damaged proteins: The main function of the 20S proteasome. *Prog. Mol. Biol. Transl. Sci.* **2012**, *109*, 227–248. [CrossRef]

25. Costa, L.; Vicente, A.R.; Civello, P.M.; Chaves, A.R.; Martínez, G.A. UV-C treatment delays postharvest senescence in broccoli florets. *Postharvest Biol. Technol.* **2006**, *39*, 204–210. [CrossRef]

26. Xie, Z.; Fan, J.; Charles, M.T.; Charlebois, D.; Khanizadeh, S.; Rolland, D.; Roussel, D.; Zhang, Z. Preharvest ultraviolet-C irradiation: Influence on physicochemical parameters associated with strawberry fruit quality. *Plant Physiol. Biochem.* **2016**, *108*, 337–343. [CrossRef]

27. Yannam, S.K.; Patras, A.; Pendyala, B.; Vergne, M.; Ravi, R.; Gopisetty, V.; Sasges, M. Effect of UV-C irradiation on the inactivation kinetics of oxidative enzymes, essential amino acids and sensory properties of coconut water. *J. Food Sci. Technol.* **2020**, *57*, 3564–3572. [CrossRef]

28. Lei, J.; Li, B.; Zhang, N.; Yan, R.; Guan, W.; Brennan, C.S.; Peng, B. Effects of UV-C treatment on browning and the expression of polyphenol oxidase (PPO) genes in different tissues of Agaricusbisporus during cold storage. *Postharvest Biol. Technol.* **2018**, *139*, 99–105. [CrossRef]

29. Mohammadi, N.; Mohammadi, S.; Abdossi, V.; Akbar-Boojar, M.A. Effect of UV-C radiation on antioxidant enzymes in strawberry fruit (*Fragaria* x *ananassa* cv. Camarosa). *J. Agric. Biol. Sci.* **2012**, *7*, 860–864.

30. Kim, J.; Marshall, M.R.; Wei, C. Polyphenoloxidase. In *Seafood Enzymes Utilization and Influence on Postharvest Seafood Quality*; Haard, N.F., Simpson, B.K., Eds.; Marcel Dekker: New York, NY, USA, 2000; pp. 271–315.

31. Gonzalez-Aguilar, G.; Wang, C.Y.; Buta, G.J. UV-C irradiation reduces breakdown and chilling injury of peaches during cold storage. *J. Sci. Food Agric.* **2004**, *84*, 415–422. [CrossRef]

32. Akgün, M.P.; Ünlütürk, S. Effects of ultraviolet light emitting diodes (LEDs) on microbial and enzyme inactivation of apple juice. *Int. J. Food Microbiol.* **2017**, *260*, 65–74. [CrossRef]

33. Teoh, L.S.; Lasekan, O.; Adzahan, N.M.; Hashim, N. The effect of ultraviolet treatment on enzymatic activity and total phenolic content of minimally processed potato slices. *J. Food Sci. Technol.* **2016**, *53*, 3035–3042. [CrossRef]

34. Wang, D.; Chen, L.; Ma, Y.; Zhang, M.; Zhao, Y.; Zhao, X. Effect of UV-C treatment on the quality of fresh-cut lotus (*Nelumbo nucifera* Gaertn.) root. *Food Chem.* **2019**, *278*, 659–664. [CrossRef]

35. Winter, T.R.; Rostás, M. Ambient ultraviolet radiation induces protective responses in soybean but does not attenuate indirect defense. *Environ. Pollut.* **2008**, *155*, 290–297. [CrossRef]

36. Jagadeesh, S.L.; Charles, M.T.; Gariepy, Y.; Goyette, B.; Raghavan, G.S.V.; Vigneault, C. Influence of postharvest UV-C hormesis on the bioactive components of tomato during post-treatment handling. *Food Bioproc. Technol.* **2011**, *4*, 1463–1472. [CrossRef]

37. Bravo, S.; García-Alonso, J.; Martín-Pozuelo, G.; Gómez, V.; García-Valverde, V.; Navarro-González, I.; Periago, M.J. Effects of postharvest UV-C treatment on carotenoids and phenolic compounds of vine-ripe tomatoes. *Int. J. Food Sci.* **2013**, *48*, 1744–1749. [CrossRef]

38. Sheng, K.; Zheng, H.; Shui, S.; Yan, L.; Liu, C.; Zheng, L. Comparison of postharvest UV-B and UV-C treatments on table grape: Changes in phenolic compounds and their transcription of biosynthetic genes during storage. *Postharvest Biol. Technol.* **2018**, *138*, 74–81. [CrossRef]

39. Wu, X.; Guan, W.; Yan, R.; Lei, J.; Xu, L.; Wang, Z. Effects of UV-C on antioxidant activity, total phenolics and main phenolic compounds of the melanin biosynthesis pathway in different tissues of button mushroom. *Postharvest Biol. Technol.* **2016**, *118*, 51–58. [CrossRef]

40. Perkins-Veazie, P.; Collins, J.K.; Howard, L. Blueberry fruit response to postharvest application of ultraviolet radiation. *Postharvest Biol. Technol.* **2008**, *47*, 280–285. [CrossRef]

41. González-Aguilar, G.A.; Villegas-Ochoa, M.A.; Martínez-Téllez, M.A.; Gardea, A.A.; Ayala-Zavala, J.F. Improving Antioxidant Capacity of Fresh-Cut Mangoes Treated with UV-C. *J. Food Sci.* **2007**, *72*, S197–S202. [CrossRef]

42. Pombo, M.A.; Rosli, H.G.; Martínez, G.A.; Civello, P.M. UV-C treatment affects the expression and activity of defense genes in strawberry fruit (*Fragaria* × *ananassa*, Duch.). *Postharvest Biol. Technol.* **2011**, *59*, 94–102. [CrossRef]

43. Kim, H.J.; Fonseca, J.M.; Kubota, C.; Kroggel, M.; Choi, J.H. Quality of fresh-cut tomatoes as affected by salt content in irrigation water and post-processing ultraviolet-C treatment. *J. Sci. Food Agric.* **2008**, *88*, 1969–1974. [CrossRef]

44. Qin, G.Q.; Yan, C.L.; Wei, L.L. Effect of cadmium stress on the contents of tannin, soluble sugar and proline in *Kandeliacandel* (L.) Druce seedlings. *Acta Ecol. Sin.* **2006**, *36*, 112–123. [CrossRef]

45. Zhang, L.H.; Shao, H.B.; Ye, G.F.; Lin, Y.M. Effects of fertilization and drought stress on tannin biosynthesis of *Casuarina equisetifolia* seedlings branchlets. *Acta Physiol. Plant.* **2012**, *34*, 1639–1649. [CrossRef]

46. Rabelo, M.C.; Bang, W.Y.; Nair, V.; Alves, R.E.; Jacobo-Velázquez, D.A.; Sreedharan, S.; de Miranda, M.R.A.; Cisneros-Zevallos, L. UVC light modulates vitamin C and phenolic biosynthesis in acerola fruit: Role of increased mitochondria activity and ROS production. *Sci. Rep.* **2020**, *10*, 21972–21985. [CrossRef]

47. Meitha, K.; Pramesti, Y.; Suhandono, S. Reactive Oxygen Species and Antioxidants in Postharvest Vegetables and Fruits. *Int. J. Food Sci.* **2020**, *2020*, 8817778–8817788. [CrossRef]
48. Park, M.H.; Kim, J.G. Low-dose UV-C irradiation reduces the microbial population and preserves antioxidant levels in peeled garlic (*Allium sativum* L.) during storage. *Postharvest Biol. Technol.* **2015**, *100*, 109–112. [CrossRef]
49. Wallace, T.C.; Bailey, R.L.; Blumberg, J.B.; Burton-Freeman, B.; Chen, C.O.; Crowe-White, K.M.; Drewnowski, A.; Hooshmand, S.; Johnson, E.; Lewis, R.; et al. Fruits, vegetables, and health: A comprehensive narrative, umbrella review of the science and recommendations for enhanced public policy to improve intake. *Crit. Rev. Food Sci. Nutr.* **2020**, *60*, 2174–2211. [CrossRef]
50. Ouhibi, C.; Attia, H.; Rebah, F.; Msilini, N.; Chebbi, M.; Aarrouf, J.; Urban, L.; Lachaal, M. Salt stress mitigation by seed priming with UV-C in lettuce plants: Growth, antioxidant activity and phenolic compounds. *Plant Physiol. Biochem.* **2014**, *83*, 126–133. [CrossRef]
51. Lemoine, M.L.; Civello, P.M.; Chaves, A.R.; Martínez, G.A. Influence of a combined hot air and UV-C treatment on quality parameters of fresh-cut broccoli florets at 0 °C. *Int. J. Food Sci.* **2010**, *45*, 1212–1218. [CrossRef]
52. Martínez-Hernández, G.B.; Gómez, P.A.; Pradas, I.; Artés, F.; Artés-Hernández, F. Moderate UV-C pretreatment as a quality enhancement tool in fresh-cut Bimi®broccoli. *Postharvest Biol. Technol.* **2011**, *62*, 327–337. [CrossRef]
53. Vicente, A.R.; Pineda, C.; Lemoine, L.; Civello, P.M.; Martinez, G.A.; Chaves, A.R. UV-C treatments reduce decay, retain quality and alleviate chilling injury in pepper. *Postharvest Biol. Technol.* **2005**, *35*, 69–78. [CrossRef]
54. Molan, A.L.; Flanagan, J.; Wei, W.; Moughan, P.J. Selenium-containing green tea has higher antioxidant and prebiotic activities than regular green tea. *Food Chem.* **2009**, *114*, 829–835. [CrossRef]
55. Hosseini, F.S.; Akhavan, H.R.; Maghsoudi, H.; Hajimohammadi-Farimani, R.; Balvardi, M. Effects of a rotational UV-C irradiation system and packaging on the shelf life of fresh pistachio. *J. Sci. Food Agric.* **2019**, *99*, 5229–5238. [CrossRef]
56. Bhat, R. Impact of ultraviolet radiation treatments on the quality of freshly prepared tomato (*Solanum lycopersicum*) juice. *Food Chem.* **2016**, *213*, 635–640. [CrossRef]
57. Niu, L.; Wu, Z.; Yang, L.; Wang, Y.; Xiang, Q.; Bai, Y. Antimicrobial Effect of UVC Light-Emitting Diodes against *Saccharomyces cerevisiae* and Their Application in Orange Juice Decontamination. *J. Food Prot.* **2021**, *84*, 139–146. [CrossRef]
58. Unluturk, S.; Atilgan, M.R. Microbial Safety and Shelf Life of UV-C Treated Freshly Squeezed White Grape Juice. *J. Food Sci.* **2015**, *80*, M1831–M1841. [CrossRef]

Article

Antibacterial Activity of Nanoparticles of Garlic (*Allium sativum*) Extract against Different Bacteria Such as *Streptococcus mutans* and *Poryphormonas gingivalis*

Tuyishime Gabriel [1], Abimana Vestine [1], Ki Deok Kim [2], Seong Jung Kwon [3], Iyyakkannu Sivanesan [4] and Se Chul Chun [1,*]

[1] Department of Environmental Health Science, Konkuk University, Seoul 05029, Korea;
    tuyishime@konkuk.ac.kr (T.G.); abimanavestine@konkuk.ac.kr (A.V.)
[2] Department of Plant Biotechnology, Korea University, Seoul 02841, Korea; kidkim@korea.ac.kr
[3] Department of Chemistry, Konkuk University, Seoul 05029, Korea; sjkwon@konkuk.ac.kr
[4] Department of Bioresources and Food Science, Konkuk University, Seoul 05029, Korea; siva74@konkuk.ac.kr
*   Correspondence: scchun@konkuk.ac.kr

Abstract: To combat the threat of antimicrobial resistance, it is important to discover innovative and effective alternative antibacterial agents. Garlic has been recommended as a medicinal plant with antibacterial qualities. Hence, we conducted this study to evaluate the antibacterial activity of ultrasonicated garlic extract against *Escherichia coli*, *Staphylococcus aureus* sub. *aureus*, *Streptococcus mutans*, and *Poryphyromonas gingivalis*. Aqueous ultrasonicated garlic extract was tested against these strains, and their antibacterial activity quantified using both agar disk diffusion and agar well diffusion methods; the plate count technique was used to estimate the total viable count. Moreover, Fourier-transform infrared spectroscopy (FTIR), transmission electron microscopy (TEM), and microplate spectrophotometry were used to characterize garlic nanoparticles. The results confirmed that all tested bacteria were sensitive to both sonicated and non-sonicated garlic extracts. *Streptococcus mutans* was the most susceptible bacteria; on the other hand, *Escherichia coli* was the most resistant bacteria. Furthermore, characterization of the prepared garlic nanoparticles, showed the presence of organosulfur and phenolic compounds, carboxyl groups, and protein particles. Based on the obtained results, ultrasonicated garlic extract is a potent antibacterial agent. It can come in handy while developing novel antibiotics against bacteria that have developed resistance.

Keywords: garlic; ultrasonication; agar disk diffusion method; plate count technique; nanoparticles

Citation: Gabriel, T.; Vestine, A.; Kim, K.D.; Kwon, S.J.; Sivanesan, I.; Chun, S.C. Antibacterial Activity of Nanoparticles of Garlic (*Allium sativum*) Extract against Different Bacteria Such as *Streptococcus mutans* and *Poryphormonas gingivalis*. *Appl. Sci.* **2022**, *12*, 3491. https://doi.org/10.3390/app12073491

Academic Editors: Ionel Calin Jianu and Georgeta Pop

Received: 22 December 2021
Accepted: 28 March 2022
Published: 30 March 2022

Publisher's Note: MDPI stays neutral with regard to jurisdictional claims in published maps and institutional affiliations.

1. Introduction

The increasing mortality rate of infectious diseases is one the most challenging public health problems faced by different countries worldwide. This compromises and poses a threat to human health. Numerous synthetic antibiotic agents have always been used for the management of infectious diseases. Looking at the figures from 2000 to 2015, the statistics confirm that global antibiotics consumption levels have reached 65%, including the use of strong antibiotic agents like colistin and carbapenem [1]. Unfortunately, the wide use of antibiotics has increased the development of bacterial resistant strains to antibiotics [2], which has resulted in a reduction in the effectiveness of some of the antibacterial agents, leading to high mortality rates. Antibiotic resistance is considered one of the world's most urgent public health problems [3,4].

A literature review has provided guidelines to minimize the increase of antibiotic resistance in the management of infections. Antibiotic resistance can be reduced by making an appropriate, timely diagnosis before doing any treatment planning; proper prescription and use of antibiotics by physicians as well as patients; effective implementation of strategies to prevent the transmission of infectious diseases; and discovering new antibacterial agents might help to control microbial drug resistance [5]. Of these, invention of new antimicrobial agents has been given more attention compared to other strategies [6].

Medicinal plants have been used for many years in the treatment of a vast number of human diseases by the community, specifically in traditional medicine. They are considered the main source of new, natural, and safe drugs to be utilized in managing diseases as an effective and harmless alternative medicine [7,8], because they are not expensive and pose minimum side effects to humans. According to a report published in 2002 by the World Health Organization (WHO) in Geneva, medicinal plants have significant value and could be considered as the best source of complementary and alternative natural medicines [9].

Garlic, scientifically known as *Allium sativum*, is one of the oldest plants used as a spice in food and also used as a medicine because of its many benefits to human health and wellbeing [10,11]. Findings have shown that garlic can be used in the management of various diseases such as cardiovascular disease and hyperglycemia. Additionallygarlic has been approved to reduce the risk of cancer, boosting the immune system and protecting against inflammation as well as infectious diseases [12–14].

Results from different studies have shown that garlic extracts have the capacity to inhibit the growth of some pathogenic microorganisms [15,16]. Their antimicrobial activity has been linked to the presence of sulphur compounds [17], specifically, allicin which is the compound produced by the alliin lyase enzyme, after crushing or bruising a garlic bulb [12]. Many medical bacteria, including gram-positive and gram-negative strains, are sensitive to garlic extracts [18,19], indicating that that garlic has a reliable broad-spectrum of activity related to its chemical composition [20]. However, the amounts and types of antimicrobial substances extracted depend on diverse extraction methods. Ultrasound-assisted extraction is considered as one of the best green extraction methods for extracting bioactive compounds from various spices, including garlic. The advantages include low-temperature extraction, easy operation, less cost, time, and energy requirement, and reduced use of toxic chemicals [21,22]. Several studies have conducted ultrasonic-assisted extraction of bioactive antimicrobial substances from garlic such as allicin, essential oil, flavonoids, polyphenols, and sulfur compounds [23–27]. However, despite numerous studies done on garlic extract particles against bacteria, the antibacterial activity of probe ultrasonicated garlic extract against *Staphylococcus aureus* sub. *aureus*, *Streptococcus mutans*, *Escherichia coli*, and *Poryphyromonas gingivalis* are still not certain. This present study evaluates the antibacterial activity of probe ultrasonicated garlic extract against the four listed bacteria so that the results of this study can be utilized for the future development of novel antibacterial agents for replacing the existing antibacterial agents, against which the tested bacteria have developed resistance.

2. Materials and Methods

2.1. Source of Bacteria Strains

The tested microorganisms, *Streptococcus mutans* 11823 (ATCC 25175), *Escherichia coli* (ATCC 11234), *Staphylococcus aureus* sub. *aureus* (KCT 1928), and *Poryphyromonas gingivalis* (KCT 5352) were purchased from the Korean Culture Center for Microorganisms.

2.2. Culture Preparation

All tested bacteria were activated by re-culturing them on their specific agar growth media, *E. coli* was re-cultured on trypticase soy agar (TSA), *S. mutans*, *P. gingivalis*, and *S. aureus* sub. *aureus* were re-cultured on brain heart infusion (BHI) agar, and the agar plates were incubated for 24 h at 37 degrees Celsius in an inverted position. After 24 h, bacteria were picked from each agar plate as single colonies and then sub-cultured into their specific broth media. Using the spectrophotometer (Optizen 2120UV plus), the turbidity of the broth culture was standardized at an optical density (OD) of 0.05 at 600 nm, before being tested against a garlic extract.

2.3. Preparation of Ultrasonicated Aqueous Garlic Extract

Fresh bulbs of garlic (*Allium sativum*, obtained from a local market, Gwangjin-gu, Seoul, Korea) were washed using tap water, peeled, cut into small pieces, and then dried in the oven at 55 degrees Celsius for seven days. The dried garlic was grounded using an electric blender, and 20 g of powder was measured and mixed with 100 mL of distilled water (DW) in a conical flask by the stirring machine. Four samples of 10 mL each were taken from the aqueous garlic mix, placed in a plastic tube, sonicated (Sonopuls HD 2200 probe ultrasonicator, Bandelin GmbH and Co. KG, Berlin, Germany) at 20 kHz frequency for 10 min at 100 power (W), and placed in a shaking incubator for 24 h at 300 rpm. After incubation, all sample solutions were separated from impurities by centrifugation at $10,000 \times g$ for 5 min, the supernatants were collected, and the pellets were discarded (ultrasonicated supernatant extract). Different concentrations of 100, 80, 60, 40, 20, 10, and 5 mg/mL were prepared from the ultrasonicated extract by dilution with distilled water (DW) and then tested against all four tested bacteria. Two more samples of 40 mg/mL were prepared from the sonicated aqueous garlic mix. The ultrasonicated samples were either filtrated by using Whatman No. 1 paper filter (ultrasonicated extract without centrifugation) or centrifuged (ultrasonicated supernatant). Another sample of the same concentration was prepared without sonication (extract without sonication or unsonicated extract).

2.4. Antibacterial Activity by Agar Disc Diffusion

The agar disk diffusion method [28] was used to determine the antibacterial effect of garlic extract against *E. coli*, *S. mutans*, *P. gingivalis*, and *S. aureus* sub. *aureus*. Prepared agar plates of different nutrient agar media were inoculated with 0.1 mL of a broth culture of tested bacteria. Using a sterile L-shape spreader; the inoculums were spread over the agar surface of the plates and kept aside. Sterile paper disks of 10 mm of diameter were saturated with 0.1 mL of different concentrations (100,80,60,40,20,10, and 5 mg/mL) of ultrasonicated garlic extract or 40 mg/mL of garlic extract sonicated without centrifugation, sonicated supernatant, or without sonication to be laid onto the seeded plates. Another disk was impregnated with 25 mg/mL of streptomycin (standard) and used as a positive control. Petri dishes with disks (saturated with garlic extract and control) were incubated overnight at 37 degrees Celsius in an inverted position. After incubation, the diameters of the zone of inhibition for each respective bacteria for every prepared concentration was measured around each disk using a ruler in millimeters [29]. Each assay was repeated in triplicate.

2.5. Antibacterial Activity by Agar Well Diffusion Method

The Agar well diffusion method is a well known method that is used frequently to determine the antibacterial effect of plant extracts. Using this method, prepared sterile agar plates were seeded with 0.1 mL of standardized bacterial inoculum on agar surfaces by an L-shaped spreader, and a sterile 9 mm cork borer was used to create eight uniform wells (holes) into the agar. Using the micropipette 100 µL of each concentration (100, 80, 60, 40, 20, 10, and 5 mg/mL) of ultrasonicated garlic extract or 40 mg/mL of garlic extract sonicated without centrifugation, sonicated supernatant, or without sonication were introduced into different wells. Moreover, well number eight or four was used as a positive control, and was inoculated with streptomycin (25 mg/mL). Plates were kept on a clean bench for 1 h to facilitate full penetration of garlic extracts in seeded agar petri dishes, then all plates were incubated for 24 h at 37 degrees Celsius and then the diameter of the inhibition zone around each well was measured in millimeter [29]. Each assay was repeated in triplicate, and the mean inhibition zone was calculated and recorded as the final inhibition zone for each set.

2.6. Post Interaction Antibacterial Activity

The antibacterial effect of garlic extract was determined by using the plate count technique [18], which indicates the number of bacteria that survived (total viable count) after overnight interaction between the garlic extract and tested bacteria. Total viable count (TVC) was calculated by mixing 5 mL of the selected serially diluted bacterial broth with 150 µL of each concentration of garlic extract in a test tube. Then, the tubes were incubated for 24 h in a shaking incubator at 37 degrees Celsius at 120 rpm. After incubation,100 µL of each interacted sample was poured out of the tubes and seeded on agar plates for overnight incubation at 37 degrees Celsius. Then, the total viable count was calculated on each plate and represented as colony-forming units per milliliter (CFU/mL) [30]. Each assay was done in triplicate to minimize errors.

2.7. Characterization of Garlic Nanoparticles

Characterization of garlic nanoparticles was done using different methods. First the garlic extract was characterized using a microplate spectrophotometer (SPECTRAmax PLUS 384). Briefly, four samples of 10 mL each were taken from the mother solution and placed in a plastic tube, then sonicated using variable frequencies for differing amounts of time and at different power as indicated below: sample 1 was sonicated at 20 kHz for 5 min at 100 power (W) ultrasonic, sample 2 was sonicated at 20 kHz for 5 min at 200 power (W), sample 3 was sonicated at 20 kHz for 10 min at 100 Power (W), sample 4 at 20 kHz for 10 min at 200 power (W), and then all the samples were placed in shaking incubator for 24 h at 300 rpm. After incubation, all sample solutions were separated from impurities by centrifugation at $10,000 \times g$ for 5 min, and the supernatants were collected, and their absorbance was scanned from 200 nm to 700 nm wavelength. Three replicates for each analysis were used, and the mean value of absorbance was obtained and recorded for graphic representation.

The obtained garlic extract nanoparticles were characterized using the Fourier-transform infrared spectroscopy (FTIR) method. Four prepared samples of ultrasonicated garlic extract were dried in the oven for seven days, and then the precipitated pellets were analyzed using FTIR, and the results were recorded on an FTIR spectrometer in the range 4000–500 cm^{-1}. An additional sample without sonication was used as a garlic control.

Furthermore, the size and morphology of the ultrasonicated garlic extract particles were characterized by utilizing transmission electron microscopy (TEM), where all four prepared samples were diluted from 1/100, 1/1000 to 1/1000 dilution factors, 2 µL of each sample were placed on the carbon-coated copper grid for overnight incubation, and then all prepared copper grids were analyzed using TEM.

2.8. Statistical Analysis

All treatments were repeated two times. Data were analyzed using a SAS program, Release 9.2. The significance of differences among the means was determined using analysis of variance and Duncan's multiple range test ($p \leq 0.05$).

3. Results and Discussion

3.1. Inhibition of Nanoparticles of Garlic Extract on the Different Bacteria

The antibacterial activity of the nanoparticles of the ultrasonicated extract was assessed against two strains of gram-positive bacteria (*S. mutans* and *S. aureus* sub. *aureus*) and two strains of gram-negative bacteria (*E. coli* and *P. gingivalis*) by using both the agar disk diffusion and agar well diffusion methods. The diameters of the inhibition zones were determined, and the results are illustrated in Figure 1 and Tables 1–4. Garlic extracts have also been reported to be effective against both gram-positive and -negative bacteria such as *Bacillus cereus, Escherichia coli, Enterobacter cloacae, Klebsiella pneumoniae, Micrococcus flavus, Proteus mirabilis, Pseudomonas aeruginosa, Salmonella typhi, Salmonella typhimurium*, and *Staphylococcus aureus* [31–34].

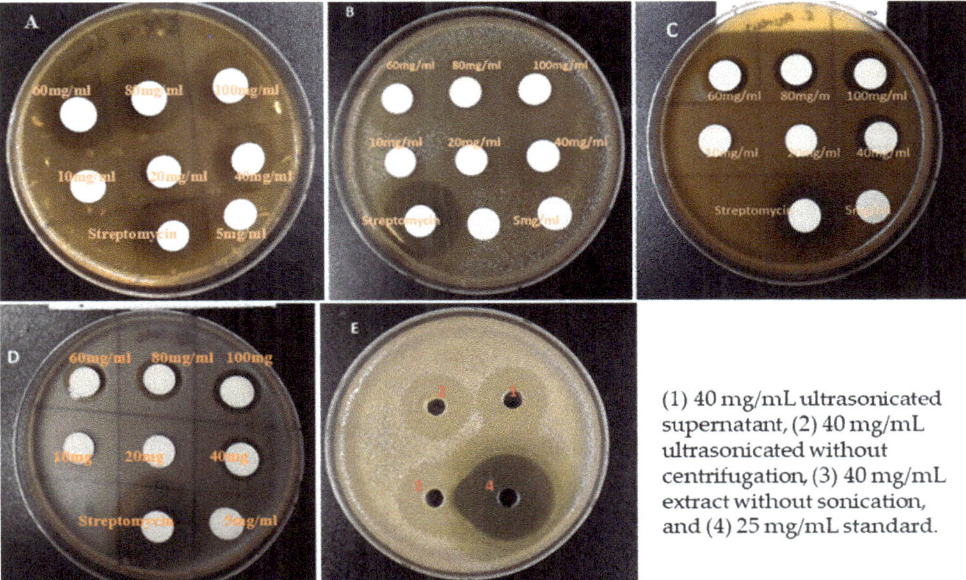

Figure 1. Antibacterial activity of different concentrations of nanoparticles (supernatant) of garlic extract against *S. mutans* (**A**), *P. gingivalis* (**B**), *S. aureus* sub. *aureus* (**C**), and *E. coli* (**D**) by agar disk diffusion method. (**E**) represents the antibacterial effect of 40 mg/mL garlic extract treated in different conditions against *P. gingivalis* by the agar well diffusion method.

Table 1. Antibacterial activity of nanoparticles (supernatant) of garlic extract against *S. mutans*, *S. aureus* sub. *aureus*, *E. coli*, and *P. gingivalis* by agar disk diffusion method.

Garlic Extract (mg/mL)	Inhibition Zone (mm)			
	S. mutans	*S. aureus* sub. *aureus*	*E. coli*	*P. gingivalis*
5	0.0 ± 0.0 g	0.0 ± 0.0 d	0.0 ± 0.0 f	0.0 ± 0.0 g
10	0.0 ± 0.0 g	0.0 ± 0.0 d	0.0 ± 0.0 f	0.0 ± 0.0 g
20	12.1 ± 0.7 f	0.0 ± 0.0 d	0.0 ± 0.0 f	11.2 ± f
40	20.2 ± 0.6 e	15.7 ± 0.6 c	14.2 ± 0.8 e	19.3 ± e
60	22.2 ± 1.0 d	16.0 ± 1.0 c	15.3 ± 0.6 d	21.5 ± d
80	24.2 ± 0.7 c	18.1 ± 0.9 b	16.4 ± 0.5 c	23.2 ± c
100	26.2 ± 0.8 b	19.1 ± 1.0 b	17.4 ± 0.5 b	25.3 ± b
Standard	33.0 ± 0.0 a	26.3 ± 0.0 a	24.1 ± 0.0 a	31.9 ± 0.0 a
R-Square	0.9982	0.9922	0.9985	0.9989
Coeff Var	3.41	8.84	3.91	2.71
Root MSE	0.59	1.05	0.43	0.45

Values are the mean ± standard deviation (S.D.) of three replicates. According to Duncan's multiple range test, S.D. within a column followed by different letters are significantly different at $p \leq 0.05$ level. Coeff Var: oefficient of variation; Root MSE: Root-mean-square deviation.

Table 2. Antibacterial activity of nanoparticles (supernatant) of garlic extract against *S. mutans*, *S. aureus* sub. *Aureus*, *E. coli*, and *P. gingivalis* by agar well diffusion method.

Garlic Extract (mg/mL)	Inhibition Zone (mm)			
	S. mutans	*S. aureus* sub. *aureus*	*E. coli*	*P. gingivalis*
5	0.0 ± 0.0 g	0.0 ± 0.0 e	0.0 ± 0.0 e	0.0 ± 0.0 g
10	0.0 ± 0.0 g	0.0 ± 0.0 e	0.0 ± 0.0 e	0.0 ± 0.0 g
20	15.6 ± 0.6 f	0.0 ± 0.0 e	0.0 ± 0.0 e	14.6 ± 0.7 f
40	22.2 ± 0.4 e	17.3 ± 0.8 d	16.2 ± 0.7 d	21.1 ± 0.4 e
60	24.5 ± 0.9 d	18.9 ± 1.5 c	17.4 ± 0.4 c	23.6 ± 0.5 d
80	26.8 ± 0.6 c	20.0 ± 1.1 c	18.1 ± 0.4 c	25.2 ± 0.9 c
100	28.7 ± 0.9 b	21.4 ± 0.5 b	19.3 ± 0.6 b	27.1 ± 0.7 b
Standard	34.1 ± 0.0 a	27.2 ± 0.0 a	25.5 ± 0.0 a	35.4 ± 0.0 a
R-Square	0.9985	0.9967	0.9990	0.9987
Coeff Var	2.94	5.60	3.05	2.82
Root MSE	0.56	0.73	0.37	0.52

Values are the mean $\pm$ standard deviation (S.D.) of three replicates. According to Duncan's multiple range test, S.D. within a column followed by different letters are significantly different at $p \leq 0.05$ level. Coeff Var: coefficient of variation; Root MSE: Root-mean-square deviation.

Table 3. Antibacterial activity of 40 mg/mL garlic extracts treated in different conditions by agar disk diffusion method.

Garlic Extract (40 mg/mL)	Inhibition Zone (mm)			
	S. mutans	*S. aureus* sub. *aureus*	*E. coli*	*P. gingivalis*
Extract without sonication	16.4 ± 0.5 d	12.7 ± 0.3 d	11.7 ± 0.4 d	15.1 ± 0.3 d
Sonicated without centrifugation	18.6 ± 0.5 c	14.7 ± 0.4 c	13.5 ± 0.3 c	17.5 ± 0.6 c
Sonicated supernatant	19.5 ± 0.5 b	15.5 ± 0.4 b	14.5 ± 0.5 b	18.7 ± 0.9 b
Standard	33.0 ± 0.0 a	26.3 ± 0.0 a	24.1 ± 0.0 a	34.2 ± 0.0 a
R-Square	0.9973	0.9976	0.9967	0.9960
Coeff Var	1.88	1.84	2.12	2.73
Root MSE	0.41	0.32	0.34	0.59

Values are the mean $\pm$ standard deviation (S.D.) of three replicates. According to Duncan's multiple range test, S.D. within a column followed by different letters are significantly different at $p \leq 0.05$ level. Coeff Var: coefficient of variation; Root MSE: Root-mean-square deviation.

The results show that different bacteria species exhibited different sensitivities against the ultrasonicated garlic extract at different concentrations of 5, 10, 20, 40, 80, and 100 mg/mL. The highest inhibition was observed against *S. mutans* at concentrations of 100 mg/mL of garlic nanoparticles and showed the zone of inhibition of 26.2 ± 0.8 mm by agar disk diffusion (Table 1) and 28.7 ± 0.9 mm by agar well diffusion method (Table 2), which is the greatest inhibition among all tested strains. *P. gingivalis* was ranked as the second most susceptible, followed by *S. aureus* sub. *Aureus,* and then *E.coli* was the bacteria to be least inhibited by nanoparticles from garlic extract in both the well and disk diffusion methods. Several studies have shown that garlic extracts possess an intense antibacterial activity against *S. mutans* [35]. The antibacterial activity of garlic is due to its phytochemicals such as allicin, flavonoids, polyphenols, and sulfur compounds [23–27].

Table 4. Antibacterial activity of 40 mg/mL garlic extracts treated in different conditions by agar well diffusion method.

Garlic Extract (40 mg/mL)	Inhibition Zone (mm)			
	S. mutans	*S. aureus* sub. *aureus*	*E. coli*	*P. gingivalis*
Extract without sonication	18.3 ± 0.3 d	14.3 ± 0.5 d	13.2 ± 0.3 d	17.6 ± 0.5 d
Sonicated without centrifugation	20.2 ± 0.5 c	16.2 ± 0.3 c	14.3 ± 0.4 c	19.3 ± 0.6 c
Sonicated supernatant	21.9 ± 0.3 b	17.2 ± 0.4 b	16.4 ± 0.6 b	20.4 ± 0.7 b
Standard	34.1 ± 0.0 a	27.2 ± 0.0 a	25.5 ± 0.0 a	35.4 ± 0.0 a
R-Square	0.9983	0.9971	0.9958	0.9961
Coeff Var	1.32	1.75	2.21	2.36
Root MSE	0.31	0.32	0.38	0.55

Values are the mean ± standard deviation (S.D.) of three replicates. According to Duncan's multiple range test, S.D. within a column followed by different letters are significantly different at $p \leq 0.05$ level. Coeff Var: coefficient of variation; Root MSE: Root-mean-square deviation.

Escherichia coli was the least susceptible and showed the minimum sensibility when tested against 40 mg/mL with a corresponding inhibition zone of 14.2 ± 0.8 mm and 16.3 ± 0.7 mm using the agar disk diffusion and agar well diffusion methods, respectively. The positive control result (streptomycin) confirmed that the bacteria were susceptible to streptomycin, the bacteria had various diameters of inhibition zones of 33.0, 31.9, 26.3, and 24.1 mm around the disk for *S. mutans*, *P. gingivalis*, *S. aureus* sub. *Aureus*, and *E. coli*, respectively, when tested by using agar disk diffusion (Table 1). Similarly, the antibacterial activity of garlic extracts was also found to be less effective against *E. coli* and *S. aureus* [36–38].

On the other hand, the present study proved that all tested bacteria were resistant to 5 mg/mL and 10 mg/mL. Moreover, both *E. coli and S. aureus* sub. *aureus* were resistant specifically to 20 mg/mL. These results are in agreement with Khashan [32], who assessed the antibacterial activity of garlic extract against *S. aureus* and found that concentrations of garlic extract ranging from 10 to 20 mg/mL were unable to inhibit the growth of *S. aureus*. Moderate growth inhibition was observed at concentrations of garlic extract ranging from 40 to 60 mg/mL, while the concentrations of 80 to 100 mg/mL showed the strongest inhibition activity against *S. aureus* [32]. Contrary to *E. coli* and *S. aureus* sub. *aureus*, during our study *S. mutans* and *P. gingivalis* were sensitive to 20 mg/mL (Tables 1 and 2).

In general, all four bacteria tested, whether gram-negative or gram-positive, were sensitive to nanoparticles of garlic extract, regardless of the concentration tested in this study. However, the growth inhibition depended on the bacterial species. These findings are consistent with previous research that looked at the antibacterial activity of south Indian spices [39], Greek garlic genotypes [33], and Chinese and Desi varieties [40] against *Aeromonas hydrophila*, *B. cereus*, *E. coli*, *E. cloacae*, *Enterococcus faecalis*, *K. pneumonia*, *Listeria monocytogenes*, *M. flavus*, *P. mirabilis*, *P. aeruginosa*, and *Salmonella* species.

The obtained results also show that all tested bacteria had higher inhibition zones when tested using the agar well diffusion method than when tested using the disk diffusion method. This could be due to the high absorption of garlic extract when introduced into the wells, whereas in the disk diffusion method, the disk is pressed on the agar surface and does not allow for the complete diffusion of garlic extract on the agar surface. Another supporting point is that when utilizing the disk diffusion approach, some extract's active components may be held within the disk's pores and limiting their ability to reach the inoculation media, preventing the extract from performing to its full potential [41].

In the present study, the diameter of the inhibition zone increased with the concentration of garlic extract; the more the concentration increased, the more the zone of inhibition increased (Tables 1 and 2). The results of this study are in line with the findings of the research done by Fatemeh et al. [42], who investigated the antibacterial effect of garlic and *Eucalyptus* extracts on oral cariogenic bacteria, and reported that both *Streptococcus mutans* and *Lactobacillus acidophilus* were sensitive to garlic extract and the association between concentration of garlic extract and the inhibition zone was proved. As the concentration increased, the diameter of the inhibition zone gradually increased [42].

The results presented in Tables 3 and 4 indicate the inhibition zones of the garlic extract on the four tested bacteria at a 40 mg/mL concentration under three different extraction conditions. The sonicated garlic extract treatment without centrifugation was the second strongest extract to inhibit the growth of all tested bacteria. Garlic extract without sonication showed the minimum inhibition zone compared to other extracts (Table 3). By using the agar well diffusion method, 40 mg/mL of the sonicated supernatant extract treatment showed high inhibition zones when compared with other types of garlic extracts. The inhibition zones of all tested bacteria were 21.9 ± 0.3, 17.2 ± 0.4, 16.4 ± 0.6, and 20.4 ± 0.7 mm on *S. mutans*, *S. aureus* sub. *aureus*, *E. coli*, and *P. gingivalis* respectively. The results were better than that of Garba et al. [43] and Liaqat et al. [44]. Inhibition zones of 24 mm and 23 mm were obtained against *E. coli* and *S. aureus*, respectively, when methanolic extract of garlic was used at 200 mg/mL [43]. On the other hand, methanolic garlic extract (200 mg/mL) exhibited higher activity against *S. aureus* (25.33 mm) followed by *E. coli* (22.33 mm) [44].

3.2. Antibacterial Activity of Nanoparticles of Garlic Extract on Tested Bacteria by Plate Count Method

Results showing the total viable counts of all tested bacteria in colony-forming units per millimeter (CFU/mL) post interaction with the garlic extract and bacteria broth culture are illustrated in Figure 2. The results proved that all tested bacteria were challenged by the garlic extract during overnight incubation and led to a high reduction in cell number of *S. mutans* at 100 mg/mL. *S. mutans* decreased from 1.77×10^4 CFU/mL (control) up to 0 CFU/mL, *P. gingivalis* showed a decline from 2.01×10^4 to 0 CFU/mL, *S. aureus* sub. *aureus* reduced from 2.05×10^5 CFU/mL to 0 CFU/mL, and the most resistant bacteria *E. coli*, decreased to 0 CFU/mL when tested with 100 mg/mL concentrations, compared to the control (2.03×10^5 CFU/mL).

In general, there was a reduction in total viable counts of all tested bacteria when bacteria interacted with the different concentrations of garlic extract from the lowest concentration of 5 mg/mL to the highest concentration of 100 mg/mL compared with the control bacteria cultures. In this study, it was discovered that increasing the concentration of garlic extract reduces the number of bacteria that survive in CFU/mL; these findings are consistent with those of Alwazni et al. [45], who compared the antibacterial effects of garlic and onion on *E. coli*, *Salmonella typhi*, *Pseudomonas aeruginosa*, and *Klebsiella pneumonia*.

3.3. Role of Sonication on Antibacterial Activity of Garlic Extract

The current findings show that 40 mg/mL of the sonicated supernatant garlic extract had the greatest inhibition on *S. mutans*. The counts reduced from 1.77×10^4 CFU/mL (control) to 0.58×104 CFU/mL, whereas 40 mg/mL the sonicated garlic extract without centrifugation reduced *S. mutans* colonies to 0.64×10^4 CFU/mL, and garlic extract without sonication did not show significant inhibition in the sonicated extracts. It reduced the bacterial counts to 0.88×10^4 CFU/mL. The second bacteria to be challenged with 40 mg/mL of the sonicated garlic extract were *P. gingivalis*; a decline from 2.01×10^4 CFU/mL to 0.64×10^4 CFU/mL was observed. The sonicated samples without centrifugation were effective against *P. gingivalis* up to 0.7×10^4 and the unsonicated garlic extract inhibited *P. gingivalis* up to 1.0×10^4 CFU/mL, *Staphylococcus aureus* followed, showing maximum effectivity at 40 mg/mL of the sonicated supernatant sample, bringing about a reduction in the bacterial count from 2.05×10^5 CFU/mL to 0.67×10^5 CFU/mL compared to 1.2×10^5 CFU/mL of the unsonicated sample (Figure 3).

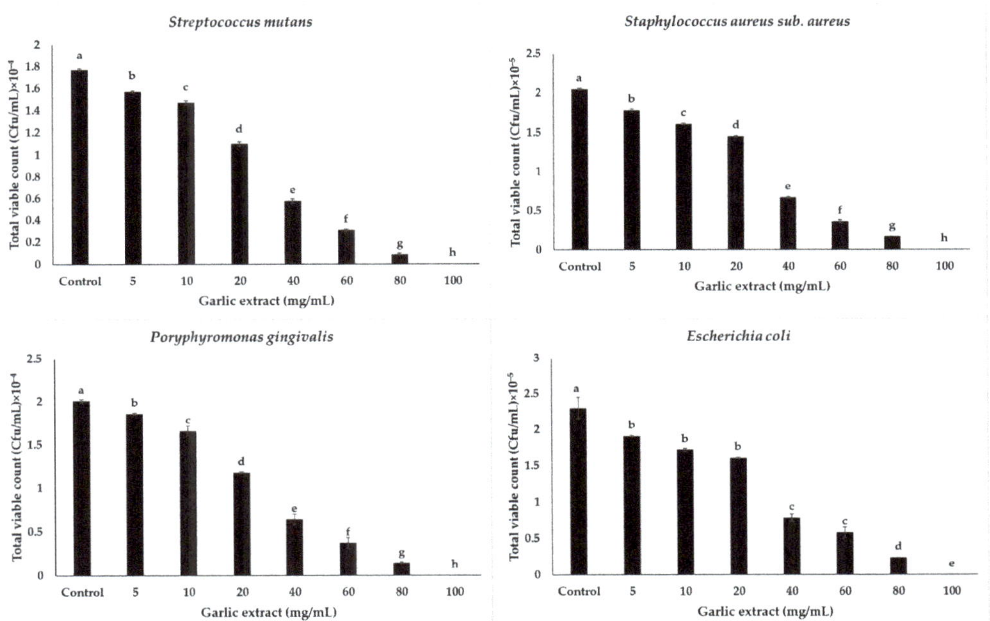

Figure 2. Graphical presentation of antibacterial activity of nanoparticles (supernatant) of garlic extract in terms of total viable counts (CFU/mL) on tested bacteria. Values are means of determinations of three replicates, and bars represent standard deviations (S.Ds.) of the means. S.Ds. followed by different letters are significantly different at $p \leq 0.05$ level by Duncan's multiple range test.

The least susceptible bacteria (*E. coli*) when treated at concentrations of 40 mg/mL showed a reduction in counts from 2.30×10^5 CFU/mL to 0.78×10^5 CFU/mL against the sonicated supernatant garlic extracts. The unsonicated garlic extracts without centrifugation decreased *E. coli* to 0.84×10^5 CFU/mL. The sample that showed the least bacterial inhibition on *E. coli* was garlic extract without sonication (1.4×10^5 CFU/mL). The above findings indicate the positive influence of sonication on increasing the capacity of garlic extract to inhibit the growth of all tested bacteria through increasing extraction of active compounds, mostly allicin, the biological compound of garlic that is responsible for many biological benefits of garlic extract, including its antibacterial property [46]. This is in accordance with another investigations based on the influence of ultrasound, microwaves, and other factors on synthetic allicin and showed that allicin production during ultrasonication extraction increases with sonication, through cavitation effects caused by ultrasound on the cell material by disrupting the cell wall structure, increasing the speed of diffusion, causing cell lysis, and ultimately releasing the cell contents as well, which indicates the high release of allicin during ultrasonication of garlic during extraction [47]. Effective extraction of garlic active compounds during sonication could be explained by the mechanism of the high solubility of garlic in water when ultrasonicated [21], leading to the breaking of garlic into nanoparticles that easily interact with biological systems.

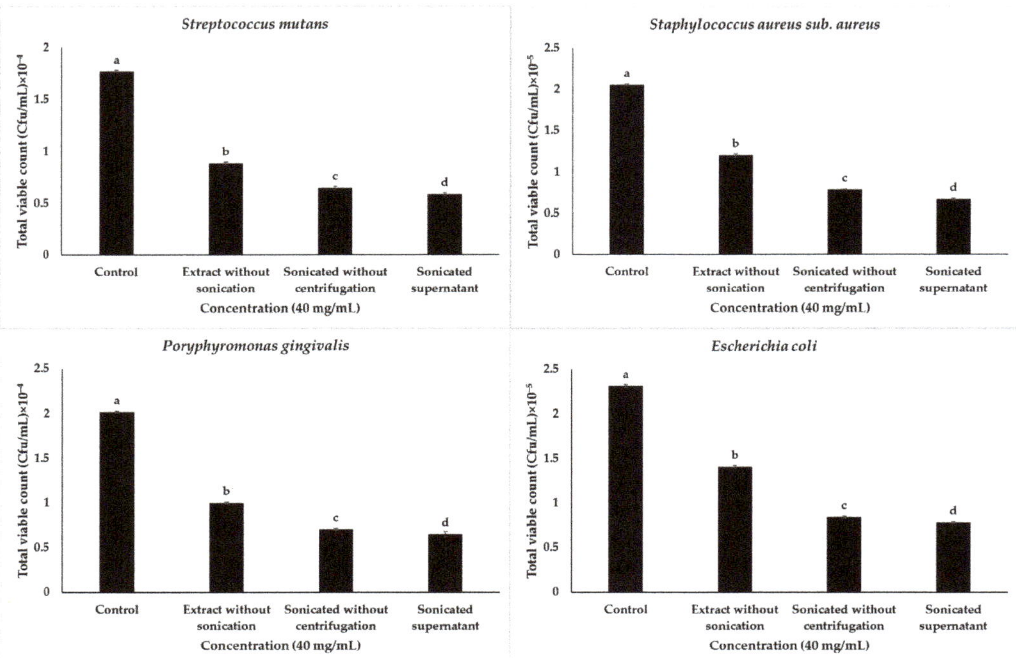

Figure 3. Graphical presentation of the total viable count of 40 mg/mL of garlic extract treated in different conditions on tested bacteria. Values are means of determinations of three replicates, and bars represent standard deviations (S.Ds.) of the means. S.Ds. followed by different letters, are significantly different at $p \leq 0.05$ level by Duncan's multiple range test.

Briefly, the effect of sonication on the antibacterial capacity of the garlic extract in the reduction of total viable counts when bacteria were tested against 40 mg/mL treated in different conditions was reported in Figure 3 and indicated that the sonicated supernatant garlic extract significantly reduced the total viable bacteria counts compared to other types of garlic extracts.

3.4. Characterization of Nanoparticles of Garlic Extract by Microplate Spectrophotometer

The UV-Vis spectrophotometric characterization was carried out using a microplate spectrophotometer, and the absorbance was scanned in the wavelength range from 200 nm to 750 nm. The spectrophotometer results showed that the maximum absorbance of nanoparticles of garlic extract was in the visible wavelength range of around 240–300 nm, which is typically the range of allicin [48], which is an organosulfur compound found in garlic and plays the vital role in the antibacterial property of garlic extract.

During this study, it was observed that the sonicated samples showed high absorbance compared to the unsonicated sample (control); however, no significant differences in the peak location or absorbance was observed among the four sonicated samples, prepared at various times and at different sonication power (Figure 4).

A slight difference was observed in sonication time which indicated that as sonication time increased, the absorbance of the four sonicated samples slightly increased. The marginal differences might be due to high extraction efficiency when the sonication time increased; however, the power (Watt) did not show an impact on the peak location or absorbance.

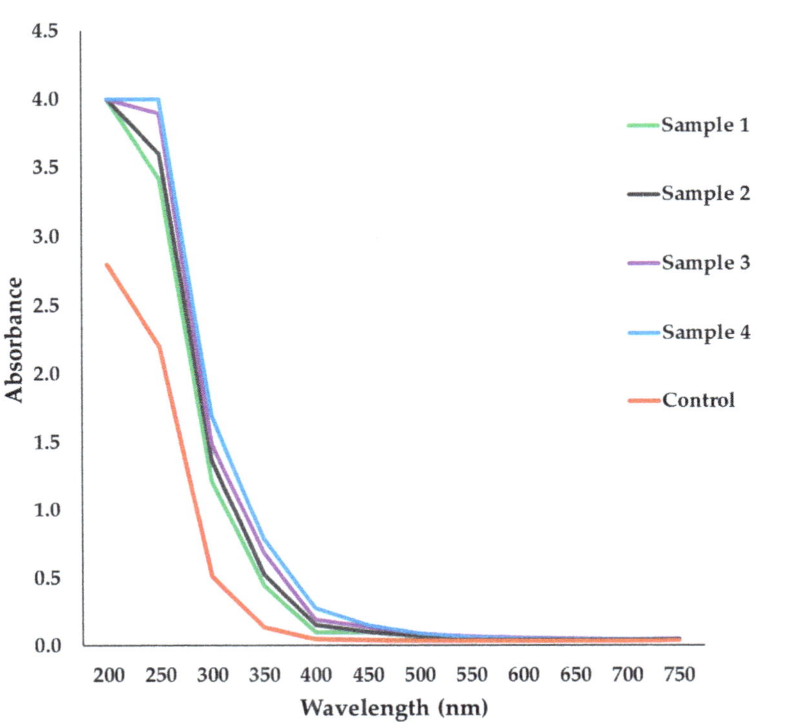

Figure 4. Microplate spectrophotometer absorption of nanoparticles of the garlic extract prepared using various sonication time and power (Sample 1, 2, 3, 4: sonicated samples, control: unsonicated sample).

3.5. Characterization of Nanoparticle of Garlic Extract by FTIR

FTIR spectroscopy was used to discover the successful extraction of garlic nanoparticles, and spectra were obtained in the wavenumber range of 500–3500 cm^{-1}. The results of the FTIR spectrum of the ultra-sonicated garlic extract showed visible peaks at around 3345.8, 2934.1, 2360.5, 1635.6, 1417.5, 1130.7, 1025, 930.61, 815.4, and 591.5 cm^{-1} wavenumbers (Figure 5).

The wide peak at 3345.8 cm^{-1} can be assigned to the presence of the O–H stretching vibration in the hydroxyl group. These results also indicated that there is asymmetric stretching in the C–H bonds at 2934.1 cm^{-1}, at 1635 cm^{-1} FTIR revealed the presence of carbonyl or carboxylic (C=O) stretching bands, the same results confirm the presence of the –O–H bend in carboxylic at 1417 cm^{-1}, at 1130.7 cm^{-1} the results revealed the presence of an S=O bond, the presence of C–N stretching vibrations in primary amines was observed at 1025 cm^{-1}, at 930.6 cm^{-1} FTIR showed the presence of a γ-C–H deformation in =CH2, at 815 cm^{-1} we observed the presence of an S–C bond which might indicate the absorption of allicin, an organosulfur compound present in the garlic extract [49], and finally these results recorded the presence of a C–H bend in the alkynes at 530 cm^{-1}.

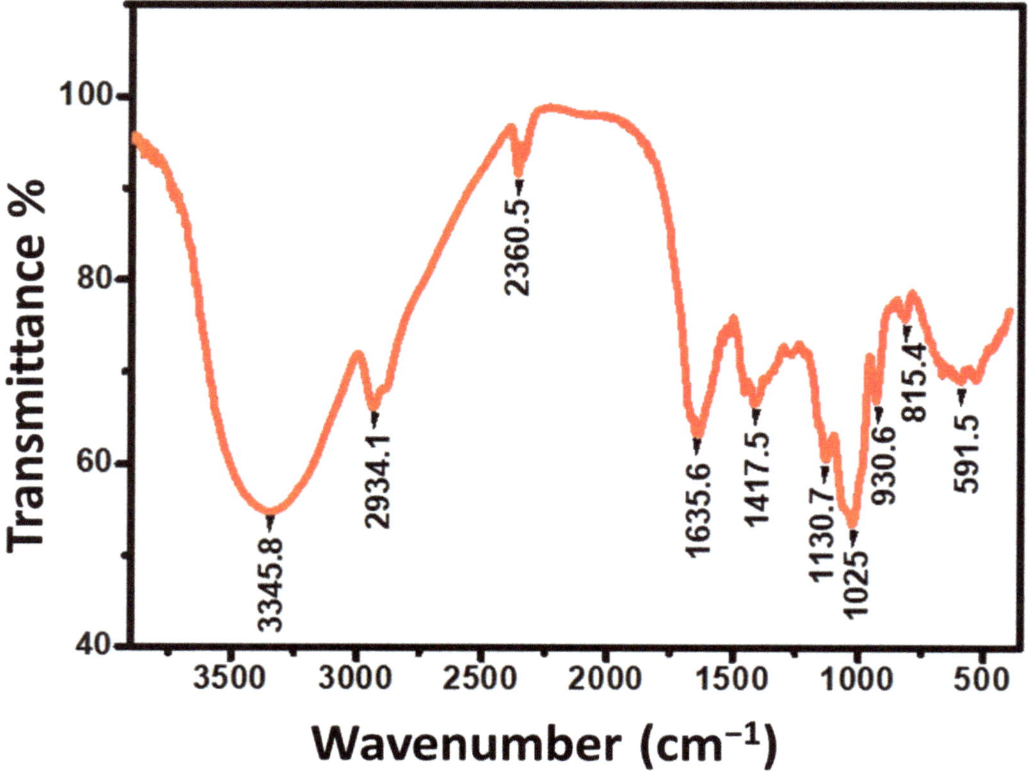

Figure 5. FTIR spectrum of nanoparticles from garlic extract.

The current ultrasonicated garlic extract FTIR spectrum matches previous research on the presence of functional groups in aqueous garlic extracts [50,51]. Based on the spectrum mentioned above, we would say that phenolic, organosulfur compounds, amino acids, carboxylic groups, and proteins are the active groups that play a key part in the antibacterial activity of ultrasonicated garlic extract. This conclusion is supported by findings of several investigations that found the same major phytochemicals in garlic extract [52,53].

3.6. Characterization of Nanoparticles of Garlic Extract by TEM

In order to confirm the nature of the nanoparticles from the ultrasonicated garlic extract, TEM was used and it revealed that the garlic nanoparticles consisted of random sized particles, a few rods, and a few spherical particles (Figure 6), which were randomly dispersed, and had small sizes (less than 50 nm). The antibacterial property of garlic extract could be attributed to its bioactive particles' small size and morphology since particle size and surface area influence the interaction between chemical compounds and biological systems. Reduction in the size of bioactive particles leads to an increased surface area coming in contact with microorganisms, enhancing their interaction and leading to antibacterial activities [54], this is in line with other studies that have been done on the antibacterial properties of nanoparticles and have indicated that small and formless particles are the most effective particles to inhibit the growth of bacteria [55,56]. Various writings have documented how nanoparticles act and highlighted that they inhibit bacterial growth by anchoring to and penetrating the bacterial cell wall. When the reach the inside of the bacteria, they start modulating cellular signaling by dephosphorylating putative key peptide substrates on tyrosine residues leading to the inhibition of bacteria growth [57].

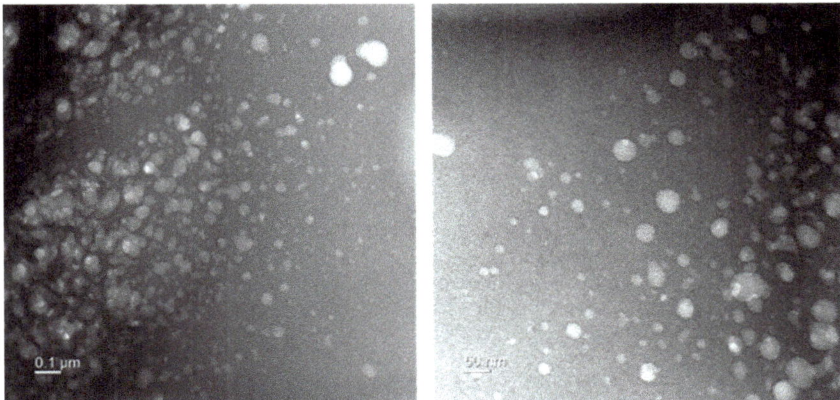

Figure 6. Transmission electron microscope image of garlic nanoparticles.

4. Conclusions

In general, the results of this present study demonstrated that the ultrasonicated garlic extracts showed the antibacterial capacity to inhibit the growth of *S. mutans*, *S. aureus* sub. *aureus*, *P. gingivalis*, and *E. coli* bacteria. The efficiency of garlic against bacteria might be related to its phenolic, organosulfur compounds, amino acids, carboxylic groups, and protein contents. Furthermore, we strongly recommend further studies to evaluate the antibacterial activity of ultrasonicated garlic extract against viruses and fungi.

Author Contributions: Conceptualization, T.G. and S.C.C.; methodology, T.G., A.V., S.J.K. and I.S.; formal analysis, T.G., A.V. and S.C.C.; investigation, T.G. and S.C.C.; writing—original draft preparation, T.G., I.S., K.D.K. and S.C.C.; writing—review and editing, I.S. and S.C.C.; supervision, S.C.C. All authors have read and agreed to the published version of the manuscript.

Funding: This research received no external funding.

Institutional Review Board Statement: Not applicable.

Informed Consent Statement: Not applicable.

Data Availability Statement: Not applicable.

Conflicts of Interest: The authors declare no conflict of interest.

Abbreviations

CFU: Colony-forming unity; FTIR: Fourier-transform infrared spectroscopy; TEM: Transmission electron microscopy; TVC: Total viable count; WHO: World Health Organization.

References

1. Klein, E.Y.; Van Boeckel, T.P.; Martinez, E.M.; Pant, S.; Gandra, S.; Levin, S.A.; Goossens, H.; Laxminarayan, R. Global increase and geographic convergence in antibiotic consumption between 2000 and 2015. *Proc. Natl. Acad. Sci. USA* **2018**, *115*, E3463–E3470. [CrossRef] [PubMed]
2. Andersson, D.I.; Hughes, D. Persistence of antibiotic resistance in bacterial populations. *FEMS Microbiol. Rev.* **2011**, *35*, 901–911. [CrossRef]
3. Schroeder, M.; Brooks, B.D.; Brooks, A.E. The Complex Relationship between Virulence and Antibiotic Resistance. *Genes* **2017**, *8*, 39. [CrossRef] [PubMed]
4. Bassoum, O.; Sougou, N.M.; Diongue, M.; Lèye, M.M.M.; Mbodji, M.; Fall, D.; Seck, I.; Faye, A.; Tal-Dia, A. Assessment of General Public's Knowledge and Opinions towards Antibiotic Use and Bacterial Resistance: A Cross-Sectional Study in an Urban Setting, Rufisque, Senegal. *Pharmacy* **2018**, *6*, 103. [CrossRef]
5. Lee, C.R.; Cho, I.H.; Jeong, B.C.; Lee, S.H. Strategies to minimize antibiotic resistance. *Int. J. Environ. Res. Public Health* **2013**, *10*, 4274–4305. [CrossRef]

6.	Álvarez-Martínez, F.J.; Barrajón-Catalán, E.; Micol, V. Tackling Antibiotic Resistance with Compounds of Natural Origin: A Comprehensive Review. *Biomedicines* **2020**, *8*, 405. [CrossRef]

7.	Vuorela, P.; Leinonen, M.; Saikku, P.; Tammela, P.; Wennberg, T.; Vuorela, H. Natural products in the process of finding new drug candidates. *Curr. Med. Chem.* **2004**, *11*, 1375–1389. [CrossRef] [PubMed]

8.	Singh, B.; Singh, B.; Kishor, A.; Singh, S.; Bhat, M.N.; Surmal, O.; Musarella, C.M. Exploring Plant-Based Ethnomedicine and Quantitative Ethnopharmacology: Medicinal Plants Utilized by the Population of Jasrota Hill in Western Himalaya. *Sustainability* **2020**, *12*, 7526. [CrossRef]

9.	WHO. *Traditional Medicine Strategy 2002–2005*; WHO: Geneva, Switzerland, 2002.

10.	Karuppiah, P.; Rajaram, S. Antibacterial effect of Allium sativum cloves and Zingiber officinale rhizomes against multiple drug resistant chemical pathogens. *Asian Pac. J. Trop. Biomed.* **2012**, *2*, 597–601. [CrossRef]

11.	Botas, J.; Fernandes, Â.; Barros, L.; Alves, M.J.; Carvalho, A.M.; Ferreira, I.C.F.R. A Comparative Study of Black and White *Allium sativum* L.: Nutritional Composition and Bioactive Properties. *Molecules* **2019**, *24*, 2194. [CrossRef]

12.	Chekki, R.Z.; Snoussi, A.; Hamrouni, I.; Bouzouita, N. Chemical composition, antibacterial and antioxidant activities of Tunisian garlic (*Allium sativum*) essential oil and ethanol extract. *Med. J. Chem.* **2014**, *3*, 947–956.

13.	Gam, D.-H.; Park, J.-H.; Kim, J.-H.; Beak, D.-H.; Kim, J.-W. Effects of *Allium sativum* Stem Extract on Growth and Migration in Melanoma Cells through Inhibition of VEGF, MMP-2, and MMP-9 Genes Expression. *Molecules* **2022**, *27*, 21. [CrossRef]

14.	Patiño-Morales, C.C.; Jaime-Cruz, R.; Sánchez-Gómez, C.; Corona, J.C.; Hernández-Cruz, E.Y.; Kalinova-Jelezova, I.; Pedraza-Chaverri, J.; Maldonado, P.D.; Silva-Islas, C.A.; Salazar-García, M. Antitumor Effects of Natural Compounds Derived from *Allium sativum* on Neuroblastoma: An Overview. *Antioxidants* **2022**, *11*, 48. [CrossRef]

15.	Yetgin, A.; Canli, K.; Altuner, E.M. Comparison of antimicrobial activity of *Allium sativum* cloves from China and Taşköprü, Turkey. *Adv. Pharmacol. Sci.* **2018**, *2018*, 9302840. [PubMed]

16.	Chen, C.; Liu, C.H.; Cai, J.; Zhang, W.; Qi, W.L.; Wang, Z.; Liu, Z.-B.; Yang, Y. Broad-spectrum antimicrobial activity, chemical composition and mechanism of action of garlic (*Allium sativum*) extracts. *Food Control* **2018**, *86*, 117–125. [CrossRef]

17.	Bhatwalkar, S.B.; Mondal, R.; Krishna, S.B.N.; Adam, J.K.; Govender, P.; Anupam, R. Antibacterial Properties of Organosulfur Compounds of Garlic (*Allium sativum*). *Front. Microbiol.* **2021**, *12*, 613077. [CrossRef]

18.	Fufa, B. Anti-bacterial and anti-fungal properties of garlic extract (*Allium sativum*): A review. *Microbiol. Res. J. Int.* **2019**, *28*, 1–5. [CrossRef]

19.	Kshirsagar, M.M.; Dodamani, A.S.; Karibasappa, G.N.; Vishwakarma, P.K.; Vathar, J.B.; Sonawane, K.R.; Khobragade, V.R. Antibacterial activity of garlic extract on cariogenic bacteria: An in vitro study. *Ayu* **2018**, *39*, 165–172. [CrossRef]

20.	Sasi, M.; Kumar, S.; Kumar, M.; Thapa, S.; Prajapati, U.; Tak, Y.; Changan, S.; Saurabh, V.; Kumari, S.; Kumar, A.; et al. Garlic (*Allium sativum* L.) Bioactives and Its Role in Alleviating Oral Pathologies. *Antioxidants* **2021**, *10*, 1847. [CrossRef]

21.	Žlabur, J.Š.; Brajer, M.; Voća, S.; Galić, A.; Radman, S.; Rimac-Brnčić, S.; Xia, Q.; Zhu, Z.; Grimi, N.; Barba, F.J.; et al. ultrasound as a promising tool for the green extraction of specialized metabolites from some culinary spices. *Molecules* **2021**, *26*, 1866. [CrossRef] [PubMed]

22.	Ranjha, M.M.A.N.; Irfan, S.; Lorenzo, J.M.; Shafique, B.; Kanwal, R.; Pateiro, M.; Arshad, R.N.; Wang, L.; Nayik, G.A.; Roobab, U.; et al. Sonication, a Potential Technique for Extraction of Phytoconstituents: A Systematic Review. *Processes* **2021**, *9*, 1406. [CrossRef]

23.	Kimbaris, A.C.; Siatis, N.G.; Daferera, D.J.; Tarantilis, P.A.; Pappas, C.S.; Polissiou, M.G. Comparison of distillation and ultrasound-assisted extraction methods for the isolation of sensitive aroma compounds from garlic (*Allium sativum*). *Ultrason Sonochem.* **2006**, *13*, 54–60. [CrossRef] [PubMed]

24.	Tomšik, A.; Pavlic, B.; Vladic, J.; Ramic, M.; Brindza, J.; Vidovic, S. Optimization of ultrasound-assisted extraction of bioactive compounds from wild garlic (*Allium ursinum* L.). *Ultrason. Sonochem.* **2016**, *29*, 502–511. [CrossRef]

25.	Mathialagan, R.; Mansor, N.; Shamsuddin, M.R.; Uemura, Y.; Majeed, Z. Optimisation of ultrasonic-assisted extraction (UAE) of allicin from garlic (*Allium sativum* L.). *Chem. Eng. Trans.* **2017**, *56*, 1747–1752.

26.	Ciric, A.R.; Krajnc, B.; Heath, D.; Ogrinc, N. Response surface methodology and artificial neural network approach for the optimization of ultrasound-assisted extraction of polyphenols from garlic. *Food Chem. Toxicol.* **2020**, *135*, 110976. [CrossRef] [PubMed]

27.	Bhattacharya, S.; Gupta, D.; Sen, D.; Bhattacharjee, C. Process intensification on the enhancement of allicin yield from *Allium sativum* through ultrasound attenuated nonionic micellar extraction. *Chem. Eng. Process. Process Intensif.* **2021**, *169*, 108610. [CrossRef]

28.	Salie, F.; Eagles, P.F.; Leng, H.M. Preliminary antimicrobial screening of four South African Asteraceae species. *J. Ethnopharmacol.* **1996**, *52*, 27–33. [CrossRef]

29.	Wikler, M.A. *Performance Standards for Antimicrobial Susceptibility Testing; Seventeenth Informational Supplement*; Part M2-A9. M100-S17; C.L.S.I. (Clinical and Laboratory Standard Institute): Pennsylvania, PA, USA, 2007.

30.	Gopal, J.; George, R.P.; Muraleedharan, P.; Khatak, H.S. Photocatalytic inhibition of microbial adhesion by anodized titanium. *Biofouling* **2004**, *20*, 167–175. [CrossRef]

31.	Durairaj, S.; Srinivasan, S.; Lakshmana-perumalsamy, P. In vitro Antibacterial Activity and Stability of Garlic Extract at Different pH and Temperature. *Electron. J. Biol.* **2009**, *5*, 5–10.

32.	Khashan, A.A. Antibacterial activity of garlic extract (*Allium sativum*) against *Staphylococcus aureus* in vitro. *GJBB* **2014**, *3*, 346–348.

33. Petropoulos, S.; Fernandes, Â.; Barros, L.; Ciric, A.; Sokovic, M.; Ferreira, I.C. Antimicrobial and antioxidant properties of various Greek garlic genotypes. *Food Chem.* **2018**, *245*, 7–12. [CrossRef] [PubMed]
34. Ismail, R.M.; Saleh, A.H.; Ali, K.S. GC-MS analysis and antibacterial activity of garlic extract with antibiotic. *J. Med. Plants Stud.* **2020**, *8*, 26–30.
35. Hoglund, K.B.; Barnett, B.K.; Watson, S.A.; Melgarejo, M.B.; Kang, Y. Activity of bioactive garlic compounds on the oral microbiome: A literature review. *Gen. Dent.* **2020**, *68*, 27–33.
36. Rees, L.P.; Minney, S.F.; Plummer, N.T.; Slater, J.H.; Skyrme, D.A. A quantitative assessment of the antimicrobial activity of garlic (*Allium sativum*). *World J. Microbiol. Biotechnol.* **1993**, *9*, 303–307. [CrossRef]
37. Hughes, B.G.; Lawson, L.D. Antimicrobial effects of *Allium sativum* L. (garlic), *Allium ampeloprasum* L. (elephant garlic), and *Allium cepa* L. (onion), garlic compounds and commercial garlic supplement products. *Phytother. Res.* **1991**, *5*, 154–158. [CrossRef]
38. Magryś, A.; Olender, A.; Tchórzewska, D. Antibacterial properties of *Allium sativum* L. against the most emerging multidrug-resistant bacteria and its synergy with antibiotics. *Arch. Microbiol.* **2021**, *203*, 2257–2268. [CrossRef]
39. Indu, M.N.; Hatha, A.; Abirosh, C.; Harsha, U.; Vivekanandan, G. Antimicrobial activity of some of the south-Indian spices against serotypes of Escherichia coli, Salmonella, Listeria monocytogenes and Aeromonas hydrophila. *Braz. J. Microbiol.* **2006**, *37*, 153–158. [CrossRef]
40. Shahid, M.; Naureen, I.; Riaz, M.; Anjum, F.; Fatima, H.; Rafiq, M.A. Biofilm Inhibition and Antibacterial Potential of Different Varieties of Garlic (*Allium sativum*) Against Sinusitis Isolates. *Dose-Response* **2021**, *19*, 15593258211050491. [CrossRef]
41. Gangoué-Piéboji, J.; Eze, N.; Djintchui, A.N.; Ngameni, B.; Tsabang, N.; Pegnyemb, D.E.; Biyiti, L.; Ngassam, P.; Koulla-Shiro, S.; Galleni, M. The in vitro antimicrobial activity of some traditionally used medicinal plants against beta-lactam-resistant bacteria. *J. Infect. Dev. Ctries* **2009**, *3*, 671–680. [CrossRef]
42. Fatemeh, A.M.; Soraya, H.; Mohammad, Y.A.; Jalal, P.; Javad, S. Antibacterial effect of *Eucalyptus globulus Labill*) and garlic (*Allium sativum*) extracts on oral Cariogenic bacteria. *J. Microbiol. Res. Rev.* **2013**, *1*, 12–17.
43. Garba, I.; Umar, A.I.; Abdulrahman, A.B.; Tijjani, M.B.; Aliyu, M.S.; Zango, U.U.; Muhammad, A. Phytochemical and antibacterial properties of garlic extracts. *Bayero J. Pure Appl. Sci.* **2013**, *6*, 45–48. [CrossRef]
44. Liaqat, A.; Zahoor, T.; Atif, M.; Muhammad, R. Characterization and antimicrobial potential of bioactive components of sonicated extract from garlic (*Allium sativum*) against foodborne pathogens. *J. Food Process. Preserv.* **2019**, *43*, e13936. [CrossRef]
45. Alwazni, D.S.; Drwesh, M.F. Comparative study of effect garlic and anion extract on the growth of some gram negative bacteria. *Mag. Al-Kufa Univ. Biol.* **2010**, *2*, 238–244.
46. Ankri, S.; Mirelman, D. Antimicrobial properties of allicin from garlic. *Microbes Infect.* **1999**, *1*, 125–129. [CrossRef]
47. Ilić, D.; Nikolić, V.; Stanković, M.; Nikolić, L.; Stanojević, L.; Mladenović-Ranisavljević, I.; Smelcerović, A. Transformation of synthetic allicin: The influence of ultrasound, microwaves, different solvents and temperatures, and the products isolation. *Sci. World J.* **2012**, *2012*, 561823. [CrossRef]
48. Wanyika, H.; Gachanja, A.; Kenji, G.; Keriko, J.; Mwangi, A. A rapid method based on uv spectrophotometry for quantitative determination of allicin in aqueous garlic extracts. *J. Agric. Sci. Technol.* **2010**, *12*, 74–82.
49. Rastogi, L.; Arunachalam, J. Green synthesis route for the size controlled synthesis of biocompatible gold nanoparticles using aqueous extract of garlic (*Allium sativum*). *Adv. Mater. Lett.* **2013**, *4*, 548–555. [CrossRef]
50. Rastogi, L.; Arunachalam, J. Sunlight based irradiation strategy for rapid green synthesis of highly stable silver nano particles using aqueous garlic (*Allium sativum*) extract and their antibacterial potential. *Mater. Chem. Phys.* **2011**, *129*, 558–563. [CrossRef]
51. Yulizar, Y.; Harits, A.A.; Abduracman, L. Green synthesis of gold nanoparticles using aqueous garlic (*Allium sativum* L.) Extract, and its interaction study with melamine. *Bull. Chem. React. Eng. Catal.* **2017**, *12*, 212. [CrossRef]
52. Beato, V.M.; Orgaz, F.; Mansilla, F.; Montaño, A. Changes in phenolic compounds in garlic (*Allium sativum* L.) owing to the cultivar and location of growth. *Plant Foods Hum. Nutr.* **2011**, *66*, 218–223. [CrossRef]
53. Stan, M.; Popa, A.; Toloman, D.; Dehelean, A.; Lung, I.; Katona, G. Enhanced photocatalytic degradation properties of zinc oxide nanoparticles synthesized by using plant extracts. *Mater. Sci. Semicond. Process.* **2015**, *39*, 23–29. [CrossRef]
54. Whitesides, G.M. The "right" size in nanobiotechnology. *Nat. Biotechnol.* **2003**, *21*, 1161–1165. [CrossRef]
55. Pal, S.; Tak, Y.K.; Song, J.M. Does the Antibacterial Activity of Silver Nanoparticles Depend on the Shape of the Nanoparticle? A Study of the Gram-Negative Bacterium *Escherichia coli*. *Appl. Environ. Microbiol.* **2007**, *73*, 1712–1720. [CrossRef]
56. Bouqellah, N.A.; Mohamed, M.M.; Ibrahim, Y. Synthesis of eco-friendly silver nanoparticles using *Allium* sp. and their antimicrobial potential on selected vaginal bacteria. *Saudi J. Biol. Sci.* **2019**, *26*, 1789–1794. [CrossRef] [PubMed]
57. Shrivastava, S.; Bera, T.; Roy, A.; Singh, G.; Ramachandrarao, P.; Dash, D. Characterization of enhanced antibacterial effects of novel silver nanoparticles. *Nanotechnology* **2007**, *18*, 225103. [CrossRef]

Article

Studies Regarding the Antibacterial Effect of Plant Extracts Obtained from *Epilobium parviflorum* Schreb

Erdogan Elvis Șachir [1,*,†], Cristina Gabriela Pușcașu [1,*,†], Aureliana Caraiane [1,†], Gheorghe Raftu [1,†], Florin Ciprian Badea [1,†], Mihaela Mociu [1,†], Claudia Maria Albu [1,†], Liliana Sachelarie [2,*,†], Loredana Liliana Hurjui [3,*,†] and Cristina Bartok-Nicolae [1,*,†]

[1] Department of Dentistry, Faculty of Dentistry, "Ovidius" University of Constanța, 900527 Constanța, Romania; drcaraiane@yahoo.com (A.C.); gheorgheraftu@yahoo.com (G.R.); florinb.md@gmail.com (F.C.B.); mihaelabacula@gmail.com (M.M.); albumariaclaudia@yahoo.com (C.M.A.)

[2] Department of Biopysics, Faculty of Dentistry, Apollonia University, 700511 Iasi, Romania

[3] Faculty of Dentistry, Grigore T. Popa University of Medicine and Pharmacy, 700115 Iasi, Romania

[*] Correspondence: erdogan_sachir@yahoo.com (E.E.Ș.); cristinap@gmb.ro (C.G.P.); lisachero@yahoo.com (L.S.); loredana.hurjui@umfiasi.ro (L.L.H.); dr.cristina_nicolae@yahoo.com (C.B.-N.); Tel.: +40-74-507-4720 (C.B.-N.)

[†] These authors contributed equally to this work.

Abstract: The present study was carried out to develop an experimental endodontic irrigant solution based on plant extracts obtained from *Epilobium parviflorum* Schreb. that largely replenish the properties of the usual antiseptics used in dentistry. Background: This study investigated the phytochemical contents of plant extracts obtained from *Epilobium parviflorum* Schreb. and their potential antibacterial activity. Methods: Identification and quantification of biologically active compounds were made by UV field photo spectrometry, adapting the Folin-Ciocalteu test method. Antibacterial activity was tested on pathological bacterial cultures collected from tooth with endodontic infections using a modified Kirby-Bauer diffuse metric method. Results: Polyphenols and flavonoids were present in all plant extracts; the hydroalcoholic extract had the highest amount of polyphenols—17.44 pyrogallol equivalent (Eq Pir)/mL and flavonoids—3.13 quercetin equivalent (Eq Qr)/mL. Plant extracts had antibacterial activity among the tested bacterial species with the following inhibition diameter: White *Staphylococcus* (16.5 mm), *Streptococcus mitis* (25 mm), *Streptococcus sanguis* (27 mm), *Enterococcus faecalis* (10 mm). Conclusions: All plant extracts contain polyphenols and flavonoids; the antibacterial activity was in direct ratio with the amount of the bioactive compounds.

Keywords: antibacterial activity; *Epilobium parviflorum* Schreb.; flavonoids; phenols; vegetable extract

Citation: Șachir, E.E.; Pușcașu, C.G.; Caraiane, A.; Raftu, G.; Badea, F.C.; Mociu, M.; Albu, C.M.; Sachelarie, L.; Hurjui, L.L.; Bartok-Nicolae, C. Studies Regarding the Antibacterial Effect of Plant Extracts Obtained from *Epilobium parviflorum* Schreb. *Appl. Sci.* **2022**, *12*, 2751. https://doi.org/10.3390/app12052751

Academic Editor: Ionel Calin Jianu

Received: 2 February 2022

Accepted: 3 March 2022

Published: 7 March 2022

Publisher's Note: MDPI stays neutral with regard to jurisdictional claims in published maps and institutional affiliations.

1. Introduction

Epilobium is a genus of perennial herbaceous plant (*Onagraceae* family). The most common species include *Epilobium parviflorum* Schreb., *Epilobium hirsutum*, *Epilobium rosmarinifolium* (*Epilobium dodonaei* Vill.), *Epilobium roseum* Schreb. and *Epilobium angustifolium* [1].

The anti-inflammatory and antiproliferative activity of *Epilobium parviflorum* Schreb. extracts was initially studied on cells in the context of benign prostatic hyperplasia; the analgesic, antioxidant, antibacterial and antifungal actions were studied in parallel [2–4]. Plant polyphenols represent a group of chemical substances ubiquitously distributed in all higher plants. These secondary metabolites possess free radical scavenging and antibacterial activity. These properties can be advantageously exploited, especially because of the abundance of polyphenols and their derivatives in various agricultural and food industry waste and by-products, and the possibility of convenient extraction by either organic or aqueous solvents [5–7].

In the oral cavity, nearly 700 species of bacteria can be found, including most of *Firmicutes*, *Bacteroidetes*, *Actinobacteria* and *Proteobacteria* [8]. There are two types of root canal infections: primary and secondary, each with specific microbiology. The primary infection results from colonization with a heterogenous group of microbes that have entered the pulp tissue, when exposed during caries or traumas; this group is dominated by Gram-negative oral anaerobic bacteria (*Prevotella* species, *Porphyromonas* species, *Fusobacterium* species, *Veillonella* species), aside from bacterial species which are also part of the commensal oral microflora: Gram-positive anaerobic bacteria (*Propionibacterium* species, *Bifidobacterium* species, anaerobic *streptococci*), treponemas or *Campylobacter* species [8,9]. The secondary infection appears inside the root canal system, after the treatment of the affected tooth has been initiated and harbor a limited number of bacterial species, with predominance of Gram-positive bacteria, namely *Enterococcus faecalis* (*E. faecalis*), oral *streptococci, lactobacilli* or *Candida albicans*; in contrast, Gram-negative bacteria are involved to a lesser extent [10]. The most common microorganism found in asymptomatic, persistent endodontic infections is *E. faecalis*. Its incidence in this type of infection varies from 24% to 77% [8,10]. *E. faecalis* possesses enzymes and constitutive structures that are able to suppress the action of lymphocytes and promote inflammation, which contributes to progression of endodontic infections [10,11]. Also, this microorganism has the ability to attach to dentin due to an inner component which is the collagen-binding protein [9,11,12].

Several studies have evaluated the antimicrobial efficacy of endodontic irrigants such as sodium hypochlorite (NaOCl), ethylene diamine tetra acetic acid (EDTA) and chlorhexidine (CHX) against *E. faecalis* biofilms [13]. In general, the aim of any disinfection strategy is to reduce the bacterial load to a subcritical level so that the patient's immune response allows healing by itself [14]. Endodontic research has always been focused on developing methods or endodontic irrigants that can completely remove the bacterial biofilm with minimal side-effects. Various endodontic irrigants are being widely used in the treatment of biofilms with varied effectiveness. Although 2% CHX has been proven to possess a high antimicrobial property, it has failed to disrupt the biofilm. NaOCl can disrupt the biofilm, disintegrating the dental pulp which is a built-in organ, but it is a well-known irritant to periapical tissues. Therefore, identification of natural products in the disinfection of root canals can be interesting [13,15,16].

In unconventional modern practice, endodontists seek the use of natural remedies as an adjunct to classical therapy, especially since the antibacterial action of plant extracts meets the phenomenon of antibiotic resistance, a problem that dominated medical practice in the second millennium and which seems to persist in an upward trend in the third millennium [17].

Plant extracts are fluid, soft or dry pharmaceutical/phytopharmaceutical preparations obtained by extracting plant products with different solvents [18].

In recent years, emphasis has been placed on the pharmaceutical and therapeutic revaluation of herbal preparations through a good knowledge of the physio-chemical and therapeutic properties of the active ingredients in medicinal plants and the development of extraction techniques and quality control means. Extractive solutions are pharmaceutical forms that contain smaller or larger proportions of active substances together with less active ones and ballast, extracted with the help of solvents. The extraction process involves the separation of the medically active portions of the plant tissues from the inactive or inert components by the use of selective solvents in standardized procedures [19].

Aqueous extractive solutions are prepared from different parts of the plant or mixtures of medicinal plants, using water as a solvent, preferably distilled or softened [20]. Description of the double maceration technique: the extracted product is first mixed with 1/2–2/3 of the total amount of solvent, after which the liquid is separated and the residue is pressed. It will contact the rest of the solvent, thus obtaining a new amount of extractive solution. The two extractive liquids will combine and filter after a 24-h rest [21].

Ultrasound-assisted extraction (UAE) is one of the most important techniques used for the extraction of valuable compounds from plant materials and is quite adaptable in the laboratory or on an industrial scale [22,23]. The method involves the use of ultrasound, with frequencies ranging from 20 kHz to 2000 kHz, which increase the permeability of cell walls and produce cell lysis, thus promoting the extraction of biologically active compounds [24].

The aim of the study is the use of plant extracts, obtained from *Epilobium parviflorum* Schreb., as a basis for alternative therapy in perspective in the treatment of endodontic pathology.

There were two distinct objectives of this study:

(1) Identification and quantification of phenols and flavonoids from the vegetable extracts obtained from *Epilobium parviflorum* Schreb.;

(2) Testing the antibacterial activity of the vegetable extracts.

2. Materials and Methods

2.1. Harvesting the Plant and Obtaining the Extracts

Plants were picked during the summer period from rural region of Dobrogea area and left to dry for one month. In the technological process, a drying yield of 20% was used.

In order to obtain an increased concentration of polyphenols in the aqueous extract, a double maceration extraction technique was used according to the European Pharmacopoeia (2016) [25].

To obtain the hydroalcoholic extract, in a borosilicate container, 50 g of dried *Epilobium parviflorum* Schreb. were added to 100 mL of 95% ethanol and 100 mL distilled water, then stored at room temperature for 7 days.

To obtain the ultrasonicated hydroalcoholic extract, in a borosilicate container, 50 g of *Epilobium parviflorum* Schreb. plant extract, 100 mL of double-distilled water and 100 mL of absolute ethyl alcohol were placed in a glass container; then it was placed in an ultrasonic bath (30,000 Hz) for 10 min a day, over 5 days.

The hydroalcoholic plant extracts were placed in a rotary evaporator (IKA-RV 10 digital V, Staufen Baden-Wurtemberg, Germany) to evaporate the solvent, at temperatures lower than 40 °C and under reduced pressure. Finally, the remaining ethanol was evaporated, placing the flask on the kiln drier until obtaining a consistent weight (three days) [26]. In order to remove the microorganisms from the vegetable extract and to obtain a sterile product, a 0.22 μm filter membrane was used, according to the specialized data from the European Pharmacopoeia—Sterility Control (2016), mounted on the vacuum filtration system. Both the filter membrane and the filtration system were mounted together and then autoclaved. To verify the absence of microorganisms in the obtained extracts, a sterility test was performed by the cultivation method, according to European Pharmacopoeia (9th ed, harmonized, Chapter 2.VI.1 Sterility, 2016). Sterility testing method: 10 mL samples were taken aseptically and transferred to 100 mL liquid medium (Bio-Merieux, Craponne, France); 3 replicates/type of environment were tested (to promote the development of aerobic and anaerobic microorganisms). The result came as a sterile sample; all test samples tested showed absent microbial growth, as well as the associated negative controls [25].

2.2. Identification and Quantification of Biologically Active Compounds

Ethanol as a solvent extraction in analytical grade was obtained from Merck in Darmstadt, Germany; pyrogallol and quercetin were obtained from Sigma-Aldrich Co. (St. Louis, MO, USA).

The Folin-Ciocalteu test method is the simplest method available for measuring the phenolic content of organic products [27]. The basic mechanism is an oxidation/reduction reaction, with the phenolic group being oxidized and the reduction of the metal ion [28]. The principle of the method is based on the reaction between Folin-Ciocalteu reagent and phosphomolybdic acid with the phenolic compounds in the sample, resulting in a mixture of blue oxides [29]. The spectrophotometric measurements were performed with a Specord M400 spectrophotometer, Carl Zeiss, Yena, Germany, using a 1 cm quartz cell.

2.3. Bacteriological Tests

- Study group

A prospective study was conducted in Constanța, on 60 adults, who required endodontic therapy, in the Endodontics Department of Dentistry Faculty, "Ovidius" University of Constanța. Ethical approval for the study was obtained from the Bioethic's committee of

Ovidius University 15547/09.11.2018. The study was conducted following the Helsinki declaration that was revised in 2013. Evaluation took place for two years, started on December 2018 and finished on November 2020. All patients with posterior teeth diagnosed with pulp necrosis (i.e., negative response to pulp sensitivity test with cold stimulus, confirmed by absence of bleeding during access cavity preparation), asymptomatic or symptomatic apical periodontitis were considered eligible and included in the study group after providing signed informed consent for the research. All personal data collected from the subject were blinded and stored with the given unique registration number.

The inclusion criteria in the study were: participants had to be over 18 years of age without allergic reactions with posterior teeth diagnosed with pulp necrosis. Moreover, the included teeth also had either symptomatic apical periodontitis, asymptomatic apical periodontitis or chronic apical abscess associated with a periapical radiolucency; absence of root fracture of the involved tooth, absence of periodontal pocket. The exclusion criteria from the study were: patients who are unable to appear for periodic check-ups, teeth with simple decay, teeth with hyperemia, poor oral hygiene, absence of enough tooth structure for rubber dam isolation, patients with the contributory medical history, patients who received antibiotic therapy during the previous 6 months.

At the beginning of the study, 60 patients with asymptomatic or symptomatic apical periodontitis were examined; only 40 of them were accepted for inclusion in the study based on acceptance criteria.

- Harvesting pathological products

The pathological product, infected dentin, was harvested using a sterile Kerr file (20 ISO, Dentsply Sirona), which was then placed in a container with culture medium (Bio-Merieux, France) and transported to the Microbiology Laboratory of the Faculty of Dentistry of the "Ovidius" University of Constanța. The products were seeded on culture medium (Columbia agar +5% sheep blood, Bio-Merieux, Craponne, France) which were then thermostated at 37 °C for 24 h (Figure 1a,b).

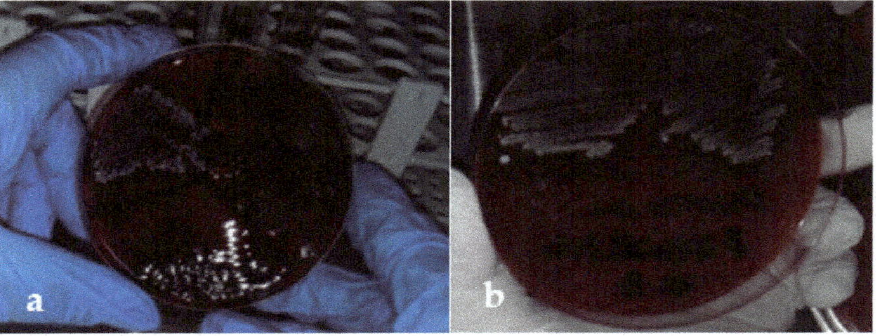

Figure 1. (**a,b**)—Bacterial cultures developed from pathological products after thermostating at 37 °C for 24 h.

- Bacteriological identification technique

After seeding on culture media and thermostating, bacterial identification was made using the API (Bio-Merieux, Craponne, France) method (Figure 2a,b).

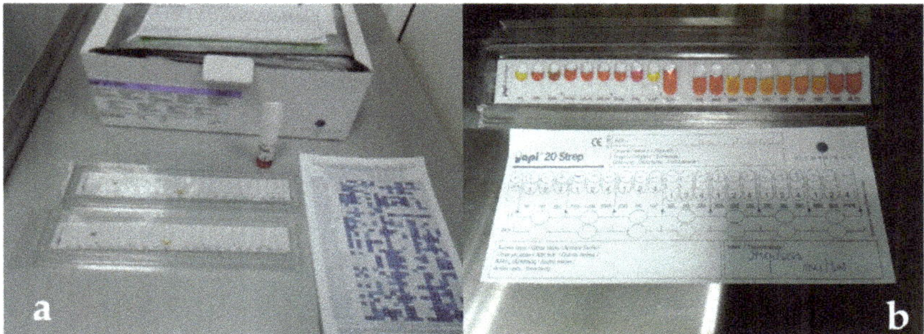

Figure 2. (**a**,**b**) API kit (Bio-Merieux, Craponne, France) for the identification of bacterial cultures.

Among the bacterial species, identified from the pathological products taken, are: White *Staphylococcus*, *Streptococcus mitis*, *Streptococcus sanguis*, *Enterococcus faecalis* and *Escherichia coli* (Figures 3 and 4).

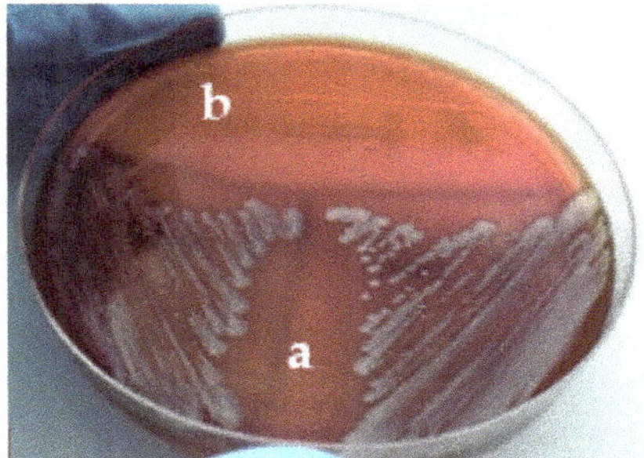

Figure 3. Bacterial cultures, (**a**) *Escherichia coli*; (**b**) *Enterococcus faecalis*.

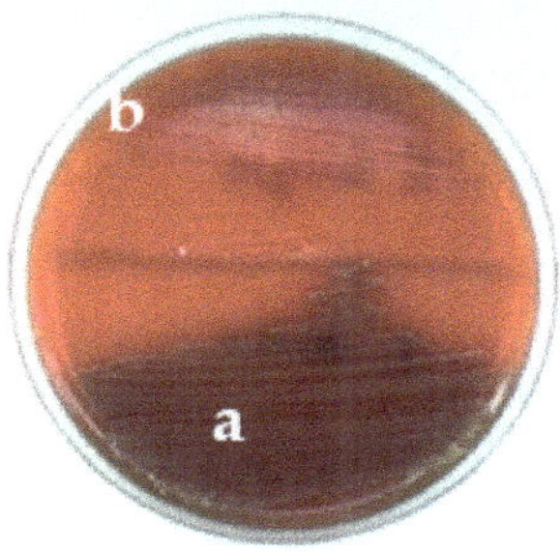

Figure 4. Bacterial cultures, (**a**) *Streptococcus mitis*; (**b**) *Streptococcus sanguis*.

- Antibacterial effect testing technique of the vegetable extracts

The antibacterial properties were tested on the following bacterial species: White *Staphylococcus*, *Streptococcus mitis*, *Streptococcus sanguis*, *Escherichia coli*, *Enterococcus faecalis*. They were sown in Mueller-Hinton and Columbia culture medium (Bio-Merieux, Craponne, France). The bacterial suspension used corresponded to a concentration of 10^7 colonies per training unit/mL micro-test, which correspond to a turbidity of the 0.5 Mac Farland tube (Figures 5 and 6) [30].

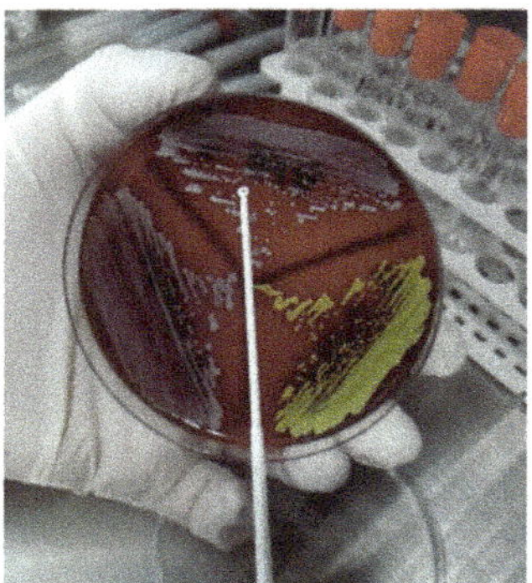

Figure 5. Sampling of bacterial colonies for later introduction into isotonic chloride suspension.

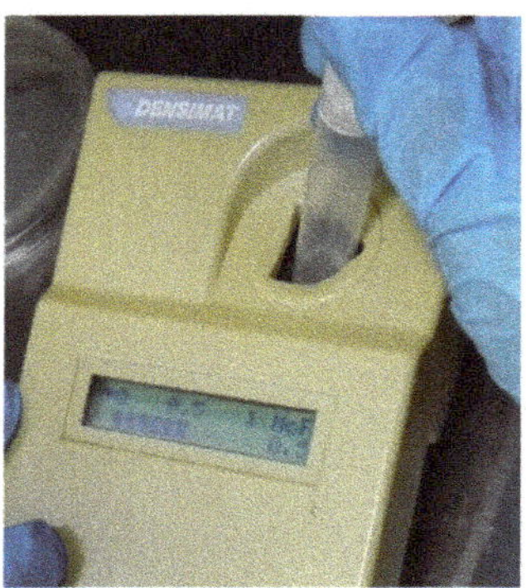

Figure 6. Measurement of the turbidity of the sodium microorganism suspension solution.

The Kirby-Bauer test for antibiotic susceptibility, known as the disc diffusion test, is a standard that has been used for years. This test determines the sensitivity or resistance of bacterial strains to antibiotics. The principle of the method relies on the direct proportionality relationship between the level of sensitivity and the size of the inhibition area of germs colonies developed around the tested substance. Bacterial sensitivity test was realized by adapting the Kirby-Bauer diffusion method in accordance with the protocol described by Romanian Microbiology Society in the Clinical microbiology guide (2017) [31,32].

The method's adaptation consisted of replacing the antibiotic discs with sterile filter paper discs, with the same size as those used in antibiograms; the discs were saturated with 50 µL of the test solution. These discs were applied on the culture medium previously seeded with the bacterial strains to be tested (Figure 7); the incubation time was 24 h at 37 °C. For comparison, similar to the test solutions we applied filter paper discs impregnated with commercial irrigants used in the practice of endodontics NaOCl 5.25% (Cerkamed, Stalowa Wola, Poland) and CHX 2% (Cerkamed, Stalowa Wola, Poland) on the surface of the sown medium (Figure 8).

2.4. Statistical Analysis

The values in the figures are expressed as mean ± standard error (SEM) of three independent experiments. When required, data sets were examined by one-way analysis of variance (ANOVA) and Dunnett's test as post analysis. A *p*-value less than 0.05 was considered statistically significant.

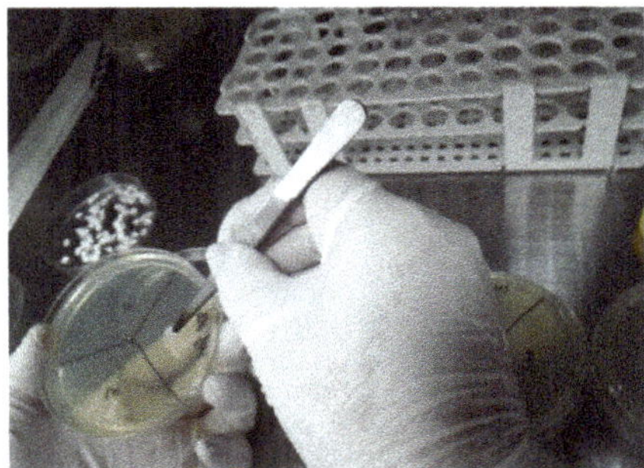

Figure 7. Applying the filter paper disc saturated with 50 µL of the test solution.

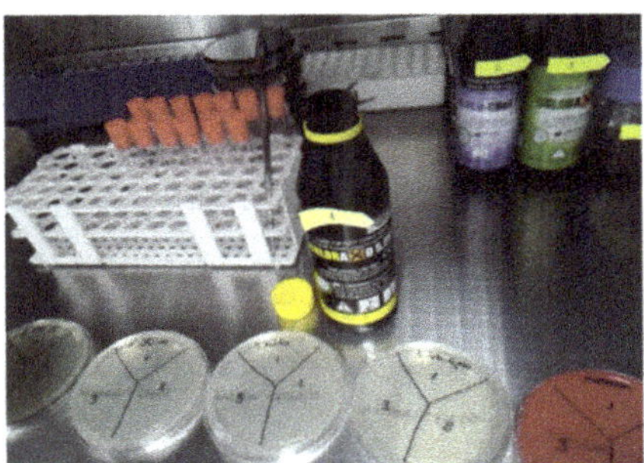

Figure 8. Commercial endodontic irrigants.

3. Results

3.1. Quantification of Total polyphenols in the Three Types of Plant Extracts

The linear regression equation was calculated and a good linear relationship was obtained between the standard concentrations (0.05–5 mg/mL) and the absorption at 760 nm of the reaction solution (Table 1), using as standard pyrogallol calibration. The total polyphenol content of the extracts was expressed as pyrogallol equivalent (EqPir) in the calibration curve (Scheme 1).

Table 1. Pyrogallol standard.

Concentration mg/mL	Absorbance at 760 nm
5.00	3.6022
2.5	2.1865
1.00	1.2698
0.5	0.9899
0.1	0.6654
0.05	0.4049

Table 2 presents the total phenolic content of the hydroalcoholic plant extract obtained from *Epilobium parviflorum* Schreb. which was determined from the regression equation (Scheme 1) of the calibration curve, ranging from 17.27 to 17.56 mg EqPir/g plant powder with an arithmetic mean of 17.44 mg EqPir/g vegetable powder; the total phenolic content of the ultrasonicated hydroalcoholic plant extract was determined from the regression equation of the calibration curve (Scheme 1), ranging from 16.90 to 17.15 mg EqPir/g vegetable powder with an average arithmetic mean of 17.05 mg EqPir/g vegetable powder; the total phenolic content of the aqueous plant extract was determined from the regression equation (Scheme 1) of the calibration curve, ranging from 5.62 to 5.65 mg EqPir/g plant powder with an arithmetic mean of 5.63 mg EqPir/g vegetable powder (Table 2). The highest polyphenol content was the hydroalcoholic plant extract with a mean of 17.44 Eq Pir/mL; at the opposite pole, the lowest polyphenol content was the aqueous plant extract with a mean of 5.63 Eq Pir/mL.

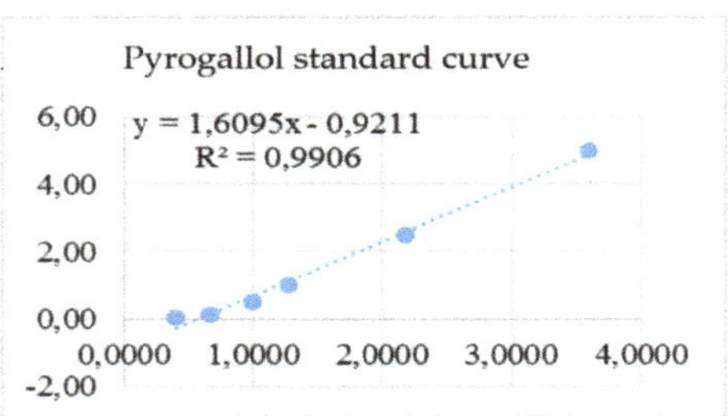

Scheme 1. Calibration curve for the pyrogallol standard.

Table 2. Results of determinations in Eq Pir/mL of total polyphenol content.

Vegetable Extract	Determinations No.	Absorbance	EqPir/mL
Hydroalcoholic	1	1.6459	17.27
	2	1.6635	17.56
	3	1.6589	17.48
	Mean/SD		17.44 ± 0.14
Ultrasonicated hydroalcoholic	1	1.6349	17.10
	2	1.6381	17.15
	3	1.6225	16.90
	Mean/SD		17.05 ± 0.13
Aqueous	1	0.9236	5.65
	2	0.9218	5.62
	3	0.9220	5.62
	Mean/SD		5.63 ± 0.01

SD = Standard deviation.

3.2. Quantification of Total Flavonoids

The linear regression equation was calculated and a good linear relationship was obtained between the standard concentrations (0.05–5 mg/mL) and the absorption of the reaction solution at 750 nm (Table 3), using as standard quercetin calibration. The total flavonoid content of the extracts was expressed as quercetin equivalent (EqQr) in the calibration curve (Scheme 2).

Table 3. Quercetin standard.

Concentration mg/mL	Absorbance at 760 nm
5	3.1443
2.5	1.6345
1	0.8510
0.5	0.4407
0.1	0.1307
0.05	0.0768

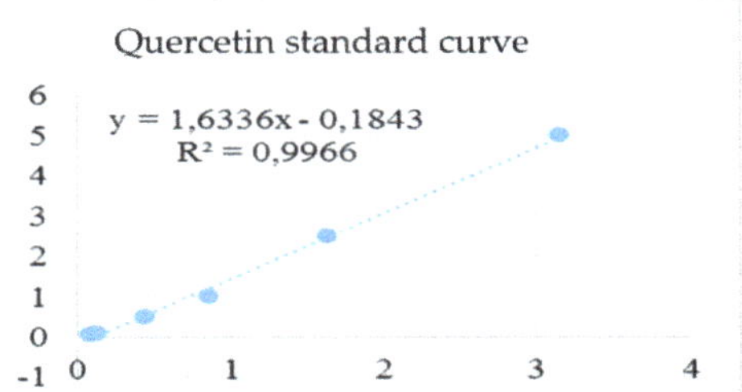

Scheme 2. Calibration curve for the quercetin standard.

As there are no significant differences in the total phenol content between the hydroalcoholic plant extract and the ultrasonicated hydroalcoholic extract, it was decided to determine the total flavonoid content only for the hydroalcoholic extract.

Table 4 presents the total flavonoid content of the hydroalcoholic plant extract obtained from *Epilobium parviflorum* Schreb. which was determined from the regression equation (Scheme 2) of the calibration curve, ranging from 2.84 to 2.9 mg EqQr/g plant powder with an arithmetic mean of 2.87 mg EqQr/g vegetable powder; the content of the aqueous plant extract was determined from the regression equation (Scheme 2) of the calibration curve, ranging from 0.78 to 0.9 mg EqQr/g plant powder with an arithmetic mean of 0.86 mg EqQr/g vegetable powder. The hydroalcoholic plant extract had the highest flavonoid content with a mean of 2.87 Eq Qr/mL, and the aqueous plant extract had the lowest level with a mean of 0.86 Eq Qr/mL.

Table 4. Results of determinations in Eq Qr/mL of total flavonoid content.

Vegetable Extract	Determinations No.	Absorbance	EqQr/mL
Hydroalcoholic	1	2.3569	2.87
	2	2.3411	2.84
	3	2.3779	2.90
	Mean/SD		2.87 ± 0.02
Aqueous	1	1.1332	0.9
	2	1.0619	0.78
	3	1.1341	0.9
	Mean/SD		0.86 ± 0.06

SD = Standard deviation.

3.3. Test Results for Antibacterial Activity of Plant Extracts

The aqueous solution lacks antibacterial activity on White *Staphylococcus*, while the two hydroalcoholic plant extracts have an antibacterial effect proportional to the concentration of phenolic compounds. The same aspect was identified on other tested bacterial species, *Enterococcus faecalis*, *Streptococcus mitis*, *Streptococcus sanguis* and *Escherichia coli*, respectively; the diameter of the areas of inhibition expressed in mm for the above-mentioned bacterial species tested are given in Table 5.

As can be seen in Scheme 3, the aqueous plant extract has no antibacterial action on any of the tested strains; on the *Escherichia coli* strain, none of the plant extracts have antibacterial action. On the rest of the tested strains, the antibacterial action is in direct proportionality to the concentration of total polyphenols from the hydroalcoholic plant extracts.

Table 5. Diameters of inhibition areas expressed in mm.

Vegetable Extract	White Staphylococcus	Streptococcus mitis	Streptococcus sanguis	Enterococcus faecalis	Escherichia coli
Hydroalcoholic	18	26.5	25.5	11.5	0
	20	26	26	12	0
	19	25.5	26.5	12.5	0
Ultrasonicated hydroalcoholic	16	25.5	26.5	9.5	0
	17	25	27	10	0
	16.5	24.5	27.5	10.5	0
Aqueous	Lacks antibacterial activity on all strains tested				

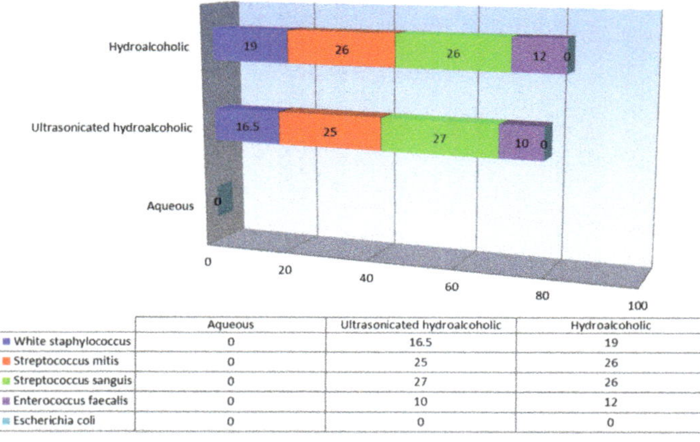

	Aqueous	Ultrasonicated hydroalcoholic	Hydroalcoholic
White staphylococcus	0	16.5	19
Streptococcus mitis	0	25	26
Streptococcus sanguis	0	27	26
Enterococcus faecalis	0	10	12
Escherichia coli	0	0	0

Scheme 3. Inhibition areas in mm of plant extracts.

3.4. Correlations between Biologically Active Compounds and the Antibacterial Effect

As can be seen in Table 6 there is a direct proportionality between the total content of biologically active compounds (polyphenols and flavonoids) in the hydroalcoholic plant extract and the antibacterial action on White *Staphylococcus*; the statistical study shows that there is a high correlation between the two parameters, expressed by a Pearson correlation index of 0.998. The same aspect can be identified for the other bacterial species except for *Escherichia coli*.

Table 6. Correlations between antibacterial action and total polyphenol and flavonoid content of plant extracts.

	Polyphenol Content		Flavonoid Content	
	Pearson Index	*p* **Value**	**Pearson Index**	*p* **Value**
White *Staphylococcus*	0.998	<0.01	0.920	<0.01
Streptococcus mitis	0.999	<0.01	0.939	<0.01
Streptococcus sanguis	0.999	<0.01	0.930	<0.01
Enterococcus faecalis	0.999	<0.01	0.924	<0.01
Escherichia coli	there is no correlation	<0.01	there is no correlation	<0.01

3.5. Antibacterial Effect of Experimental vs. Commercial Irrigants

As can be seen in Figures 9–11, the best parallelism regarding the antibacterial effect was shown to be between the test solutions and the antiseptics mentioned on the strains of

White *Staphylococcus*, *Streptococcus mitis* and *Streptococcus sanguis*. The highest antibacterial activity was had by the hydroalcoholic plant extract, which managed to inhibit the growth of the following bacterial strains: White *Staphylococcus*, *Streptococcus mitis*, *Streptococcus sanguis* and *Enterococcus faecalis* having the areas of inhibition (expressed in mm) of 19, 26, 26 and 12. The aqueous plant extract had no antibacterial activity; all bacterial strains developed on the disc saturated with the test solution.

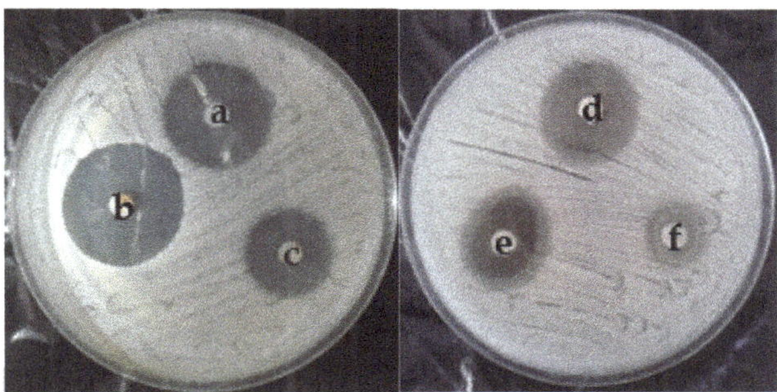

Figure 9. White *Staphylococcus*. (**a**) NaOCl 5.25%; (**b**) CHX 2%; (**c**) NaOCl 2%; (**d**) hydroalcoholic vegetable extract; (**e**) ultrasonicated hydroalcoholic vegetable extract; (**f**) aqueous vegetable extract.

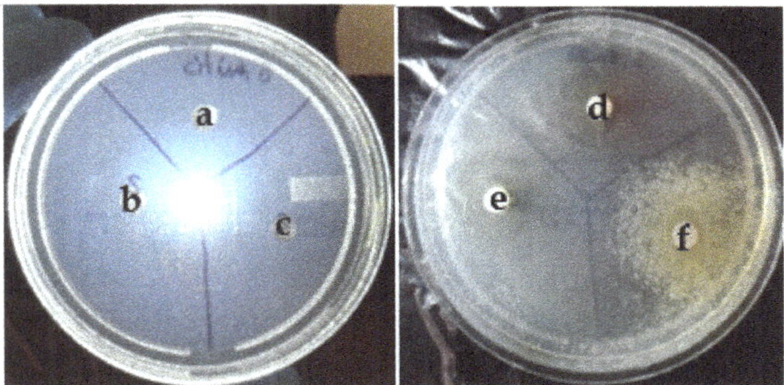

Figure 10. *Streptococcus mitis*. (**a**) NaOCl 5.25%; (**b**) CHX 2%; (**c**) NaOCl 2%; (**d**) hydroalcoholic vegetable extract (**e**) ultrasonicated hydroalcoholic vegetable extract; (**f**) aqueous vegetable extract.

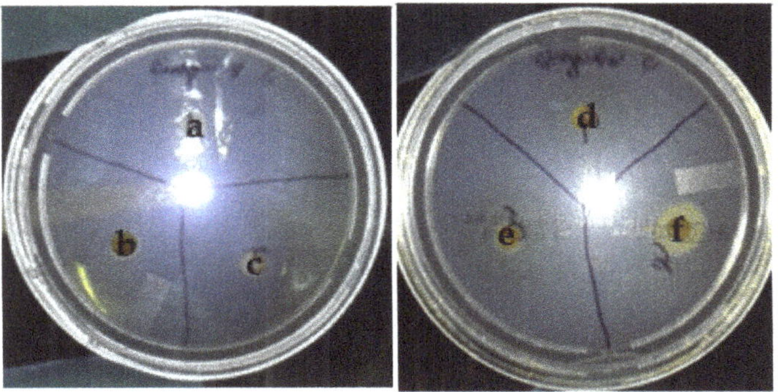

Figure 11. *Streptococcus sanguis*. (**a**) NaOCl 5.25%; (**b**) CHX 2%; (**c**) NaOCl 2%; (**d**) hydroalcoholic vegetable extract; (**e**) ultrasonicated hydroalcoholic vegetable extract; (**f**) aqueous vegetable extract.

The group of Gram-negative bacilli represented by *Escherichia coli* (Figure 12) showed resistance to all three types of extracts studied, noting that when reading was performed at an interval of 4 h after seeding the bacteria and applying test solutions, it showed an area of inhibition of 19 mm for the aqueous solution, an area that was initially maintained but in which resistant mutant colonies increased during the following hours.

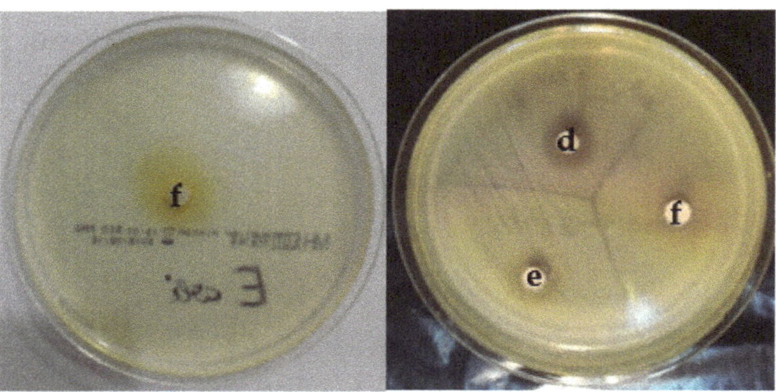

Figure 12. *Escherichia coli.* (**d**) hydroalcoholic vegetable extract; (**e**) ultrasonicated hydroalcoholic vegetable extract; (**f**) aqueous vegetable extract.

The most resistant strain was that of *E. faecalis*, with the mention that there is a proportionality between the level tested and those of antiseptics as seen in Figure 13; this aspect is justified by the established resistance of all strains of enterococci.

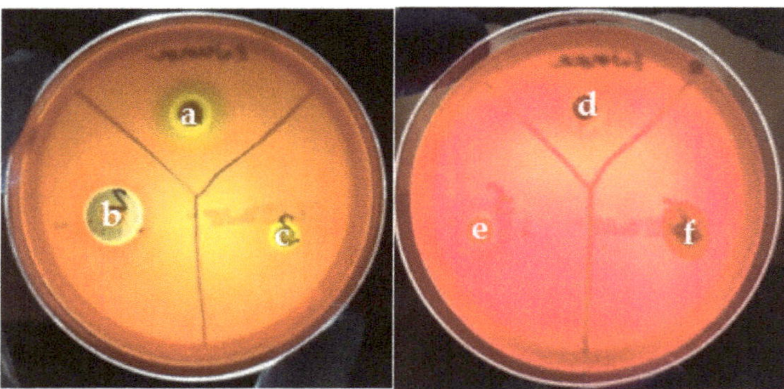

Figure 13. *Enterococcus faecalis.* (**a**) NaOCl 5.25%; (**b**) CHX 2%; (**c**) NaOCl 2%; (**d**) hydroalcoholic vegetable extract; (**e**) ultrasonicated hydroalcoholic vegetable extract; (**f**) aqueous vegetable extract.

4. Discussion

Using the same methods of determination, UV field photo spectrometry, Remmel et al. (2012) analyzed the content of polyphenols and flavonoids of an aqueous extract of concentration 110 mg/mL of *Epilobium parviflorum* Schreb. with a total content of 10.45 Eq. chlorogenic acid/g vegetable powder polyphenols and 31.9 mg EqPir/g vegetable powder flavonoids. Although difficult to compare in terms of the amount of polyphenols, because different controls were used for calibration, the data in our study show the presence of polyphenols in a higher amount in hydroalcoholic extracts than in aqueous ones because the alcohol mixed with water acquires a power of higher dissolution and ensures a better yield. Regarding the amount of flavonoids, this is higher in the study carried out by Remmel compared to the level determined in the present study from the two extracts, both aqueous (0.86 EqQr/g vegetable powder) and hydroalcoholic (2.87 EqQr/g vegetable powder); one of the possible reasons why it is found in larger quantities would be the temperature at which the extraction was carried out (40 °C) and the crushing of the plant to the powder stage, thus increasing the contact surface between the plant and the solvent [33]. A very recent study (Merighi et al., 2021) used UV field photo spectrometry to analyze the content of polyphenols and flavonoids of a 40% ethanolic extract of *Epilobium parviflorum* Schreb. Although difficult to compare in terms of the amount of polyphenols and flavonoids, because different controls were used for calibration, the data in our study show the presence of polyphenols (16.79 μg/μL Gallic Acid equivalent) and flavonoids (2.87 μg/μL Catechin equivalent) above the one founded by Merighi et al. [34].

The constituent compounds of a plant extract obtained from *Epilobium parviflorum* Schreb. by the method of hydrodistillation was determined in a study by Bajera et al. (2017), among which were 40% alcohols and 13.8% polyphenols, with the remaining percentages being occupied by esters, aldehydes, ketones, aromatic hydrocarbons. Although the method for determining the biologically active compounds is completely different from that of the present study, the results obtained compete with those found by Bajera who demonstrated the existence of polyphenols in the extract of *Epilobium parviflorum* Schreb. The results of the present study show that extracts obtained from *Epilobium parviflorum* Schreb. are effective on strains isolated from pathological products taken from patients with endodontic pathology, being in competition with those obtained by Bajera, although both antibacterial activity and the type of bacterial strains were achieved by different methods in the two studies [35].

The evaluation of the level of efficacy shows that there are differences between Gram-positive and Gram-negative bacteria, so the strains of White *Staphylococcus*, *Streptococcus mitis* and *Streptococcus sanguis* proved to be the most sensitive to these extracts, maintaining that the hydroalcoholic ones were the most effective. Of all the strains tested, the hydroalcoholic extracts were the most effective, compared to the aqueous plant extracts. The lowest level of sensitivity to the tested solutions had the bacteria from the group of Gram-positive shells, specifically *Enterococcus faecalis*, a bacterial species with established resistance to antibiotics. There were no significant differences in antibacterial action between the hydroalcoholic and ultrasonicated extracts.

The results obtained on antibacterial susceptibility are consistent with the content of polyphenols and flavonoids, with a direct proportionality between biologically active compounds and antibacterial activity.

Although it is difficult to compare, the data in our study are consistent with those in the literature; both ethanolic extracts have antibacterial activity in direct proportion to the concentration of polyphenols and flavonoids.

The limitations of the study are given by the final number of the study group and the method for testing the antibacterial activity; to overcome these limitations for future perspectives, it is necessary to conduct a study with a larger number of patients and to extend the research using microscopy method for the antibacterial activity observations with the same high standards used in the studies conducted by Imani et al. (2020) and Li et al. (2022) [36,37].

5. Conclusions

All plant extracts contain polyphenols and flavonoids. Hydroalcoholic plant extract obtained from *Epilobium parviflorum* Schreb. had the highest quantity of polyphenols, flavonoids and antibacterial activity. The most sensitive bacterial species were those belonging to the Gram-positive group.

Author Contributions: Conceptualization, E.E.Ș., C.G.P. and C.B.-N.; methodology, G.R. and E.E.Ș.; validation, C.G.P. and C.B.-N.; formal analysis, L.L.H., C.G.P., C.B.-N., C.M.A. and M.M.; investigation, E.E.Ș.; resources, E.E.Ș.; data curation, G.R. and C.B.-N.; writing—original draft preparation, E.E.Ș. and C.G.P.; writing—review and editing, E.E.Ș., C.G.P., L.S., F.C.B., A.C., G.R., C.B.-N. and C.G.P.;

visualization, C.G.P., E.E.Ș., C.B.-N. and M.M.; supervision L.S., F.C.B., A.C. and C.M.A. All authors have read and agreed to the published version of the manuscript.

Funding: This research received no external funding.

Institutional Review Board Statement: The study was conducted according to the guidelines of the Declaration of Helsinki, and approved by the Ethics Committee of "Ovidius" University (15547/09.11.2018).

Informed Consent Statement: Informed consent was obtained from all subjects involved in the study.

Conflicts of Interest: The authors declare no conflict of interest.

References

1. Vitalone, A.; Allkanjari, O. *Epilobium* spp: Pharmacology and phychemistry. *Phytother. Res.* **2018**, *32*, 1229–1240. [CrossRef] [PubMed]
2. Gafner, S. Herbal Drugs and Phytopharmaceuticals: A Handbook for Practice on a Scientific Basis. *J. Nat. Prod.* **2004**, *67*, 1774–1775. [CrossRef]
3. Heves, B.T.; Houghton, P.J.; Habtemariam, S.; Kéry, A. Antioxidant and antiinflammatory effect of *Epilobium parviflorum* Schreb. *Phytother Res.* **2009**, *23*, 719–724. [CrossRef] [PubMed]
4. Kim, J.H.; Park, K.M.; Lee, J.A. Herbal medicine for benign prostatic hyperplasia: A protocol for a systematic review of controlled trials. *Medicine* **2019**, *98*, e14023. [CrossRef]
5. Kodeš, Z.; Vrublevskaya, M.; Kulišová, M.; Jaroš, P.; Paldrychová, M.; Pádrová, K.; Lokočová, K.; Palyzová, A.; Maťátková, O.; Kolouchová, I. Composition and Biological Activity of *Vitis vinifera* Winter Cane Extract on Candida Biofilm. *Microorganisms* **2021**, *9*, 2391. [CrossRef]
6. Tsao, R. Chemistry and Biochemistry of Dietary Polyphenols. *Nutrients* **2010**, *2*, 1231–1246. [CrossRef]
7. Squillaci, G.; Giorio, L.A.; Cacciola, N.A.; La Cara, F.; Morana, A. Effect of temperature and time on the phenolic extraction from grape canes. In *Wastes-Solutions, Treatments and Opportunities, Wastes III*; Routledge: London, UK, 2019.
8. Nagy-Bota, M.C.; Man, A.; Santacroce, L.; Brinzaniuc, K.; Pap, Z.; Pacurar, M.; Pribac, M.; Ciurea, C.N.; Pintea-Simon, I.A.; Kovacs, M. Essential Oils as Alternatives for Root-Canal Treatment and Infection Control against Enterococcus faecalis—A Preliminary Study. *Appl. Sci.* **2021**, *11*, 1422. [CrossRef]
9. Lee, D.; Im, J.; Na, H.; Ryu, S.; Yun, C.-H.; Han, S.H. The Novel *Enterococcus* Phage VB_EfaS_HEf13 Has Broad Lytic Activity against Clinical Isolates of *Enterococcus faecalis*. *Front. Microbiol.* **2019**, *10*, 2877. [CrossRef]
10. Alghamdi, F.; Shakir, M. The Influence of *Enterococcus faecalis* as a Dental Root Canal Pathogen on Endodontic Treat-ment: A Systematic Review. *Cureus* **2020**, *12*, e7257.
11. Amaral, R.R.; Oliveira, A.G.G.; Braga, T.; Reher, P.; de Macedo Farias, L.; Magalhães, P.P.; Ferreira, P.G.; de Souza Côrtes, M.I. Quantitative Assessment of the Efficacy of Two Different Single-File Systems in Reducing the Bacterial Load in Oval-Shaped Canals: A Clinical Study. *J. Endod.* **2020**, *46*, 1228–1234. [CrossRef]
12. Pușcașu, C.G.; Caraiane, A.; Dumea, E.; Șachir, E.E.; Raftu, G.; Pușcașu, R.A.; Bartok-Nicolae, C.; Cernei, E.R.; Sachelarie, L.; Hurjui, L.L. Measurement of the Clinical Effects of a Marine Fish Extract on Periodontal Healing—A Preliminary Clinical Interventional Study. *Appl. Sci.* **2022**, *12*, 677. [CrossRef]
13. Parolia, A.; Kumar, H.; Ramamurthy, S.; Madheswaran, T.; Davamani, F.; Pichika, M.R.; Mak, K.-K.; Fawzy, A.S.; Daood, U.; Pau, A. Effect of Propolis Nanoparticles against *Enterococcus faecalis* Biofilm in the Root Canal. *Molecules* **2021**, *26*, 715. [CrossRef] [PubMed]
14. Song, W.; Ge, S. Application of Antimicrobial Nanoparticles in Dentistry. *Molecules* **2019**, *24*, 1033. [CrossRef] [PubMed]
15. Singh, M.; Singh, S.; Salgar, A.R.; Prathibha, N.; Chandrahari, N.; Swapna, L.A. An in vitro comparative evaluation of antimicrobial efficacy of propolis, Morindacitrifolia Juice, sodium hypochlorite and chlorhexidine on *Enterococcus faecalis* and *Candida albicans*. *J. Contemp. Dent. Pract.* **2019**, *20*, 40–45.
16. Pușcașu, C.G.; Ștefănescu, C.L.; Murineanu, R.M.; Grigorian, M.; Petcu, L.C.; Dumea, E.; Sachelarie, L.; Pușcașu, R.A. Histological Aspects Regarding Dental Pulp of Diabetic Patients. *Appl. Sci.* **2021**, *11*, 9440. [CrossRef]
17. Jamshidi-Kia, F.; Lorigooini, Z.; Amini-Khoei, H. Medicinal plants: Past history and future perspective. *J. Herbmed. Pharmacol.* **2018**, *7*, 1–7. [CrossRef]
18. Jayaraj, A.J.; Uchimahali, J.; Gnanasundaram, T.; Thirumal, S. Evaluation of antimicrobial activity and phytochemicals analysis of whole plant extract of Vinca rosea. *Evaluation* **2019**, *12*, 132–136.
19. Arguelles-Peña, K.; Olguín-Rojas, J.A.; Acosta-Osorio, A.A.; Carrera, C.; Barbero, G.F.; García-Alvarado, M.Á.; Rodríguez-Jimenes, G.d.C. An Evaluation of the Equilibrium Properties in Hexane and Ethanol Extractive Systems for *Moringa oleifera* Seeds and Fatty Acid Profiles of the Extracts. *Separations* **2021**, *8*, 217. [CrossRef]
20. Darwish, W.S.; Khadr, A.E.S.; Kamel, M.A.E.N.; Abd Eldaim, M.A.; El Sayed, I.E.T.; Abdel-Bary, H.M.; Ullah, S.; Ghareeb, D.A. Phytochemical Characterization and Evaluation of Biological Activities of Egyptian Carob Pods (*Ceratonia siliqua* L.) Aqueous Extract: In Vitro Study. *Plants* **2021**, *10*, 2626. [CrossRef]

21. Świątek, Ł.; Sieniawska, E.; Mahomoodally, M.F.; Sadeer, N.B.; Wojtanowski, K.K.; Rajtar, B.; Polz-Dacewicz, M.; Paksoy, M.Y.; Zengin, G. Phytochemical Profile and Biological Activities of the Extracts from Two Oenanthe Species (*O. aquatica* and *O. silaifolia*). *Pharmaceuticals* **2022**, *15*, 50. [CrossRef]

22. Vilkhu, K.; Mawson, R.; Simons, L.; Bates, D. Applications and opportunities for ul-trasound assisted extraction in the food industry—A review. *Innov. Food Sci. Emerg. Technol.* **2008**, *9*, 161–169. [CrossRef]

23. Vinatoru, M. An overview of the ultrasonically assisted extraction of bioactive principles from herbs. *Ultrason. Sonochem.* **2001**, *8*, 303–313. [CrossRef]

24. Hromàdkovà, Z.; Ebringerovà, A. Ultrasonic extraction of plant materials–investigation of hemicellulose release from buckwheat hulls. *Ultrason. Sonochem.* **2003**, *10*, 127–133.

25. EDQM. *European Pharmacopoeia*, 9th ed.; European Pharmacopoeia, Council of Europe, B.P. 907, F-67029; EDQM: Strasbourg, France, 2016.

26. Shami, A.M.M.; Philip, K.; Muniady, S. Synergy of antibacterial and antioxidant activities from crude extracts and peptides of selected plant mixture. *BMC Complementary Altern. Med.* **2013**, *13*, 360. [CrossRef] [PubMed]

27. El-Mahrouk, M.E.; Dewir, Y.H. Physico-Chemical Properties of Compost Based Waste-Recycling of Grape Fruit as Nursery Growing Medium. *Am. J. Plant Sci.* **2016**, *7*, 48–54. [CrossRef]

28. Gabriel, A.; Joe, V.; Patrick, E.D. Folin-Ciocalteu Reagent for Polyphenolic Assay. *IJFS* **2014**, *3*, 147–156.

29. Adamiak, K.; Kurzawa, M.; Sionkowska, A. Physicochemical Performance of Collagen Modified by Melissa officinalis Extract. *Cosmetics* **2021**, *8*, 95. [CrossRef]

30. Gaire, U.; Thapa Shrestha, U.; Adhikari, S.; Adhikari, N.; Bastola, A.; Rijal, K.R.; Ghimire, P.; Banjara, M.R. Antibiotic Susceptibility, Biofilm Production, and Detection of *mec*A Gene among *Staphylococcus aureus* Isolates from Different Clinical Specimens. *Diseases* **2021**, *9*, 80. [CrossRef]

31. Contreras-Alvarado, L.M.; Zavala-Vega, S.; Cruz-Córdova, A.; Reyes-Grajeda, J.P.; Escalona-Venegas, G.; Flores, V.; Alcázar-López, V.; Arellano-Galindo, J.; Hernández-Castro, R.; Castro-Escarpulli, G.; et al. Molecular Epidemiology of Multidrug-Resistant Uropathogenic Escherichia coli O25b Strains Associated with Complicated Urinary Tract Infection in Children. *Microorganisms* **2021**, *9*, 2299. [CrossRef]

32. Buiuc, D.; Neguţ, M. *Tratat de Microbiologieclinică*, 3rd ed.; EdituraMedicală: Bucharest, Romania, 2017.

33. Remmel, I.; Lauri, V.; Lauri, T.; Vallo, M.; Ain, R. Phenolic Compounds in Five Epilobium Species Collected from Estonia. *Nat. Prod. Commun.* **2012**, *7*, 1323–1324. [CrossRef]

34. Merighi, S.; Travagli, A.; Tedeschi, P.; Marchetti, N.; Gessi, S. Antioxidant and Antiinflammatory Effects of *Epilobium parviflorum*, *Melilotus officinalis* and *Cardiospermum halicacabum* Plant Extracts in Macrophage and Microglial Cells. *Cells* **2021**, *10*, 2691. [CrossRef] [PubMed]

35. Bajer, T.; Šilha, D.; Karel, V.; Bajerová, P. Composition and antimicrobial activity of the essential oil, distilled aromatic water and herbal infusion from *Epilobium parviflorum* Schreb. *Ind. Crops Prod.* **2017**, *100*, 95–105. [CrossRef]

36. Li, Y.; Liu, Y.; Yao, B.; Narasimalu, S.; Dong, Z. Rapid preparation and antimicrobial activity of polyurea coatings with RE-Doped nano-ZnO. *Microb. Biotechnol.* **2022**, *15*, 548–560. [CrossRef] [PubMed]

37. Imani, S.M.; Ladouceur, L.; Marshall, T.; Maclachlan, R.; Soleymani, L.; Didar, T.F. Antimicrobial Nanomaterials and Coatings: Current Mechanisms and Future Perspectives to Control the Spread of Viruses Including SARS-CoV-2ACS. *Nano* **2020**, *14*, 12341–12369. [CrossRef] [PubMed]

Article

Phytochemical Profile, Antioxidant and Wound Healing Potential of Three *Artemisia* Species: In Vitro and In Ovo Evaluation

Daliana Minda [1,2,†], Roxana Ghiulai [2,3,†], Christian Dragos Banciu [4,*], Ioana Zinuca Pavel [1,2,*], Corina Danciu [1,2], Roxana Racoviceanu [2,3], Codruta Soica [2,3], Oana Daniela Budu [4], Delia Muntean [5], Zorita Diaconeasa [6], Cristina Adriana Dehelean [2,7] and Stefana Avram [1,2]

[1] Department of Pharmacognosy, Faculty of Pharmacy, "Victor Babes" University of Medicine and Pharmacy Timisoara, Eftimie Murgu Square No. 2, RO-300041 Timişoara, Romania; daliana.minda@umft.ro (D.M.); corina.danciu@umft.ro (C.D.); stefana.avram@umft.ro (S.A.)

[2] Research Center for Pharmaco-Toxicological Evaluations, Faculty of Pharmacy, "Victor Babes" University of Medicine and Pharmacy Timisoara, Eftimie Murgu Square No. 2, RO-300041 Timişoara, Romania; roxana.ghiulai@umft.ro (R.G.); babuta.roxana@umft.ro (R.R.); codrutasoica@umft.ro (C.S.); cadehelean@umft.ro (C.A.D.)

[3] Department of Pharmaceutical Chemistry, Faculty of Pharmacy, "Victor Babes" University of Medicine and Pharmacy Timisoara, Eftimie Murgu Square No. 2, RO-300041 Timişoara, Romania

[4] Department of Internal Medicine IV, Faculty of Medicine, "Victor Babes" University of Medicine and Pharmacy Timisoara, Eftimie Murgu Square No. 2, RO-300041 Timişoara, Romania; oanadaniela.budu@yahoo.ro

[5] Department of Microbiology, Faculty of Medicine, "Victor Babes" University of Medicine and Pharmacy Timisoara, Eftimie Murgu Square No. 2, RO-300041 Timişoara, Romania; muntean.delia@umft.ro

[6] Department of Food Science and Technology, Faculty of Food Science and Technology, University of Agricultural Science and Veterinary Medicine, Calea Manastur, 3-5, RO-400372 Cluj-Napoca, Romania; zorita.diaconeasa@gmail.com

[7] Department of Toxicology, Faculty of Pharmacy, "Victor Babes" University of Medicine and Pharmacy Timisoara, Eftimie Murgu Square No. 2, RO-300041 Timişoara, Romania

[*] Correspondence: banciu.christian@umft.ro (C.D.B.); ioanaz.pavel@umft.ro (I.Z.P.)

[†] These authors contributed equally to this work.

Citation: Minda, D.; Ghiulai, R.; Banciu, C.D.; Pavel, I.Z.; Danciu, C.; Racoviceanu, R.; Soica, C.; Budu, O.D.; Muntean, D.; Diaconeasa, Z.; et al. Phytochemical Profile, Antioxidant and Wound Healing Potential of Three *Artemisia* Species: In Vitro and In Ovo Evaluation. *Appl. Sci.* **2022**, *12*, 1359. https://doi.org/10.3390/app12031359

Academic Editor: Catarina Guerreiro Pereira

Received: 11 January 2022
Accepted: 25 January 2022
Published: 27 January 2022

Publisher's Note: MDPI stays neutral with regard to jurisdictional claims in published maps and institutional affiliations.

Abstract: Skin injuries, and especially wounds of chronic nature, can cause a major negative impact on the quality of life. New efficient alternatives are needed for wound healing therapy and herbal products are being investigated due to a high content of natural compounds with promising healing activity. For this purpose, we investigated three *Artemisia* species, *Artemisia absinthium* L. (AAb), *Artemisia dracunculus* L. (ADr) and *Artemisia annua* L. (AAn). Ethanolic extracts, containing different polyphenolic compounds, elicited strong antioxidant activities in the DPPH assay, comparable to ascorbic acid. Human ketratinocyte proliferation was stimulated and wound closure was enhanced by all three extracts at concentrations of 100 µg/mL. The *Artemisia* extracts modulated angiogenesis by increasing vessel formation, especially following treatment with *A. annua* and *A. dracunculus*, extracts with a significantly higher content of chlorogenic acid. Good tolerability and anti-irritative effects were also registered in ovo, on the chorioallantoic membrane (CAM). The three *Artemisia* species represent promising low-cost, polyphenol-rich, antioxidant, safe alternatives for wound care treatment.

Keywords: *Artemisia annua*; *Artemisia dracunculus*; *Artemisia absinthium*; polyphenols; wound healing; keratinocytes; CAM assay

1. Introduction

The skin is our largest organ, an essential barrier towards the outer environment, a key protector of our organism from dehydration, pathogens, toxic chemicals, thermal deregulation. Its exposure and vulnerability are the source of frequent injuries. Repair and regeneration are extraordinary functions, being activated through unique cross-talk mechanisms of numerous cells, growth factors and cytokines [1]. The process of cutaneous wound healing can be divided into four overlapping phases: hemostasis, inflammation, proliferation/migration, and remodeling, involving vasoconstriction, platelet aggregation, antimicrobial effects, vascular leakiness, fibroblast proliferation, angiogenesis, collagen remodeling [2].

Unfortunately, the burden of worldwide vascular diseases, metabolic syndrome and aging has an impact on the rising number of patients with dysregulated healing wounds that can cause a major negative influence on the quality of life and on healthcare systems [3]. Despite important advances in wound care therapies, focused on skin repair and regeneration, there are still major limitations such as bacterial resistance and, especially, high costs. New efficient alternatives are desirable for wound healing therapy; traditional medicine is an important source of inspiration, since natural products were used as healing remedies from ancient times, and are still dominant therapeutic approaches in Asia, Africa or Latin America [1,2].

As a result of scientific progress, a large number of new active principles have been discovered, leading to an extensive knowledge of medicinal plants uses [4,5]. *Artemisia* species are included in the genus *Artemisia*, belonging to the *Asteraceae* family. These are aromatic, medicinal plants and culinary herbs [6]. *Artemisia* species are widespread in Asia, Europe and North America, in temperate, subtropical and cold regions [7]. *Artemisia* comprises about 500 species, including the three species that were selected for the present study: *A. annua* L. (AAn), *A. absinthium* L. (AAb), *A. dracunculus* L. (ADr). These species represent traditional remedies, being available in herbal shops.

The major bioactive types of phytocompounds that are described for *Artemisia* species are terpenoids, flavonoids, coumarins, acetylenes, caffeoylquinic acids and sterols [8–10]. *A. annua*, known as sweet wormwood is used as anti-malarial, anti-ulcer, anti-hyperlipidemic, anti-plasmodial, anti-convulsant, anti-inflammatory and anti-microbial remedy [11–13]. *A. dracunculus*, also called tarragon, a well-known culinary herb, was shown to possess antidiabetic effects due to the flavonoids in the composition [14]. Tarragon is also known as an analgesic, hypnotic, antiepileptic, anti-inflammatory and antipyretic, anticoagulant, antibacterial agent; more recent studies have proven its antioxidant, immunomodulating, anti-tumor activities, as well as hepatoprotective and hypoglycemic effects [15–17]. *A. absinthium*, or wormwood, has also been shown to possess several therapeutic effects: digestive, antiprotozoal, anthelmintic, antimicrobial, anti-inflammatory, cytotoxic, antioxidative, neuroprotective [18–22].

More recently, especially due to their polyphenolic content, *Artemisia* species were associated with healing potential, by reducing the number of inflammatory cells in the wounded area, with a positive impact on the progress of wound healing, improving the proliferation of human keratinocytes [23]. Still, there are not many studies that investigate the effect of the three *Artemisia* species regarding the underlaying mechanisms involved in the process of wound repair.

One important step in the process of tissue regeneration during the proliferative phase is represented by angiogenesis, the development of vessels from pre-existing ones, orchestrated by a sophisticated communication between cells, angiogenic factors and the surrounding tissues. Unfortunately, the process is mostly dysregulated in vascular diseases. Efforts are being directed toward the exploration of potential natural agents that can modulate the impaired reparative angiogenesis [24–27].

An experimental approach for this purpose is the chorioallantoic membrane (CAM) assay, considered an in vivo experimental alternative to animal models with advantages such as time, accessibility and costs. The method can be used to assess the angiogenic effects, but also to estimate the biocompatibility and the irritation potential of natural products [28,29].

In the present study, we investigated three *Artemisia* species available on the herbal market in Romania, *A. annua*, *A dracunculus*, *A absinthium*, regarding the phenolic content and profile of ethanolic extracts, next to in vitro antioxidant assessment, healthy keratinocyte viability and migration, as well as in vivo angiogenic and anti-irritative potential.

2. Materials and Methods

2.1. Plant Materials and Extraction

The plant material of the three *Artemisia* species in our study were purchased from herbal shops in Timisoara, Romania. All three species were represented by aerial parts of the plants, *Artemisiae annuae herba* (AAn), *Dracunculi herba* (ADr), *Absinthii herba* (AAb). The plant material of AAn and AAb were collected from Romania, while ADr, from Greece.

The dried plant material was grounded and stored in amber glass containers. Powdered dry plant material was extracted in ethanol 80% (*v/v*), (10 g dry weight/100 mL) by maceration for 15 min at room temperature, followed by ultrasound assisted extraction using the ultrasonic bath (Falc LCD series), for 30 min at 50 °C, 800 W and 40 KHz.

All samples were filtered and extracts were concentrated to dryness in a rotary vacuum evaporator (Heidolph Laborota 4000, Schwalbach, Germany) at 50 °C. Dried extracts (d.e.) were stored at −20 °C prior to use.

2.2. Chemicals

Methanol (99.9% purity, CAS No. 67-56-1) and acetic acid (99.9% purity, CAS No. 64-19-7) were purchased from Merck (Darmstadt, Germany) and used without further purification. Standard polyphenols: rosmarinic acid (CAS No. 20283-92-5), caftaric acid (CAS No. 67879-58-7), gentisic acid (CAS No. 4955-90-2), chlorogenic acid (CAS No. 327-97-9), caffeic acid (CAS No. 331-39-5), p-coumaric acid (CAS No. 501-98-4), ferulic acid (CAS No. 537-98-4), sinapic acid (CAS No. 530-59-6), hyperoside (CAS No. 482-36-0), isoquercitrin (CAS No. 482-35-9), rutin (CAS No. 153-18-4), myricetin (CAS No. 529-44-2), fisetin (CAS No. 345909-34-4), quercitrin (CAS No. 522-12-3), quercetol (CAS No. 117-39-5), luteolin (CAS No. 491-70-3), kaempferol (CAS No. 520-18-3) and apigenin (CAS No. 520-36-5) were purchased from Sigma-Aldrich (Germany). Folin–Ciocâlteu reagent (FC), gallic acid (GA, CAS No. 149-91-7), 2,2-diphenyl-1-picrylhydrazyl (DPPH, CAS No. 1898-86-4), ascorbic acid (AA, CAS No. 50-81-7), indometacin (CAS No. 53-86-1) and sodium dodecyl sulfate (SDS, CAS No. 151-21-3) were purchased from Sigma-Aldrich (Steinheim, Germany). Sodium carbonate anhydrous (CAS No. 497-19-8) was obtained from VWR International bvba (Leuven, Belgium). All used reagents were of analytical grade.

2.3. Cell Culture

The immortalized human keratinocytes (HaCaT cells) were kindly provided by the University of Debrecen, Hungary. The cells were cultured in high glucose Dulbecco's Modified Eagle's Medium (DMEM; Sigma-Aldrich, Taufkirchen, Germany) supplemented with antibiotic mixture to avoid contamination (Penicillin/Streptomycin 10,000 IU/mL; Sigma-Aldrich, Taufkirchen, Germany) and fetal bovine serum 10% (FCS; Sigma-Aldrich, Taufkirchen, Germany). The cells were kept in standard conditions-5% CO_2, at a temperature of 37 °C.

2.4. Total Phenolic Content Determination

Total phenolic content (TPC) from *Artemisia* extracts was assessed using Folin-Ciocâlteu reagent [30–32] using an adapted method. The samples, represented by the diluted solutions of the dried extracts (1000 µg/mL), were pipetted into test tubes containing previously diluted (1:10) Folin Ciocâlteu's phenol reagent and, after 5 min at room temperature, sodium carbonate solution (Na_2CO_3 75 g/L) was added. The mixture was vortexed for 15 s and then left to stand at room temperature for 2 h in the dark. Absorbance was measured at 760 nm using a UV-VIS spectrophotometer (T80+, PG Instruments Ltd., Lutterworth, UK). The calibration curve was established using gallic acid (0–200 µg/mL). Estimation of the phenolic content was carried out in triplicate. Total phenolic content was expressed as mg of gallic acid equivalents (GAE)/g of dry extract (d.e.).

2.5. Antioxidant Activity In Vitro

The evaluation of the antioxidant potential of the *Artemisia* extracts (AAn, AAb, ADr) was performed by the DPPH method [31,33,34], with slight modifications.

The assay is based on the DPPH reduction by the hydrogen donating antioxidants, leading to the discoloration and, subsequently, to the decrease of solution absorbance. Various concentrations (50–1000 μg/mL) of the *Artemisia* extracts or the pure ascorbic acid (AA) as control, in volumes of 0.2 mL, were added to 1.8 mL of freshly prepared 0.1 mM DPPH in ethanol. The mixture was incubated in the dark for 30 min, at room temperature. Absorbance was measured against blank samples, at 517 nm, using an UV–VIS spectrophotometer (T80+, PG Instruments Ltd., Lutterworth, UK). The decrease in the registered absorbance indicates a free radical scavenging activity. The antioxidant activity (AOA) was calculated as the scavenging capacity of free DPPH radical (in percentages) using the formula:

$$\text{AOA } (\%) = \left[\frac{A_0 - A_s}{A_0} \right] \times 100$$

where: A_0 = absorbance of the blank sample and A_s = absorbance of the tested samples.

2.6. LC-MS

All vegetal extracts were analyzed by an LC-MS analytical method that enables the simultaneous screening and quantification of 18 polyphenols (rosmarinic acid, caftaric acid, gentisic acid, chlorogenic acid, caffeic acid, p-coumaric acid, ferulic acid and sinapic acid, hyperoside, isoquercitrin, rutin, myricetin, fisetin, quercitrin, quercetol, luteolin, kaempferol and apigenin), separated on a reverse phase Zorbax Eclipse Plus C18 column (3.0 × 100 mm × 3.5 μ) as previously described [35,36]. Briefly, the mobile phase consisted in a mixture of 0.1% acetic acid and methanol in gradient elution. The LC-MS parameters were the following: flow rate 1 mL/min, injection volume 10 μL, column temperature 40 °C, UV detection at 330 and 370 nm; the elution of all components was achieved in about 40 min. MS detection was achieved by electrospray ionization (ESI) in the single ion monitoring mode (SIM) in the negative ion mode, at the following parameters: capillary voltage 3500 V, dry gas flow 12 L/min at 350 °C, nebulizer pressure 55 psig and fragmentor 70. Calibration curves were conducted for the quantification of polyphenols by the external standard method in the 0.05–2 μg/mL range for a six-point plot for each compound. The m/z scale of the mass spectrum was calibrated by use of an external calibration standard ESI Tuning Mix from Agilent (Santa Clara, CA, USA).

HPLC/LC-MS experiments were conducted on a 6120 LC-MS analytical system from Agilent (Santa Clara, CA, USA) consisting of 1260 Infinity HPLC equipped with G1322A degasser, G1311B quaternary pump, G1316A column thermostat, G1365C MWD detector and G7129A autosampler; the Quadrupolar (Q) mass spectrometer is equipped with an electrospray ionization source (ESI). The LC-MS system is controlled by OpenLAB CDS ChemStation Workstation software.

Sample solutions were diluted with methanol, homogenized with a WisdVM-10 vortex mixer (Witeg Labortechnik, Wertheim, Germany) and centrifuged for 2 min at 10,000 rpm in a ThermoMicro CL17microcentrifuge (Thermo Fisher Scientific, Boston, MA, USA). The supernatant was collected and submitted to LC-MS analysis.

2.7. Cell Viability Assessment by Alamar Blue Assay

The technique was performed in order to establish cells viability following stimulation with various concentrations of the extracts. The principle of the technique consists in the ability of viable cells to reduce resazurin to the fluorescent form, resorufin. The assay was performed as previously described by Cosarca et al. [37]. Briefly, 1×10^4 cells/well were cultured in 96-well plates and allowed to adhere. The second day, the cells were stimulated with different concentrations of the extracts (10, 25, 50, 100, 250, 500, 750 and 1000 µg/mL) for a period of 72 h. After the 72 h stimulation, 20 µL/well of Alamar blue (10% of the volume of cell culture medium) was added and the cells were further incubated at 37 °C for 3 h. Then, the absorbance was measured at 570 and 600 nm using the xMark Microplate spectrophotometer (BioRad, xMark^{TM} Microplate, Serial No. 10578, Tokyo, Japan).

2.8. Cell Cytotoxicity Assessment by LDH Assay

Lactate dehydrogenase (LDH) assay (CyQUANT, Thermo Fisher Scientific, Boston, MA, USA) was performed to determine the cytotoxic effect of the *Artemisia* extracts on HaCaT cells. Different concentrations were used, namely 10, 25, 50, and 100 µg/mL. The assay was performed as previously described by Ghitu et al. [38]. For the experiment, 5×10^3 cells/well were cultured in 96-well culture plates and left to adhere. On the second day, the cells were stimulated with the above-mentioned concentrations and incubated for 72 h. After the 72-h stimulation, a volume of 50 µL was transferred from each well into a new 96-well culture plate, mixed with 50 µL/well of the reaction mixture and further incubated at room temperature for 30 min. Then, 50 µL of stop solution was added to each well. The level of LDH release in the medium was determined at two wavelengths (490 and 680 nm) using the xMark Microplate spectrophotometer (BioRad, xMark^{TM} Microplate, Serial No. 10578, Tokyo, Japan).

2.9. Wound Healing Technique by Scratch Assay

The method called "wound healing assay" is the most common in vitro assay for cell migration ability, useful for describing the ability of keratinocytes to migrate and to restore the epidermal barrier affected by injury, a first step in the healing process. The assay was performed as previously described [19,39]. The migration capacity of HaCaT cells following stimulation with 100 µg/mL of each *Artemisia* extract was evaluated. In brief, 2×10^5 cells/well were cultured in 12-well plates and allowed to adhere. When a 90% confluence was reached, we used a sterile pipette tip in order to draw scratches in the center of the wells. Cells that detached following the procedure were removed by washing with phosphate-buffered saline (PBS) before stimulation. Then, the cells were stimulated with the extractive solutions (100 µg/mL). Pictures of the cells were taken at 0 and 24 h post-stimulation using an inverted microscope (Olympus IX73) provided with DP74 camera (Olympus, Tokyo, Japan). Image analysis for cell migration was performed by means of ImageJ (ImageJ Version 1.53k, https://imagej.nih.gov/ij/index.html, accessed on 4 November 2021) and GIMP software (GIMP v 2.10.24, https://www.gimp.org/, accessed on 6 January 2022).

The wound closure rate was calculated using the formula described by Felice et al. [40]:

$$\text{Wound closure rate } (\%) = \left[\frac{A_{t0} - A_t}{A_{t0}}\right] \times 100$$

where: A_{t0} is the scratch area at time 0; A_t is the scratch area at 24 h.

2.10. The Chorioallantoic Membrane Assay

The evaluation of the active potential of *Artemisia* sp. was also performed in vivo, in order to assess the modulatory effect on angiogenesis with an important role in tissue healing and toxicity [40]. The in ovo chorioallantoic membrane (CAM) assay was performed. The fertilized chicken (*Gallus gallus domesticus*) eggs were incubated in controlled humidified atmosphere at 37 °C. On the third day of incubation, 4–5 mL of egg white were removed so that the developing chorioallantoic membrane could detach from the inner shell. The following day, a window was cut on the upper shells and resealed to avoid dehydration; the incubation continued until the experimental process [41,42].

To investigate the potential influence on the angiogenesis process of *Artemisia* extracts, samples were prepared in concentrations of 100 µg/mL in 0.5% DMSO and volumes of

5 μL were applied inside plastic rings previously placed on CAMs on the 8th day of incubation. All tested specimens were daily monitored by stereomicroscopy (ZEISS SteREO Discovery.V8, Göttingen, Germany). Images were registered and processed by Axiocam 105 color, AxioVision SE64. Rel. 4.9.1 Software, (ZEISS Göttingen, Germany), ImageJ (ImageJ Version 1.53k, https://imagej.nih.gov/ij/index.html, accessed on 4 November 2021) and GIMP software (GIMP v 2.10.24, https://www.gimp.org/, accessed on 6 January 2022).

2.11. The Anti-Irritant Effect In Ovo by the HET CAM Method

Using the same biological material as for the angiogenic evaluation, the anti-irritant effect of the *Artemisia* extracts was assessed by applying a modified protocol of the HET-CAM test (hen egg test chorioallantoic membrane assay), which is known as an alternative protocol for evaluating the potential irritative effect of compounds intended for ophthalmic and dermatologic products, or as a method of assessing in vivo biocompatibility [43–45]. The protocol used here was an adapted alternative to the HET-CAM method as recommended by ICCVAM [46]. The anti-irritative approach was used after Wilson et al. [47] to assess the potential beneficial preventive effects of the extracts when in contact with an irritant agent.

The fertilized eggs prepared as described above were incubated until the 9th day of incubation. At this point, by stereomicroscope assistance, 0.3 mL of the control or test samples were applied to the membrane; then, after 20 min, the irritating solution of sodium dodecyl sulfate (SDS) was applied (0.5%). The tested membranes were monitored over a period of 300 s, by means of a stereomicroscope; the time for the occurrence of the selected parameters (hemorrhage, H; vascular lysis, L; coagulation, C) was recorded in seconds. The irritation score (IS) was then calculated using the followig equation:

$$IS = 5 \times \left[\frac{301 - \text{Sec H}}{300} \right] + 7 \times \left[\frac{301 - \text{Sec L}}{300} \right] + 9 \times \left[\frac{301 - \text{Sec C}}{300} \right]$$

where, hemmorrhage (Sec H) = start of observation (in seconds) of bleeding reactions on the membrane, lysis time (Sec L) = start of observation (in seconds) of lysis of the vessel on the membrane, coagulation time (Sec L) = start of observation (in seconds) of the formation of coagulation on the membrane.

The positive control was considered treatment only with sodium dodecyl sulfate (SDS), while the negative control was treatment with just distilled water. DMSO in a concentration of 0.5% was also tested as solvent control; indometacin represented the anti-inflammatory control.

2.12. Statistical Analysis

Statistical analysis was performed with GraphPad Prism 5 software (San Diego, CA, USA). For the in vitro results; comparison among the groups was performed using the one-way ANOVA followed by Dunnett's multiple comparison test (* $p < 0.05$; ** $p < 0.01$; *** $p < 0.001$). Data were represented as Mean ± SD.

3. Results

3.1. Total Polyphenolic Content

The obtained experimental data allowed the calculation of the equation resulted from the standard curve, $y = 0.009x + 0.1865$, $R^2 = 0.9798$.

The total phenolic content, expressed in mg GAE/g dry extract, as indicated in Table 1, reached the highest value of 193.61 ± 2.36 the case of AAb extract. The value of the total phenolic content for the AAn and ADr extract had lower values of 129.28 ± 2.09 and 144.28 ± 1.87, respectively.

Table 1. Total phenolic content of *Artemisia* extracts.

Extract	Total Phenolic Content mg GAE/g Dry Extract
AAn	129.28 ± 2.09
ADr	144.28 ± 1.87
AAb	193.61 ± 2.36

3.2. Good Radical Scavenging Activity of Artemisia Species

The antioxidant activity of the samples was evaluated using the DPPH assay.

In Figure 1 is shown the AOA% (antioxidant activity %) of the studied extracts next to the standard compound, ascorbic acid 50 µg/mL. It can be observed that the antioxidant activity is directly proportional to the concentration of the extracts.

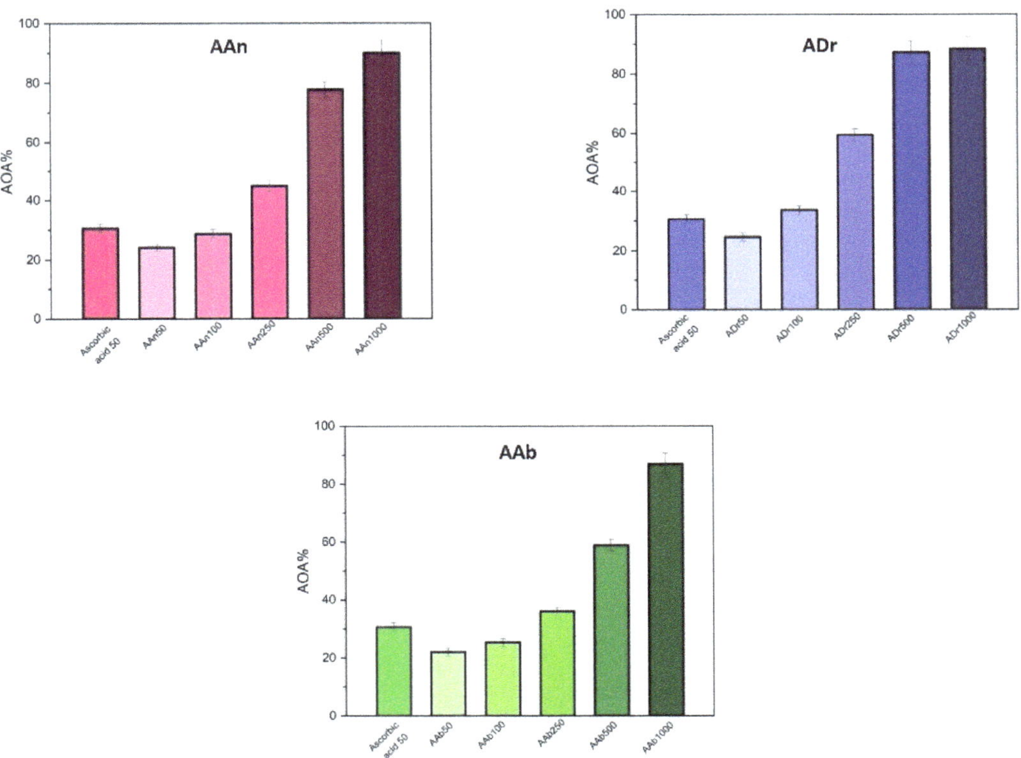

Figure 1. Antioxidant activity of the AAn, ADr and AAb extracts.

Artemisiae annue herba—AAn induced an antioxidant activity of 24.14 ± 0.6% for the lowest tested concentration AAn at 50 µg/mL and goes up to 90.04 ± 2.25% for AAn at 1000 µg/mL. The AOA% values for the AAn extracts show a comparable antioxidant activity with the one of the standard-ascorbic acid, starting with AAn 100 µg/mL. At higher concentrations, 250, 500 and 1000 µg/mL a much intense AOA as compared to the standard antioxidant can be observed.

The antioxidant activity of *A. dracunculus* ranged from 24.66 ± 0.63% (concentration 50 µg/mL) to 88.37 ± 2.25% (concentration of 1000 µg/mL). In addition, we determined in this study the AOA% for the *A. absinthium* extract and the values started at 22.03 ± 0.52%

for the lowest concentration of 50 μg/mL and increased to 86.87 ± 2.15% for the highest concentration of the extract.

All three tested extracts induced an important antioxidant effect, above 100 μg/mL, comparable to the effect of 50 μg/mL AA. At this particular concentration, of 100 μg/mL, *A. dracunculus* had a higher antioxidant capacity (33.65 ± 1.98%) than the standard AA (30.57 ± 1.02%).

Although the antioxidant capacity of the three *Artemisia* species showed similar effects, by comparing the values obtained for concentrations of 1000 μg/mL, the AOA capacity fall in the following order: AAn > ADr > AAb. Still, interestingly, above the concentration of 100 μg/mL up to 1000 μg/mL, the most potent extract is the one obtained from tarragon (ADr).

3.3. Polyphenols and Phenolic Acids in Artemisia Species

LC-MS analysis of sample extracts of *Artemisia dracunculus* L. (ADr), *Artemisia annua* L. (AAn) and *Artemisia absinthium* L. (AAb) were conducted under identical solution and instrumental conditions. The obtained results revealed the identification and in some cases quantification, according to their R_t and m/z values, of a total of 11 polyphenols in all analyzed extracts combined with some differences in terms of their expression in each extract (Table 2). Identified polyphenols fall into polyphenolic acids group, cinnamic acid derivatives, and flavonoids, flavones and flavonols respectively. Identified phytocompounds were gentisic acid, chlorogenic acid, caffeic acid, ferulic acid, isoquercitrin, rutin, quercitrin, quercetol, luteolin, kaempferol and apigenin, expressed as μg/mg d.e. The most important polyphenols identified in all extracts, consistent with their concentration, were chlorogenic acid, rutin and quercetol, the most abundant one being chlorogenic acid. Meanwhile, gentisic acid, caffeic acid, ferulic acid, isoquercitrin, quercitrin, luteolin, and apigenin were identified in smaller concentrations, or even in traces, falling below the limit of quantification in some cases. Some differences regarding the expression of polyphenols in the three types of extracts were spotted out, such as quercitrin and isoquercitrin, identified in AAb and AAn extract, but not in ADr extract, concurrently, kaempferol, which was found only in ADr extract, in a substantial amount. The expression of the dominant compound, chlorogenic acid respectively, was almost equivalent in AAn and ADr extracts, while AAb exhibited a concentration almost four times lower. The richest extract consistent with the concentration of identified compounds was ADr, while AAb contained the lowest amounts.

Table 2. Identified polyphenols by LC-MS.

No.	Compound Name	R_t (min)	$[M − H^+]^+$ (*m/z*)	AAn (μg/g d.e.)	AAb (μg/g d.e.)	ADr (μg/g d.e.)
1.	Gentisic acid	2.67	153	ND	NQ	NQ
2.	Chlorogenic acid	6.45	353	12.4	3.15	11.77
3.	Caffeic acid	6.97	179	0.06	0.009	NQ
4.	Ferulic acid	13.91	193	ND	ND	NQ
5.	Isoquercitrin	22.50	463	0.5	0.15	ND
6.	Rutin	23.01	609	0.4	0.33	2.87
7.	Quercitrin	26.18	447	0.9	0.73	ND
8.	Quercetol	30.38	301	0.11	0.07	5.54
9.	Luteolin	32.78	285	NQ	NQ	ND
10.	Kaempferol	35.63	285	ND	ND	4.44
11.	Apigenin	36.91	269	NQ	NQ	NQ

Notes: ND—not detected, below the limit of detection; NQ—not quantified, below the limit of quantification.

3.4. Human Keratinocyte Viability and Cytotoxicity

The effect of the *Artemisia* extracts—*Artemisia annua* L. (AAn) *Artemisia absinthium* L (AAb), *Artemisia dracunculus* L. (ADr), was evaluated on HaCaT keratinocytes after a stimulation period of 72 h (Figure 2). The Control group is represented by cells treated with the solvent dimethyl sulfoxide (the DMSO concentration did not exceed 0.5%). In Figure 2a, it can be observed that AAn, at concentrations ranging from 10 to 250 µg/mL, produced a significant increase in cells viability. Only at higher concentrations, 500–1000 µg/mL, was a decrease in cells viability noticed. A similar effect was obtained for AAb (Figure 2b) and ADr extracts (Figure 2c).

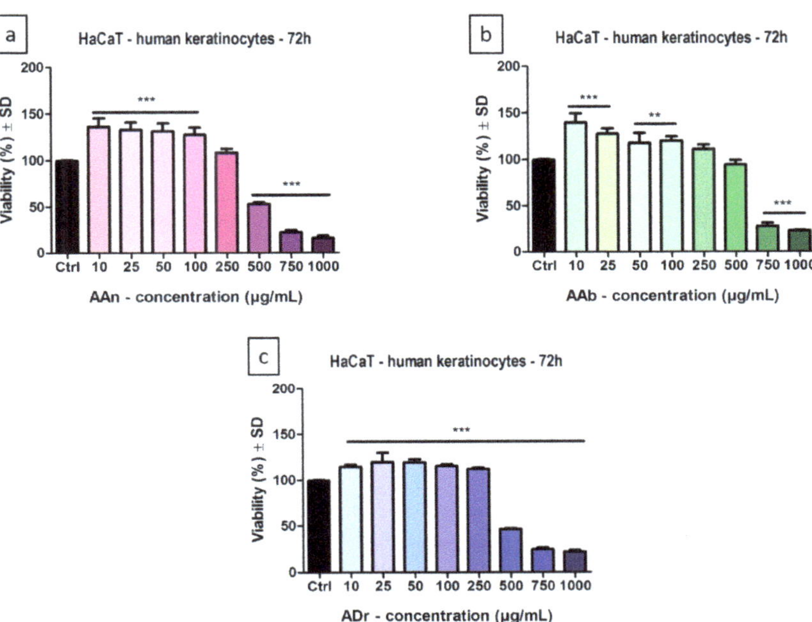

Figure 2. HaCaT cells viability 72 h after stimulation with the extracts (a—AAn, b—AAb, c—ADr) at different concentrations (10, 25, 50, 100, 250, 500, 750 and 1000 µg/mL). Data are expressed as mean ± SD. One-way ANOVA and Dunnett's multiple comparison post-test were used for comparison among groups (** $p < 0.01$, *** $p < 0.001$ vs. Control).

The highest increase in keratinocytes viability was induced by the sweet wormwood, followed by the wormwood extract and last by the tarragon extract (AAn > AAb > ADr). In terms of decreasing the cells viability, the most significant reduction was obtained at the highest concentration, 1000 µg/mL, for AAn—cells viability was 15.8 ± 2.6% vs. Control; for AAb—cells viability was 22.6 ± 0.4% vs. Control; for ADr—cells viability was 21.6 ± 1.4% vs. Control.

Figure 3 depicts the effect of the *Artemisia* extracts on LDH release. At the tested concentrations, for all samples, no cytotoxic effect was observed compared to Control. Moreover, a decrease in LDH release was obtained following stimulation with the extracts. We determined the cytotoxic effect at these concentrations by analyzing the viability assay results, where an increase in cells viability was obtained and we intended to add a proof that the human keratinocytes were not affected by these extracts.

3.5. Wound-Healing Effect In Vitro on Human Keratinocyte

This experimental model was performed using healthy keratinocyte cells, in order to explore the ability of the three *Artemisia* extracts to stimulate cells migration, a process that can be considered a first phase of lesion regeneration and can be correlated with wound closure, thus with the reduction of the open and vulnerable surface.

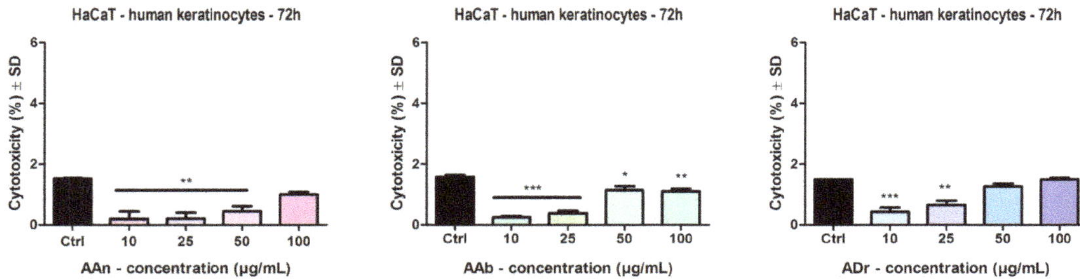

Figure 3. The cytotoxic effect of the extracts (AAn, AAb, ADr) at different concentrations (10, 25, 50, and 100 µg/mL). Data are expressed as mean ± SD. One-way ANOVA and Dunnett's multiple comparison post-test were used for comparison among groups (* $p < 0.05$, ** $p < 0.01$, *** $p < 0.001$ vs. Control).

Representative images of HaCaT cells are presented in Figure 4. The Control group is represented by cells treated with the culture medium, while the DMSO group represents cells treated with the solvent dimethyl sulfoxide. As shown in Figure 4, AAn and AAb extracts produced, at 100 µg/mL, almost a complete restoration of the scratch area, with wound closure rates of 98.55 ± 0.64% and 97.84 ± 1.98%, respectively, thus, indicating that the samples stimulated keratinocytes migration and wound closure. The ADr extract, at 100 µg/mL, increased cell migration with a wound closure rate of 89.65 ± 2.18%. The scratch closure percentage was lowest for the solvent used, DMSO (71.13 ± 5.69%).

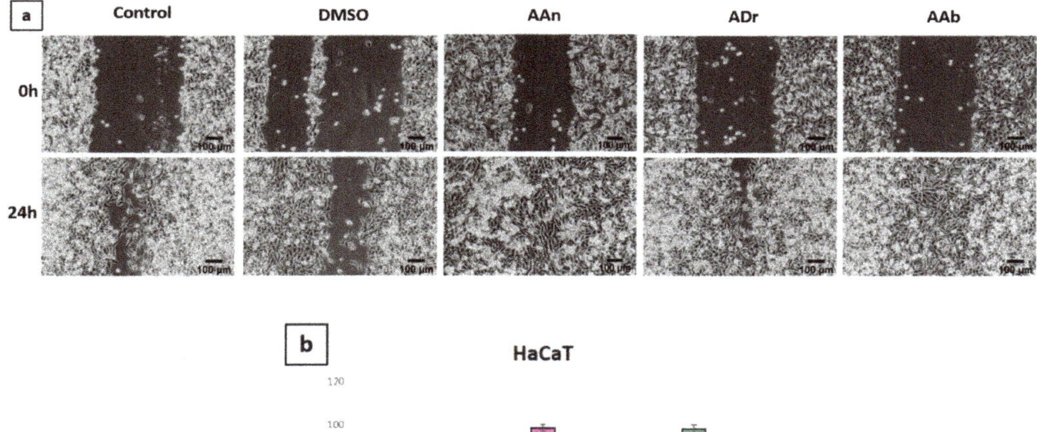

Figure 4. (**a**) Microscopic images representing the HaCaT cells treated with the three extracts of *Artemisia* species (100 µg/mL), initially at 0 h and at 24 h, respectively, visualized by light microscopy at 10× magnification; (**b**) Results were expressed as wound closure percentage after 24 h compared to the initial scratch length. Mean values ± SD of three independent experiments ($n = 3$).

3.6. Angiogenesis Modulation on CAM Assay

For the in vivo evaluation of the potential influence upon the wound healing process, we selected the 100 µg/mL concentration for the three *Artemisia* extract solutions, based on the fact that, at this concentration, the antioxidant effect was significant; when tested on the HaCaT cell viability, at this concentration, there was an increase in the viability of the cells, compared to higher concentrations that induced a reduction in cell viability.

The samples were assessed by evaluating the effect on the physiologic angiogenic process, starting with the 8th day of incubation. The rapid growth of the vessels took place from day 7 up to day 11 of incubation [48].

The effect was evaluated 24 h and 72 h after treatment, as represented in Figure 5. After one dose at 24 h post treatment, no relevant modifications were observed concerning the normal developing vessel architecture. Visible changes were noted after 3 doses at 72 h after treatment; different effects were observed for the three *Artemisia* extracts. An increase in the number of small vessels was induced by the AAn extracts, while the ADr also increased the number of bigger vessels inside the ring with a spokes–wheel pattern, compared to the control specimens. The vascular network was less affected by the AAb at 100 µg/mL concentration, inducing after 72 h post-treatment a lower increase in vessel growth; still, a significant number of small new forming capillary branches, derived from main vessels, were present.

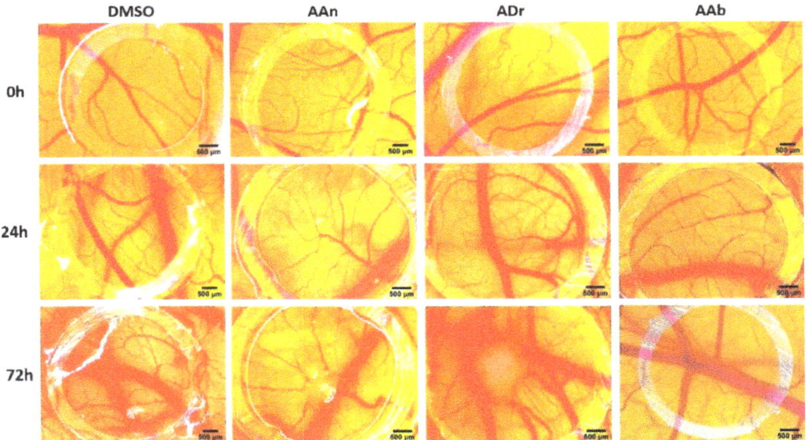

Figure 5. Stereomicroscopic images showing the effects produced by the *Artemisia* extracts on the CAM; images are represented initially at 0 h, 24 h and 72 h post-treatment, for the solvent control DMSO and the three *Artemisia* extracts; scale bars represent 500 µm.

With potential pro-angiogenesis effects at the tested concentration, extracts of sweet wormwood and tarragon may be especially beneficial in the vascular phase of wound healing. All the solutions showed good tolerability, with no influence upon embryo viability.

3.7. Anti-Irritant Effect Evaluated Using the HET-CAM Assay

Using a modified version of the HET-CAM protocol for the irritant effect, we assessed here the ability of the tested extracts to exert preventive and anti-irritative effects upon exposure to the irritant agent sodium dodecyl sulfate (SDS 0.5%). As shown in Figure 6, the results obtained after stereomicroscopic monitoring for 5 min reflected the degree of irritation reduction compared to the sample treated only with SDS. The study also included, as anti-inflammatory control, indometacin.

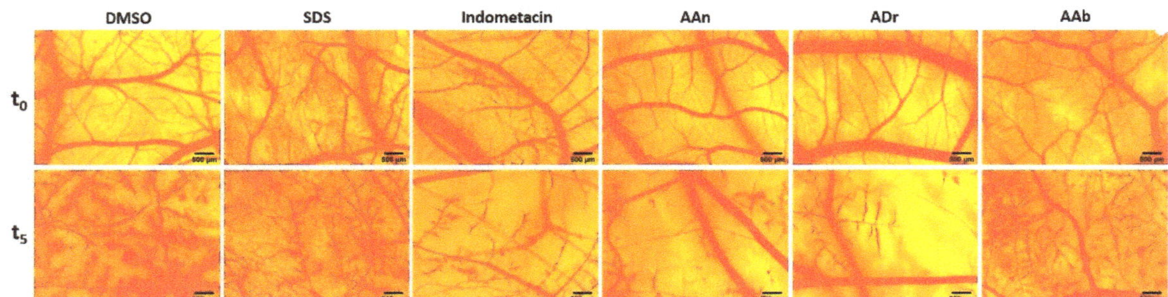

Figure 6. Stereomicroscopic images showing the effects induced by the *Artemisia* extracts on the CAM using the anti-irritative HET-CAM assay; images represent the initial moment, before application of the irritant SDS 0.5% (t_0), and 5 min after application of the irritant (t_5); DMSO represents the solvent control and indometacin, the anti-inflammatory control; scale bars represent 500 μm.

Comparing the irritative scores (Table 3) of the plant extracts at the concentration of 100 μg/mL with the score obtained by SDS alone, it is noticeable that a reduction of the IS was registered for all three extracts. The IS values calculated for the *Artemisia* extracts were also reduced compared to the solvent control; yet, the irritation scores were still higher when compared to the anti-inflammatory agent (IS = 16.69). According to the irritation scale recommended by Luepke [49], the irritation scores can be classified as following: 0–0.9—non-irritant, 1–4.9 weak irritant, 5–8.9 moderate irritant, 9–21 strong irritant.

Table 3. The irritative scores of the analyzed samples in the anti-irritative HET-CAM assay.

Samples	Irritative Score
AAn	18.38
ADr	19.01
AAb	19.69
DMSO	20.13
SDS	20.36
Indometacin	16.74

Within this experimental setting, for the evaluated concentration of the extracts, a similar anti-irritative effect was noticed for all three extracts, with AAn inducing the strongest effect (IS = 18.38), followed by ADr (IS = 19.01) and AAb (IS = 19.69).

4. Discussions

The management of the complicated disruptive wound healing processes is a real challenge that is confronted with limitations of the health systems. Thus, effort is made to identify low costs alternatives, especially of natural origin, efficient activators of the complex process of wound healing. Medicinal plants, with high content of polyphenolic compounds, are an important approach to reduce the high levels of oxidative stress that, in most cases, affects the healing process. Hence, in our study, we focused on the phenolic profile of our extracts, in correlation with their potential antioxidant activities. Likewise, the DPPH radical scavenging activity assay was performed in order to evaluate the antioxidant capacity of the three *Artemisia* extracts. A great number of natural compounds are also essential as antimicrobial agents, next to stimulating skin cell proliferation, migration or modelling the disruptive vascular process, thus, acting as multiple targeting active agents in the wound microenvironment [50–52].

In the current study, we selected three *Artemisia* species (*A. annua*, *A dracunculus* and *A. absinthium*) with a background in traditional skin ailment therapy and available on the herbal market of our country [53]. Confirmatory studies of the potential benefits in wound healing and regenerative medicine are still lacking; therefore, we intended to contribute with data concerning the phytochemical and bioactive potential of ethanolic extracts of the three *Artemisia* species.

Based on previous studies [54–56] that showed important content in polyphenolic compounds and lacking potential topical toxicity due to residual solvent remanence after drying, we selected to explore here the content in phenolic compounds of the 80% ethanolic extracts. The influence of the solvent on the antioxidant potential and chemical composition of *Artemisia annua* leaves extract was investigated by Iqbal et al. [54]. In terms of TPC, they concluded that the most efficient solvent for the extraction of phenolics from the leaves of AAn are methanol and water, and the less efficient were chloroform and hexane. In the case of the methanolic extract, the TPC was of 134.50 ± 4.37 mg/g. The extract of *Artemisia annuae herba* tested in our study contains 129.28 ± 2.09 mg GAE/g dry extract, using an ethanolic extract and the whole aerial part of the plant.

Kozlowska et al. [55] evaluated the antioxidant potential and the phenolics found in various dried and fresh plants, tarragon (*Artemisia dracunculus* L.) being one of them. For the extracts preparation the aerial part of the plant and 70% (v/v) ethanol were used and the TPC was 32.91 ± 0.68 g GAE/kg extract in case of fresh plant material, while, in the case of dried tarragon, 42.53 ± 0.93 g GAE/kg extract was obtained. Others obtained TPC values for methanolic tarragon extract in the range 97.2–253.5 mg GA/g, with a mean value of 151.6 mg GA/g, while for the water extract the range was 59.5–198.3 mg GA/g, and the mean value 102.8 mg GA/g [56]. Therefore, it can be said that the methanolic extract has the closest value to the one obtained by us for the *A. dracunculus* total phenolic content.

The regional area of *A. absinthium* L. collection influenced the total polyphenolic content of the extract, according to Msaada et al. [57]. The maximal value of TPC was 99.89 ± 3.30 mg GAE/g dried weight, obtained with vegetal material collected from Kairouan region. In the other regions, the TPC (expressed as mg GAE/g dried weight) had lower values, such as 83.70 ± 1.31 in Bou Salem, 72.05 ± 1.83 in Boukornine and 49.39 ± 2.20 in Jerissa.

A total phenolic content of 194 ± 9.7 mg gallic acid equivalent/g extract was reported by Ebrahimzadeh et al. [58] for *Artemisia absinthium* L. The extract was prepared using the aerial parts of the plants collected from Iran and methanol as extracting agent. The value reported for the TPC by Ebrahimzadeh et al. is similar to our results (193.61 ± 2.36 mg GAE/g dry extract). In addition, Bora and Sharma presented in their work a value of TPC of 123 ± 0.82 mg GA equivalents per gram of extract, for *A. absinthium* (methanolic extract); the extract was made from aerial parts and was procured from Himalaya Herbs Stores, India [59].

Therefore, from all the results presented above it is noticeable that the TPC is influenced by factors like the nature of the extracting agent, the region of the plant collection, the part of the plant used. The highest TPC was obtained for the wormwood extract, followed by the tarragon and then by the sweet wormwood ethanolic extract. The TPC established in our work for the three *Artemisia* extracts indicates important polyphenolic content of the dry medicinal products available on the market in Romania, being comparable to literature data.

Investigating the free radical scavenging capacity of the three extracts in a concentration range from 50 to 1000 µg/mL, our results indicated important antioxidant activities. Extracts in concentrations as high as 100 µg/mL had comparable AOA% with ascorbic acid (50 µg/mL) used as standard antioxidant agent. For higher concentrations, the antioxidant activities reached values over 90%, in a dose-related manner for all three species. Kim et al. [60] investigated the properties of *Artemisia annua* L. extracts obtained using different extractants such as water, 80% methanol, 80% ethanol and 80% acetone by the DPPH radical scavenging method. In the study conducted by Kim, the antioxidant activity at a concentration of 0.6 mg/mL ethanolic extract was 57.0 ± 1%, slightly lower than 77.71 ± 1.94% obtained by us for a concentration of 500 µg/mL (0.5 mg/mL).

A. dracunculus leaves and inflorescence in a methanolic extract, used in the work of Khezrilu Bandli and Heidari showed similar DPPH (%) values for the leaves and inflorescence extracts (86.43 ± 0.15% and 92.03 ± 0.11%, respectively) [61]. Lower values were obtained for *A. absinthium* extract prepared from leaves and stems using ethanol 80% as extractant; for a concentration of 1.4 mg/mL the leaves extract had 49.47 ± 0.015% and the stem extract had 56.84 ± 0.026% inhibition [19].

The most numerous studies regarding the three species consider their terpenoid-rich composition; however, their antioxidant effects were moderate. However, the extracts in polar solvents from plant products of the three species of *Artemisia* have been investigated, regarding the antioxidant effect. Studies have shown that the most concentrated compounds in tarragon are phenolic acids, especially gallic acid, synaptic acid, syringic acid, and the strong antioxidant activity can be accounted for by the high number of hydroxyl groups and the presence of type 2-carboxyl groups [61]. Wormwood is more concentrated in flavonoids next to phenolic acids, and values obtained in other studies indicate an antioxidant potential similar to sweet wormwood or tarragon [62]. Sweet wormwood is studied in most cases due to the presence of the sesquiterpene lactone artemisinin. We observed in the HPLC evaluation (data not shown), that artemisinin was not present in any of our extracts, as previously reported with significant degradability over storage [63]. Thus, the *Artemia* species tested here, as dried plant material available on the market, are valuable sources of polyphenols and antioxidant products.

The three *Artemisia* ethanolic extracts are concentrated in phenolic acids, mainly chlorogenic acid, with some differences between species. The highest content in chlorogenic acid was obtained for sweet wormwood, followed by a similar concentration in tarragon, while wormwood was three times less concentrated. As shown by others [64], this may explain the angiogenic activity reported here using the chorioallantoic membrane assay. Chlorogenic acid and flavonoids such as quercetin, kempferol and rutin were also identified in relevant concentrations in the tarragon extract only. *A. annua* polyphenolic profile was thoroughly investigated by Yi Song et al., who analyzed extracts from flowers, leaves, stems and roots by LC/MS/MS that revealed a rich polyphenolic content, out of which the most abundant compounds were caffeic acid derivatives [65]. Consistent with our findings, the group of Mumivand et al. investigated the polyphenolic content extracted from *A. dracunculus*, the LC analysis revealing a similar profile as the one identified by our group, phenolic acids out of which chlorogenic acid being the dominant one, accompanied by flavonoids [66]. In a similar approach, the polyphenolic profile of *A. absinthium* extract disclosed a vast phytocompound content, phenolic acids, in particular chlorogenic acid and p-coumaric acid being accountable as major components [57].

Great variability in the phytocompound profile and concentration are distinguished from the available literature data, and significant differences are accounted on the climate, soil, part of the plant, collection variables and extraction protocols. Once again, it reflects the importance of extract standardization worldwide.

The viability of healthy keratinocytes was assessed after exposure to the *Artemisia* extracts. Exposed to concentration of 100 μg/mL and 250 μg/mL the keratinocyte viability was not hindered, displaying even stimulatory effects in the Alamar blue assay 72 h following treatment. Wormwood extract at the concentration of 500 μg/mL also did not influence the viability of HaCaT cells; however, sweet wormwood and tarragon extracts at concentrations of 500 μg/mL and above decreased cells viability; thus, the recommended concentrations for wound healing purposes are up to 250 μg/mL. The effects of various *Artemisia* species were previously reported on different skin cells [67,68]. A study conducted on *A annua* in concentration of 100 μg/mL showed no inhibitory effects on the viability of normal fibroblasts [68]. Oh et al. indicated that *Artemisia princeps* Pampanini extract did not display a cytotoxic effect on HaCaT cells at concentrations up to 500 μg/mL (62.5, 125, 250 or 500 μg/mL) [67]; in addition, *Artemisia apiacea* showed no cytotoxicity towards HaCaT in concentration up to 200μg/mL [69].

Artemisia absinthium from Serbia [70] was also evaluated and, generally, concentrations below 400 µg/mL showed no cytotoxic effect on keratinocytes, while a Romanian cultivated wormwood extract showed a decrease in HaCaT cells, only at 1000 µg/mL [19]. Another study tested collagen scaffolds containing Romanian *A. absinthium* extract rich in luteolin and quercetin, and results showed, 5 days following treatment, an improve in fibroblast and keratinocyte proliferation to a higher extent than collagen alone. [71].

Our data regarding the three *Artemisia* species, AAn, AAb and ADr, indicated no toxicity towards HaCaT cells at doses up to 250 µg/mL. We further investigated the potential cytotoxicity through the LDH assay obtaining a confirmation of the lack of cytotoxicity for the healthy keratinocytes upon treatment with all three *Artemisia* species up to 100 µg/mL.

Interestingly *Artemisia capillaris* ethanolic extract from Korea, in concentration of 37.5 µg/mL induced an antiproliferative effect on healthy keratinocytes and, thus, considered as a potential anti-psoriatic agent [72].

The in vitro scratch assay exhibited wound closure capacity in healthy human keratinocytes after treatment with 100 µg/mL for all three *Artemisia* species, indicative for a pro-migratory and proliferative effect of keratinocytes during a wound healing process. Extracts in concentration of 100 µg/mL induced after the drawing of the scratch line, 24 h following treatment, a cellular density higher than the control. There are fewer studies that investigate in vitro the migratory potential of healthy keratinocytes for *Artemisia* species. Moaca et al. investigated the migratory potential of Romanian wormwood, and up to 250 µg/mL for the leaves and 500 µg/mL for the stems a pro-migratory effect was reported, while only higher concentrations induced a decrease in the migratory capacity oh HaCaT cells [19]. *A. asiatica* in concentrations as low as 5 µg/mL restored the proliferative and migratory potential of keratinocytes after cisplatin induced inhibition [73].

Other studies involving the wound healing potential of *Artemisia* species use animal models. *A. absinthium* from Algeria evaluated as an ointment on a rat model improved wound contraction comparable to the control allantoin ointment [74]. *A. dracunculus* with chitosan nanoparticle biofilm induced significant improvement of wound contraction on rats compared to the other treatment groups [75].

With a high TPC and a potent radical scavenging activity, the effects on the viability and migration potential of healthy keratinocytes of the wormwood extract investigated here, although less concentrated in the phenolic compounds screened in this study, may be related to other phytocompounds such as chalcones, cinnamic acid, artemetin as previously shown by others [74].

The skin wound healing process involves a cascade of complex events including angiogenesis. The angiogenic activity was assessed by the CAM model and a visible increase of new vessel formation was observed 72 h after treatment with concentrations of 100 µg/mL, compared to Control specimens. Slight differences between the three tested extracts were noted, with *A. dracunculus* and *A. annua* inducing the more pronounced effect. Limited studies investigated the angiogenic potential of the three *Artemisia* species. Several data indicate the anti-angiogenic effect, but for different species, such as *A. sieberi* [76], *A. capillaris* in a polyherbal anti-obesity preparation [77] or *A. herba-alba* essential oil [78]. Other studies involving *A. annua*, showed that mostly sesquiterpenoid artemisinins are responsible for the anti-angiogenic effect [79,80]. However, the pro-angiogenic effect induced in ovo using the CAM assay can be explained based on the dominant phenolic compound in all three extracts, chlorogenic acid, and its hormetic behavior reported as a proangiogenic agent at small concentrations of 10 µg/mL, significantly increasing endothelial cell migration and stimulating capillary tube formation [64,81,82].

Finally, using the in ovo chorioallantoic membrane, and a modified HET-CAM protocol, we assessed the anti-irritative potential of the three extracts in concentrations of 100 µg/mL. The assay allowed observing the potential preventive and healing capacities of the extracts when exposed to a highly irritant agent, sodium dodecyl sulfate. Although the results did not show a major improvement of the damaged vascular area, limitative effects can be underlined as the irritation score obtained for the three extracts indicated a superior effect as compared to the solvent control alone. *A. annua* and *A dracunculus* showed a slightly higher anti-irritative effects.

The three *Artemisia* species available on the herbal market in Romania investigated as 80% ethanolic extracts displayed a high content in phenolic compounds, with different phytochemical profile following LC-MS analysis. All extracts showed chlorogenic acid as dominant polyphenol; *A. annua* and *A. dracunculus* revealed to be almost three times more concentrated than *A. absinthium*. Still, accounting for high TPC values; similar biological effects were registered by the three species, in terms of in vitro radical scavenging activity, proliferative and pro-migratory effects on keratinocytes. The angiogenic process was also promoted by all three extracts in concentration of 100 µg/mL, with a different, more pronounced impact in the case of *A. annua* and *A. dracunculus.* Moreover, potentially beneficial in alleviating skin irritation, the three *Artemisia* extracts are safe to use on cutaneous and mucosal tissues.

5. Conclusions

We demonstrated the effective promotion of normal keratinocyte proliferation and migration by the three *Artemisia* extracts. Up to a concentration of 100 µg/mL the extracts enhanced cell viability, restored wound closure and slightly reduced the irritation induced by SDS in ovo. Angiogenesis was activated by increasing new vessel formation, especially by *A. annua* and *A. dracunculus* extracts with significantly higher content of chlorogenic acid. The high total phenolic content and the significant radical scavenging activities of the three *Artemisia* extracts are important features potentially improving the wound healing dysregulated microenvironment. The three *Artemisia* species, easily available on the herbal market, may represent low-cost alternatives, benefitting from a multi-targeted mechanism and a safe profile, hence applicable in the design of novel wound dressing preparations for wound care management.

Author Contributions: Conceptualization, R.G., D.M. (Daliana Minda), I.Z.P. and S.A.; extract preparation, D.M. (Daliana Minda); antioxidant activity, D.M. (Daliana Minda), R.R. and D.M. (Delia Muntean); physicochemical screening, Z.D.; R.G., C.D.B. and O.D.B.; in vitro assays, I.Z.P. and O.D.B.; angiogenesis, S.A.; Formal analysis, R.G. and C.D.B.; funding acquisition, I.Z.P.; resources, C.S., C.A.D., I.Z.P. and S.A.; writing—original draft preparation, R.G., C.D.B., I.Z.P., C.D., R.R., Z.D. and S.A.; writing—review and editing, C.S., C.A.D. and S.A.; validation, C.A.D., C.S. and S.A.; visualization, R.G., C.D.B., I.Z.P., R.R. and S.A.; supervision, S.A. All authors have read and agreed to the published version of the manuscript.

Funding: This research was funded by the Project PN-III-P1-1.1-PD-2019-1231, No. PD 206/2020, Project Director Ioana Zinuca Pavel.

Institutional Review Board Statement: Not applicable.

Informed Consent Statement: Not applicable.

Data Availability Statement: The data supporting the findings of the study are available within the article.

Conflicts of Interest: The authors declare no conflict of interest.

References

1. Pereira, R.F.; Bártolo, P.J. Traditional Therapies for Skin Wound Healing. *Adv. Wound Care* **2016**, *5*, 208–229. [CrossRef] [PubMed]
2. Shedoeva, A.; Leavesley, D.; Upton, Z.; Fan, C. Wound Healing and the Use of Medicinal Plants. *Evid.-Based Complement. Altern. Med.* **2019**, *2019*, 2684108. [CrossRef] [PubMed]
3. Tottoli, E.M.; Dorati, R.; Genta, I.; Chiesa, E.; Pisani, S.; Conti, B. Skin Wound Healing Process and New Emerging Technologies for Skin Wound Care and Regeneration. *Pharmaceutics* **2020**, *12*, 735. [CrossRef] [PubMed]
4. Colalto, C. What phytotherapy needs: Evidence-based guidelines for better clinical practice. *Phytother. Res.* **2018**, *32*, 413–425. [CrossRef]

5. Danciu, C.; Zupko, I.; Bor, A.; Schwiebs, A.; Radeke, H.; Hancianu, M.; Cioanca, O.; Alexa, E.; Oprean, C.; Bojin, F.; et al. Botanical Therapeutics: Phytochemical Screening and Biological Assessment of Chamomile, Parsley and Celery Extracts against A375 Human Melanoma and Dendritic Cells. *Int. J. Mol. Sci.* **2018**, *19*, 3624. [CrossRef]
6. Watson, L.E.; Bates, P.L.; Evans, T.M.; Unwin, M.M.; Estes, J.R. Molecular phylogeny of Subtribe *Artemisiinae* (*Asteraceae*), including *Artemisia* and its allied and segregate genera. *BMC Evol. Biol.* **2002**, *2*, 17. [CrossRef]
7. Lou, H.; Huang, Y.; Ouyang, Y.; Zhang, Y.; Xi, L.; Chu, X.; Wang, Y.; Wang, C.; Zhang, L. *Artemisia annua* sublingual immunotherapy for seasonal allergic rhinitis: A randomized controlled trial. *Allergy* **2020**, *75*, 2026–2036. [CrossRef]
8. Bora, K. Phytochemical and pharmacological potential of *Artemisia absinthium* Linn. and *Artemisia asiatica* Nakai: A Review. *J. Pharm. Res.* **2010**, *3*, 325–328.
9. Taleghani, A.; Emami, S.A.; Tayarani-Najaran, Z. *Artemisia*: A promising plant for the treatment of cancer. *Bioorg. Med. Chem.* **2020**, *28*, 115180. [CrossRef]
10. Czechowski, T.; Larson, T.R.; Catania, T.M.; Harvey, D.; Wei, C.; Essome, M.; Brown, G.D.; Graham, I.A. Detailed Phytochemical Analysis of High- and Low Artemisinin-Producing Chemotypes of *Artemisia annua*. *Front. Plant Sci.* **2018**, *9*, 641. [CrossRef]
11. Messaili, S.; Colas, C.; Fougère, L.; Destandau, E. Combination of molecular network and centrifugal partition chromatography fractionation for targeting and identifying *Artemisia annua* L. antioxidant compounds. *J. Chromatogr. A* **2019**, *1615*, 460785. [CrossRef]
12. Mohammadi, S.; Jafari, B.; Asgharian, P.; Martorell, M.; Sharifi-Rad, J. Medicinal plants used in the treatment of Malaria: A key emphasis to *Artemisia, Cinchona, Cryptolepis,* and *Tabebuia* genera. *Phytother. Res.* **2020**, *34*, 1556–1569. [CrossRef] [PubMed]
13. Foglio, M.A.; Dias, P.C.; Antônio, M.A.; Possenti, A.; Rodrigues, R.A.F.; da Silva, É.F.; Rehder, V.L.; de Carvalho, J.E. Antiulcerogenic Activity of Some Sesquiterpene Lactones Isolated from *Artemisia annua*. *Planta Med.* **2002**, *68*, 515–518. [CrossRef] [PubMed]
14. Yu, Y.; Simmler, C.; Kuhn, P.; Poulev, A.; Raskin, I.; Ribnicky, D.; Floyd, Z.E.; Pauli, G.F. The DESIGNER Approach Helps Decipher the Hypoglycemic Bioactive Principles of *Artemisia dracunculus* (Russian Tarragon). *J. Nat. Prod.* **2019**, *82*, 3321–3329. [CrossRef] [PubMed]
15. Durić, K.; Kovac Besovic, E.E.; Niksic, H.; Muratovic, S.; Sofic, E. Anticoagulant activity of some *Artemisia dracunculus* leaf extracts. *Bosn. J. Basic Med. Sci.* **2015**, *15*, 9–14. [CrossRef]
16. Kordali, S.; Kotan, R.; Mavi, A.; Cakir, A.; Ala, A.; Yildirim, A. Determination of the Chemical Composition and Antioxidant Activity of the Essential Oil of *Artemisia dracunculus* and of the Antifungal and Antibacterial Activities of Turkish *Artemisia absinthium, A. dracunculus, Artemisia santonicum,* and *Artemisia spicigera* Essential Oils. *J. Agric. Food Chem.* **2005**, *53*, 9452–9458. [CrossRef]
17. Ekiert, H.; Świątkowska, J.; Knut, E.; Klin, P.; Rzepiela, A.; Tomczyk, M.; Szopa, A. *Artemisia dracunculus* (Tarragon): A Review of Its Traditional Uses, Phytochemistry and Pharmacology. *Front. Pharmacol.* **2021**, *12*, 653993. [CrossRef]
18. Szopa, A.; Pajor, J.; Klin, P.; Rzepiela, A.; Elansary, H.O.; Al-Mana, F.A.; Mattar, M.A.; Ekiert, H. *Artemisia absinthium* L.—Importance in the History of Medicine, the Latest Advances in Phytochemistry and Therapeutical, Cosmetological and Culinary Uses. *Plants* **2020**, *9*, 1063. [CrossRef]
19. Moacă, A.-E.; Pavel, I.Z.; Danciu, C.; Crăiniceanu, Z.; Minda, D.; Ardelean, F.; Antal, D.S.; Ghiulai, R.; Cioca, A.; Derban, M.; et al. Romanian Wormwood (*Artemisia absinthium* L.): Physicochemical and Nutraceutical Screening. *Molecules* **2019**, *24*, 3087. [CrossRef]
20. Asghar, M.N.; Khan, I.U.; Bano, N. In vitro antioxidant and radical-scavenging capacities of *Citrullus colocynthes* (L) and *Artemisia absinthium* extracts using promethazine hydrochloride radical cation and contemporary assays. *Food Sci. Technol. Int.* **2011**, *17*, 481–494. [CrossRef]
21. Guarrera, P.M. Traditional antihelmintic, antiparasitic and repellent uses of plants in Central Italy. *J. Ethnopharmacol.* **1999**, *68*, 183–192. [CrossRef]
22. Kharoubi, O.; Slimani, M.; Aoues, A. Neuroprotective effect of wormwood against lead exposure. *J. Emerg. Trauma Shock* **2011**, *4*, 82. [CrossRef]
23. Carvalho, A.R.; Diniz, R.M.; Suarez, M.A.M.; Figueiredo, C.S.S.e.S.; Zagmignan, A.; Grisotto, M.A.G.; Fernandes, E.S.; da Silva, L.C.N. Use of Some *Asteraceae* Plants for the Treatment of Wounds: From Ethnopharmacological Studies to Scientific Evidences. *Front. Pharmacol.* **2018**, *9*, 784. [CrossRef] [PubMed]
24. Mastrullo, V.; Cathery, W.; Velliou, E.; Madeddu, P.; Campagnolo, P. Angiogenesis in tissue engineering: As nature intended? *Front. Bioeng. Biotechnol.* **2020**, *8*, 188. [CrossRef]
25. Morbidelli, L.; Genah, S.; Cialdai, F. Effect of Microgravity on Endothelial Cell Function, Angiogenesis, and Vessel Remodeling During Wound Healing. *Front. Bioeng. Biotechnol.* **2021**, *9*, 720091. [CrossRef]
26. Morbidelli, L.; Terzuoli, E.; Donnini, S. Use of Nutraceuticals in Angiogenesis-Dependent Disorders. *Molecules* **2018**, *23*, 2676. [CrossRef] [PubMed]
27. Matusz, P.; Dragoslav Miclăuş, G.; Dragoş Banciu, C.; Sas, I.; Joseph, S.C.; Cornel Pirtea, L.; Shane Tubbs, R.; Loukas, M. Congenital Solitary Kidney with Multiple Renal Arteries: Case Report Using MDCT Angiography. *Rom. J. Morphol. Embryol.* **2015**, *56*, 823–826. [PubMed]
28. Ribatti, D.; Annese, T.; Tamma, R. The use of the chick embryo CAM assay in the study of angiogenic activity of biomaterials. *Microvasc. Res.* **2020**, *131*, 104026. [CrossRef]

29. Moreno-Jiménez, I.; Hulsart-Billstrom, G.; Lanham, S.A.; Janeczek, A.A.; Kontouli, N.; Kanczler, J.M.; Evans, N.D.; Oreffo, R.O.C. The Chorioallantoic Membrane (CAM) Assay for the Study of Human Bone Regeneration: A Refinement Animal Model for Tissue Engineering. *Sci. Rep.* **2016**, *6*, 32168. [CrossRef]

30. Singleton, V.L.; Orthofer, R.; Lamuela-Raventós, R.M. Analysis of total phenols and other oxidation substrates and antioxidants by means of Folin-Ciocalteu reagent. *Methods Enzymol.* **1999**, *299*, 152–178. [CrossRef]

31. Gupta, D.; Gupta, R.K. Bioprotective properties of Dragon's blood resin: In vitro evaluation of antioxidant activity and antimicrobial activity. *BMC Complement. Altern. Med.* **2011**, *11*, 13. [CrossRef] [PubMed]

32. Moacă, E.A.; Farcaş, C.; Ghiţu, A.; Coricovac, D.; Popovici, R.; Cărăba-Meiţă, N.L.; Ardelean, F.; Antal, D.S.; Dehelean, C.; Avram, Ş.A. Comparative Study of *Melissa officinalis* Leaves and Stems Ethanolic Extracts in terms of Antioxidant, Cytotoxic, and Antiproliferative Potential. *Evid. Based Complement. Altern. Med.* **2018**, *2018*, 7860456. [CrossRef] [PubMed]

33. Santos, U.P.; Campos, J.F.; Torquato, H.F.V.; Paredes-Gamero, E.J.; Carollo, C.A.; Estevinho, L.M.; de Picoli Souza, K.; Dos Santos, E.L. Antioxidant, antimicrobial and cytotoxic properties as well as the phenolic content of the extract from *Hancornia speciosa* gomes. *PLoS ONE* **2016**, *11*, e0167531. [CrossRef] [PubMed]

34. Ghiulai, R.; Avram, S.; Stoian, D.; Pavel, I.Z.; Coricovac, D.; Oprean, C.; Vlase, L.; Farcas, C.; Mioc, M.; Minda, D.; et al. Lemon Balm Extracts Prevent Breast Cancer Progression In Vitro and In Ovo on Chorioallantoic Membrane Assay. *Evid. Based Complement. Altern. Med.* **2020**, *2020*, 6489159. [CrossRef] [PubMed]

35. Sipos, S.; Moacă, E.A.; Pavel, I.Z.; Avram, Ş.; Creţu, O.M.; Coricovac, D.; Racoviceanu, R.M.; Ghiulai, R.; Pană, R.D.; Şoica, C.M.; et al. *Melissa officinalis* L. Aqueous Extract Exerts Antioxidant and Antiangiogenic Effects and Improves Physiological Skin Parameters. *Molecules* **2021**, *26*, 2369. [CrossRef]

36. Baiceanu, E.; Vlase, L.; Baiceanu, A.; Nanes, M.; Rusu, D.; Crisan, G. New polyphenols identified in *Artemisiae abrotani herba* extract. *Molecules* **2015**, *20*, 11063–11075. [CrossRef]

37. Coşarcă, S.L.; Moacă, E.A.; Tanase, C.; Muntean, D.L.; Pavel, I.Z.; Dehelean, C.A. Spruce and Beech Bark Aqueous Extracts: Source of Polyphenols, Tannins And Antioxidants Correlated to In Vitro Antitumor Potential on Two Different Cell Lines. *Wood Sci. Technol.* **2019**, *53*, 313–333. [CrossRef]

38. Ghiţu, A.; Schwiebs, A.; Radeke, H.H.; Avram, S.; Zupko, I.; Bor, A.; Pavel, I.Z.; Dehelean, C.A.; Oprean, C.; Bojin, F.; et al. A Comprehensive Assessment of Apigenin as an Antiproliferative, Proapoptotic, Antiangiogenic and Immunomodulatory Phytocompound. *Nutrients* **2019**, *11*, 858. [CrossRef]

39. Farcas, C.G.; Dehelean, C.; Pinzaru, I.A.; Mioc, M.; Socoliuc, V.; Moaca, E.-A.; Avram, S.; Ghiulai, R.; Coricovac, D.; Pavel, I.; et al. Thermosensitive Betulinic Acid-Loaded Magnetoliposomes: A Promising Antitumor Potential for Highly Aggressive Human Breast Adenocarcinoma Cells under Hyperthermic Conditions. *Int. J. Nanomed.* **2020**, *15*, 8175–8200. [CrossRef]

40. Felice, F.; Zambito, Y.; Belardinelli, E.; Fabiano, A.; Santoni, T.; Di Stefano, R. Effect of different chitosan derivatives on in vitro scratch wound assay: A comparative study. *Int. J. Biol. Macromol.* **2015**, *76*, 236–241. [CrossRef]

41. Ribatti, D. The chick embryo chorioallantoic membrane in the study of tumor angiogenesis. *Rom. J. Morphol. Embryol.* **2008**, *49*, 131–135. [PubMed]

42. Caunii, A.; Oprean, C.; Cristea, M.; Ivan, A.; Danciu, C.; Tatu, C.; Paunescu, V.; Marti, D.; Tzanakakis, G.; Spandidos, D.; et al. Effects of Ursolic and Oleanolic on SK-MEL-2 Melanoma Cells: In Vitro and In Vivo Assays. *Int. J. Oncol.* **2017**, *51*, 1651–1660. [CrossRef] [PubMed]

43. Scheel, J.; Kleber, M.; Kreutz, J.; Lehringer, E.; Mehling, A.; Reisinger, K.; Steiling, W. Eye Irritation Potential: Usefulness of the HET-CAM under the Globally Harmonized System of Classification and Labeling of Chemicals (GHS). *Regul. Toxicol. Pharmacol.* **2011**, *59*, 471–492. [CrossRef]

44. Szuhanek, C.A.; Watz, C.G.; Avram, Ş.; Moacă, E.-A.; Mihali, C.V.; Popa, A.; Campan, A.A.; Nicolov, M.; Dehelean, C.A. Comparative Toxicological In Vitro and In Ovo Screening of Different Orthodontic Implants Currently Used in Dentistry. *Materials* **2020**, *13*, 5690. [CrossRef] [PubMed]

45. Coricovac, D.; Farcas, C.; Nica, C.; Pinzaru, I.; Simu, S.; Stoian, D.; Soica, C.; Proks, M.; Avram, S.; Navolan, D.; et al. Ethinylestradiol and Levonorgestrel as Active Agents in Normal Skin, and Pathological Conditions Induced by UVB Exposure: In Vitro and In Ovo Assessments. *Int. J. Mol. Sci.* **2018**, *19*, 3600. [CrossRef]

46. Interagency Coordinating Committee on the Validation of Alternative Methods (ICCVAM), ICCVAM Recommended Test Method Protocol: Hen's Egg Test–Chorioallantoic Membrane. 2010. Available online: http://iccvam.niehs.nih.gov/ (accessed on 6 December 2021).

47. Wilson, T.D.; Steck, W.F. A Modified HET-CAM Assay Approach to the Assessment of Anti-Irritant Properties of Plant Extracts. *Food Chem. Toxicol.* **2000**, *38*, 867–872. [CrossRef]

48. Ribatti, D. The Chick Embryo Chorioallantoic Membrane (CAM). A Multifaceted Experimental Model. *Mech. Dev.* **2016**, *141*, 70–77. [CrossRef]

49. Luepke, N.P. Hen's Egg Chorioallantoic Membrane Test for Irritation Potential. *Food Chem. Toxicol.* **1985**, *23*, 287–291. [CrossRef]

50. Lordani, T.V.A.; de Lara, C.E.; Ferreira, F.B.P.; de Souza Terron Monich, M.; da Silva, C.M.; Lordani, C.R.F.; Bueno, F.G.; Teixeira, J.J.V.; Lonardoni, M.V.C. Therapeutic Effects of Medicinal Plants on Cutaneous Wound Healing in Humans: A Systematic Review. *Mediat. Inflamm.* **2018**, *2018*, 7354250. [CrossRef]

51. Noroozi, R.; Sadeghi, E.; Yousefi, H.; Taheri, M.; Sarabi, P.; Dowati, A.; Ayatallahi, S.A.; Noroozi, R.; Ghafouri-Fard, S. Wound Healing Features of *Prosopis farcta*: In Vitro Evaluation of Antibacterial, Antioxidant, Proliferative and Angiogenic Properties. *Gene Rep.* **2019**, *17*, 100482. [CrossRef]
52. Sharma, A.; Khanna, S.; Kaur, G.; Singh, I. Medicinal Plants and Their Components for Wound Healing Applications. *Future J. Pharm. Sci* **2021**, *7*, 1–13. [CrossRef]
53. Bora, K.S.; Sharma, A. The Genus *Artemisia*: A Comprehensive Review. *Pharm. Biol.* **2011**, *49*, 101–109. [CrossRef] [PubMed]
54. Iqbal, S.; Younas, U.; Chan, K.W.; Zia-Ul-Haq, M.; Ismail, M. Chemical Composition of *Artemisia annua* L. Leaves and Antioxidant Potential of Extracts as a Function of Extraction Solvents. *Molecules* **2012**, *17*, 6020–6032. [CrossRef] [PubMed]
55. Kozlowska, M.; Scibisz, I.; Przybyl, J.L.; Ziarno, M.; Zbikowska, A.; Majewska, E. Phenolic Contents and Antioxidant Activity of Extracts of Selected Fresh and Dried Herbal Materials. *Pol. J. Food Nutr. Sci.* **2021**, *71*, 269–278. [CrossRef]
56. Ulewicz-Magulska, B.; Wesolowski, M. Total Phenolic Contents and Antioxidant Potential of Herbs Used for Medical and Culinary Purposes. *Plant Foods Hum. Nutr.* **2019**, *74*, 61–67. [CrossRef]
57. Msaada, K.; Salem, N.; Bachrouch, O.; Bousselmi, S.; Tammar, S.; Alfaify, A.; Al Sane, K.; Ben Ammar, W.; Azeiz, S.; Haj Brahim, A.; et al. Chemical Composition and Antioxidant and Antimicrobial Activities of Wormwood (*Artemisia absinthium* L.) Essential Oils and Phenolics. *J. Chem.* **2015**, *2015*, 804658. [CrossRef]
58. Ebrahimzadeh, M.A.; Nabavi, S.F.; Nabavi, S.M.; Pourmorad, F. Nitric Oxide Radical Scavenging Potential of Some Elburz Medicinal Plants. *Afr. J. Biotechnol.* **2010**, *9*, 5212–5217. [CrossRef]
59. Bora, K.; Sharma, A. Pharmaceutical Biology Evaluation of Antioxidant and Free-Radical Scavenging Potential of *Artemisia absinthium*. *Pharm. Biol.* **2011**, *49*, 1216–1223. [CrossRef]
60. Kim, W.S.; Choi, W.J.; Lee, S.; Kim, W.J.; Lee, D.C.; Sohn, U.D.; Shin, H.S.; Kim, W. Anti-Inflammatory, Antioxidant and Antimicrobial Effects of Artemisinin Extracts from *Artemisia annua* L. *Korean J. Physiol. Pharmacol.* **2015**, *19*, 21–27. [CrossRef]
61. Khezrilu Bandli, J.; Heidari, R. The Evaluation of Antioxidant Activities and Phenolic Compounds in Leaves and Inflorescence of *Artemisia dracunculus* L. by HPLC. *J. Med. Plants* **2014**, *13*, 41–50.
62. Emami, S.A.; Asili, J.; Mohagheghi, Z.; Hassanzadeh, M.K. Antioxidant activity of leaves and fruits of Iranian conifers. *Evid. Based Complement. Altern. Med.* **2007**, *4*, 313–319. [CrossRef] [PubMed]
63. Ivanescu, B.; Vlase, L.; Corciova, A.; Lazar, M.I. Artemisinin Evaluation in Romanian *Artemisia annua* Wild Plants Using a New HPLC/MS Method. *Nat. Prod. Res.* **2011**, *25*, 716–722. [CrossRef] [PubMed]
64. Moghadam, S.E.; Ebrahimi, S.N.; Salehi, P.; Farimani, M.M.; Hamburger, M.; Jabbarzadeh, E. Wound Healing Potential of Chlorogenic Acid and Myricetin-3-o-β-Rhamnoside Isolated from *Parrotia persica*. *Molecules* **2017**, *22*, 1501. [CrossRef]
65. Song, Y.; Desta, K.T.; Kim, G.S.; Lee, S.J.; Lee, W.S.; Kim, Y.H.; Jin, J.S.; Abd El-Aty, A.M.; Shin, H.C.; Shim, J.H.; et al. Polyphenolic profile and antioxidant effects of various parts of *Artemisia annua* L. *Biomed. Chromatogr.* **2016**, *30*, 588–595. [CrossRef] [PubMed]
66. Mumivand, H.; Babalar, M.; Tabrizi, L.; Craker, L.E.; Shokrpour, M.; Hadian, J. Antioxidant properties and principal phenolic phytochemicals of Iranian tarragon (*Artemisia dracunculus* L.) accessions. *Hortic. Environ. Biotechnol.* **2017**, *58*, 414–422. [CrossRef]
67. Oh, C.T.; Jang, Y.J.; Kwon, T.R.; Im, S.; Kim, S.R.; Seok, J.; Kim, G.Y.; Kim, Y.H.; Mun, S.K.; Kim, B.J. Effect of isosecotanapartholide isolated from *Artemisia princeps* Pampanini on IL-33 production and STAT-1 activation in HaCaT keratinocytes. *Mol. Med. Rep.* **2017**, *15*, 2681–2688. [CrossRef]
68. Kim, E.J.; Kim, G.T.; Kim, B.M.; Lim, E.G.; Kim, S.Y.; Kim, Y.M. Apoptosis-induced effects of extract from *Artemisia annua* Linné by modulating PTEN/p53/PDK1/Akt/ signal pathways through PTEN/p53-independent manner in HCT116 colon cancer cells. *BMC Complement. Altern. Med.* **2017**, *17*, 236. [CrossRef]
69. Yang, J.H.; Lee, E.; Lee, B.; Cho, W.K.; Ma, J.Y.; Park, K.I. Ethanolic Extracts of *Artemisia apiacea* Hance Improved Atopic Dermatitis-Like Skin Lesions In Vivo and Suppressed TNF-Alpha/IFN-Gamma-Induced Proinflammatory Chemokine Production In Vitro. *Nutrients* **2018**, *10*, 806. [CrossRef]
70. Ivanov, M.; Gašić, U.; Stojković, D.; Kostić, M.; Mišić, D.; Soković, M. New Evidence for *Artemisia absinthium* L. Application in Gastrointestinal Ailments: Ethnopharmacology, Antimicrobial Capacity, Cytotoxicity, and Phenolic Profile. *Evid. Based Complement. Altern. Med.* **2021**, *2021*, 9961089. [CrossRef]
71. Gaspar-Pintiliescu, A.; Seciu, A.M.; Miculescu, F.; Moldovan, L.; Ganea, E.; Craciunescu, O. Enhanced Extracellular Matrix Synthesis Using Collagen Dressings Loaded with *Artemisia absinthium* Plant Extract. *J. Bioact. Compat. Polym.* **2018**, *33*, 516–528. [CrossRef]
72. Lee, S.Y.; Nam, S.; Hong, I.K.; Kim, H.; Yang, H.; Cho, H.J. Antiproliferation of Keratinocytes and Alleviation of Psoriasis by the Ethanol Extract of *Artemisia capillaris*. *Phytother. Res.* **2018**, *32*, 923–932. [CrossRef] [PubMed]
73. Chang, J.W.; Hwang, H.S.; Kim, Y.S.; Kim, H.J.; Shin, Y.S.; Jittreetat, T.; Kim, C.H. Protective Effect of *Artemisia asiatica* (Pamp.) Nakai Ex Kitam Ethanol Extract against Cisplatin-Induced Apoptosis of Human HaCaT Keratinocytes: Involvement of NF-Kappa B- and Bcl-2-Controlled Mitochondrial Signaling. *Phytomedicine* **2015**, *22*, 679–688. [CrossRef] [PubMed]
74. Boudjelal, A.; Smeriglio, A.; Ginestra, G.; Denaro, M.; Trombetta, D. Phytochemical Profile, Safety Assessment and Wound Healing Activity of *Artemisia absinthium* L. *Plants* **2020**, *9*, 1744. [CrossRef] [PubMed]
75. Ranjbar, R.; Yousefi, A. *Artemisia Dracunculus* in Combination with Chitosan Nanoparticle Biofilm Improves Wound Healing in MRSA Infected Excisional Wounds: An Animal Model Study. *EurAsian J. Biosci.* **2018**, *12*, 219–226.
76. Abdolmaleki, Z.; Arab, H.A.; Amanpour, S.; Muhammadnejad, S. Anti-Angiogenic Effects of Ethanolic Extract of *Artemisia sieberi* Compared to Its Active Substance, Artemisinin. *Rev. Bras. Farmacogn.* **2016**, *26*, 326–333. [CrossRef]

77. Yoon, M.; Kim, M.Y. The Anti-Angiogenic Herbal Composition Ob-X from *Morus alba*, *Melissa officinalis*, and *Artemisia capillaris* Regulates Obesity in Genetically Obese Ob/Ob Mice. *Pharm. Biol.* **2011**, *49*, 614–619. [CrossRef]
78. Jaouadi, I.; Tansu Koparal, A.; Beklem Bostancıoğlu, R.; Tej Yakoubi, M.; el Gazzah, M. The Anti-Angiogenic Activity of *Artemisia herba-alba*'s Essential Oil and Its Relation with the Harvest Period. *AJCS* **2014**, *8*, 1395–1401.
79. Zhu, X.X.; Yang, L.; Li, Y.J.; Zhang, D.; Chen, Y.; Kostecká, P.; Kmoníéková, E.; Zídek, Z. Effects of Sesquiterpene, Flavonoid and Coumarin Types of Compounds from *Artemisia annua* L. on Production of Mediators of Angiogenesis. *Pharmacol. Rep.* **2013**, *65*, 410–420. [CrossRef]
80. Feng, X.; Cao, S.; Qiu, F.; Zhang, B. Traditional Application and Modern Pharmacological Research of *Artemisia annua* L. *Pharmacol. Ther.* **2020**, *216*, 107650. [CrossRef]
81. Majewska, I.; Gendaszewska-Darmach, E. Proangiogenic Activity of Plant Extracts in Accelerating Wound Healing—A New Face of Old Phytomedicines. *Acta Biochim. Pol.* **2011**, *58*, 449–460. [CrossRef]
82. Tsakiroglou, P.; Vandenakker, N.E.; del Bo', C.; Riso, P.; Klimis-Zacas, D. Role of Berry Anthocyanins and Phenolic Acids on Cell Migration and Angiogenesis: An Updated Overview. *Nutrients* **2019**, *11*, 1075. [CrossRef] [PubMed]

Article

Chemical Profile of *Ruta graveolens*, Evaluation of the Antioxidant and Antibacterial Potential of Its Essential Oil, and Molecular Docking Simulations

Călin Jianu [1], Ionuț Goleț [2], Daniela Stoin [1], Ileana Cocan [1], Gabriel Bujancă [1], Corina Mișcă [1,*], Marius Mioc [3,4], Alexandra Mioc [3,4], Codruța Șoica [3,4], Alexandra Teodora Lukinich-Gruia [5,*], Laura-Cristina Rusu [6,7], Delia Muntean [8,9] and Delia Ioana Horhat [8]

[1] Faculty of Food Engineering, Banat's University of Agricultural Sciences and Veterinary Medicine "King Michael I of Romania", 300645 Timișoara, Romania; calin.jianu@gmail.com (C.J.); danielastoin@usab-tm.ro (D.S.); ileanacocan@usab-tm.ro (I.C.); gabrielbujanca@usab-tm.ro (G.B.)

[2] Faculty of Economics and Business Administration, West University of Timișoara, 300233 Timișoara, Romania; ionut.golet@e-uvt.ro

[3] Faculty of Pharmacy, "Victor Babeș" University of Medicine and Pharmacy, 300041 Timișoara, Romania; marius.mioc@umft.ro (M.M.); alexandra.mioc@umft.ro (A.M.); codrutasoica@umft.ro (C.Ș.)

[4] Research Center for Pharmacotoxicological Evaluations, Faculty of Pharmacy, "Victor Babeș" University of Medicine and Pharmacy, 300041 Timișoara, Romania

[5] OncoGen Centre, County Hospital "Pius Branzeu", 300736 Timișoara, Romania

[6] Faculty of Dental Medicine, "Victor Babeș" University of Medicine and Pharmacy, 300041 Timișoara, Romania; laura.rusu@umft.ro

[7] Multidisciplinary Center for Research, Evaluation, Diagnosis and Therapies in Oral Medicine, "Victor Babeș" University of Medicine and Pharmacy, 300173 Timișoara, Romania

[8] Faculty of Medicine, "Victor Babeș" University of Medicine and Pharmacy, 300041 Timișoara, Romania; muntean.delia@umft.ro (D.M.); horhat.ioana@umft.ro (D.I.H.)

[9] Multidisciplinary Research Center on Antimicrobial Resistance, "Victor Babeș" University of Medicine and Pharmacy, 300041 Timișoara, Romania

[*] Correspondence: corinamisca@usab-tm.ro (C.M.); alexandra.gruia@gmail.com (A.T.L.-G.)

Citation: Jianu, C.; Goleț, I.; Stoin, D.; Cocan, I.; Bujancă, G.; Mișcă, C.; Mioc, M.; Mioc, A.; Șoica, C.; Lukinich-Gruia, A.T.; et al. Chemical Profile of *Ruta graveolens*, Evaluation of the Antioxidant and Antibacterial Potential of Its Essential Oil, and Molecular Docking Simulations. *Appl. Sci.* **2021**, *11*, 11753. https://doi.org/10.3390/app112411753

Academic Editor: Antony C. Calokerinos

Received: 10 November 2021
Accepted: 7 December 2021
Published: 10 December 2021

Publisher's Note: MDPI stays neutral with regard to jurisdictional claims in published maps and institutional affiliations.

Abstract: The research aimed to investigate the chemical composition and antioxidant and antibacterial potential of the essential oil (EO) isolated from the aerial parts (flowers, leaves, and stems) of *Ruta graveolens* L., growing in western Romania. *Ruta graveolens* L. essential oil (RGEO) was isolated by steam distillation (0.29% v/w), and the content was assessed by gas chromatography-mass spectrometry (GC-MS). Findings revealed that 2-Undecanone (76.19%) and 2-Nonanone (7.83%) followed by 2-Undecanol (1.85%) and 2-Tridecanone (1.42%) are the main detected compounds of the oil. The RGEO exerted broad-spectrum antibacterial and antifungal effects, *S. pyogenes*, *S. aureus*, and *S. mutans* being the most susceptible tested strains. The antioxidant activity of RGEO was assessed by peroxide and thiobarbituric acid value, 1,1-diphenyl-2-picrylhydrazyl radical (DPPH), and β-carotene/linoleic acid bleaching testing. The results indicated moderate radical scavenging and relative antioxidative activity in DPPH and β-carotene bleaching tests. However, between the 8th and 16th days of the incubation period, the inhibition of primary oxidation compounds induced by the RGEO was significantly stronger ($p < 0.001$) than butylated hydroxyanisole (BHA). Molecular docking analysis highlighted that a potential antimicrobial mechanism of the RGEO could be exerted through the inhibition of D-Alanine-d-alanine ligase (DDl) by several RGEO components. Docking analysis also revealed that a high number RGEO components could exert a potential in vitro protein-targeted antioxidant effect through xanthine oxidase and lipoxygenase inhibition. Consequently, RGEO could be a new natural source of antiseptics and antioxidants, representing an option for the use of synthetic additives in the food and pharmaceutical industry.

Keywords: *Ruta graveolens* L.; essential oil; 2-undecanone; 2-nonanone; antimicrobial activity; antioxidant activity; molecular docking

1. Introduction

Food spoilage may be caused by physical, chemical, or microbiological mechanisms. Microbial spoilage is frequently due to spoilage bacteria, yeasts, or moulds' growth and/or metabolism [1,2]. The chemical spoilage typically occurs when food exposure to oxygen triggers a chain of several chemical reactions involving lipids, fatty acids, and pigments and generates chemical compounds with undesirable biochemical properties, such as toxicity and unpleasant smell, taste, and colour [3]. These changes make foodstuff unacceptable or undesirable for consumption and, finally, generate food loss and waste [2]. The real economic losses generated by food spoilage are challenging to estimate. However, these losses represent a substantial financial burden assessed at 1.3 billion tonnes per year by FAO [4].

Consequently, to increase foodstuff quality, safety, and shelf-life without any adverse effect on their nutritional or sensorial properties, food additives such as preservatives and antioxidants have become indispensable for the food industry, mainly synthetic ones. In recent decades, there has been significant scientific progress concerning pharmacological studies of aromatic plants to identify and valorise natural extracts [5,6]. This trend is concurrent with an increasing interest in identifying new sources of preservatives and antioxidants to replace synthetic food additives because of their potential carcinogenicity [7,8]. A large plethora of plant extracts are also well known and researched for their antimicrobial potential. These extracts contain various compound classes such as terpenes, polyphenolic compounds, flavonoids, various adelhydes, and ketones or alkaloids that disrupt bacteria activity by various mechanisms [9]. These mechanisms include key enzymes that play important roles in bacterial survival and proliferation and are frequently used as targets for novel antimicrobial drug design or for the determination of active antimicrobial agents' mechanisms of action using computational methods [10]; *Rutaceae* family have been recognized for their economic value and also for the cultivated citrus fruits, timber, and essential oils (EOs), indicating a potential source of natural active principles [11–13]. One of the genera of *Rutaceae* family plants investigated is the genus *Ruta* [13]. The genus *Ruta* includes about 40 species of perennial shrubs and herbs distributed along the Mediterranean coast, the Balkan Peninsula, and Crimea [14]. In Romania, the *Ruta* genus is represented by *Ruta graveolens* L., *Ruta suaveolens* D.C., and *Dictamnus albus* L. [15]. Among the family members, *R. graveolens* L. stands out for EO production [6]. Several studies report that oxygenated compounds (e.g., aldehydes, alcohols, and esters) are predominant in the *R. graveolens* EOs (RGEO) isolated from leaves, fruits, flowers, stems, and roots [16]. In contrast, other investigations mention aliphatic compounds, especially ketones (2-undecanone and 2-nonanone), representing more than 50% of the total composition of RGEO [6,14,17]. These differences in the phytochemical profile of *R. graveolens* may explain the anti-rheumatic, anti-diarrheic, anti-inflammatory, anti-febrile, antiulcer, anti-diabetics, and antimicrobial properties reported in the recent pharmacological trials [6,18,19]. To our knowledge, no investigation of the antioxidant properties of *R. graveolens* has been previously reported. However, several studies report the in vitro antioxidant properties of the *Ruta montana* and *Ruta chalepensis*. Still, no investigations report the *Ruta* genus members' antioxidant activity in food systems.

This research aimed to investigate: (i) the chemical composition of the EO isolated from the aerial parts of *R. graveolens* cultivated in western Romania by using the GC-MS technique; (ii) the antioxidant and antimicrobial activities of the oil; and (iii) the mechanisms of interaction between RGEO chemical components and target proteins corelated with antibacterial activity and intracellular antioxidant mechanisms, thus aiming for its potential application in food and pharmaceutic industries as a green preservative and/or antioxidant.

2. Materials and Methods

2.1. Plant Material and RGEO Isolation Procedure

The fresh plant material was harvested manually, during the flowering phase in July 2019, from the experimental fields of the Didactic Station "Tinerii Naturaliști"/Banat's

University of Agricultural Sciences and Veterinary Medicine "King Michael I of Romania" in Timisoara, Romania. After identification, a voucher specimen (VSNH.BUASTM–109/1) was deposited in the Herbarium of Agricultural Technologies Department, Faculty of Agriculture, Banat's University of Agricultural Sciences and Veterinary Medicine "King Michael I of Romania" from Timisoara, Romania. The fresh plant material (flowers, leaves, and stems) was manually chopped into parts approximately 1.5 cm long and immediately submitted to steam distillation [20] in a Craveiro apparatus for 4 h. A water-cooled EO receiver was used to reduce the formation of artifacts due to overheating, which may occur during the isolation of RGEO. After separating the RGEO by decantation (yielding at 0.29% *v/w*), the oil was dried using anhydrous sodium sulphate and stored until use at $-18\,^{\circ}\text{C}$.

2.2. Gas Chromatography Coupled to Mass Spectrometry Method

The RGEO was analyzed using a sensitive and qualitative gas chromatographic technique performed on an HP6890 gas chromatograph coupled with an HP5973 mass spectrometer. The sample, diluted 1:1000 in hexane, was injected in a splitless mode in a heated inlet at 230 °C, and run through a Bruker Br-5MS column (30 m × 0.25 mm; film thickness 0.25 μm) (Agilent Technologies, Santa Clara, CA, USA), carrier gas: helium, flow rate 1.0 mL/min. The gas chromatograph oven temperature was set up to 50 °C for 5 min, raised to 300 °C at a temperature rate of 6 °C/min, and kept there for 5 min. The HP5973 mass spectrometer operating parameters were as follows: ionization potential, 70 eV; mass analyzer quadrupole 150 °C; solvent delay 3.0 min; mass range 50 to 550 amu. The NIST0.2 spectral library (USA National Institute of Science and Technology software) was employed to identify the compounds (similarity indexes > 90 %), followed by a comparison of the retention index (RI), calculated based on the n-alkanes C_8–C_{20} homologous series, with the values reported in the literature [21].

2.3. Effect of RGEO on Cold-Pressed Sunflowers Oil Oxidation

RGEO and synthetic antioxidants, butylated hydroxyanisole (BHA) and butylated hydroxytoluene (BHT), used for comparison were added at 200 mg/L concentrations separately to 10 mL of cold-pressed sunflower oil purchased from the local market. Oxidation was periodically evaluated by measuring peroxide value (PV) at the 0th, 4th, 8th, 12th, 16th, 20th, and 24th days of storage according to the potentiometric end-point determination method described by ISO 27107:2010 [22]. In addition, the thiobarbituric acid (TBA) value was analysed to measure secondary oxidation products in the cold-pressed sunflower oil at the same days of storage, according to the previous investigation described by Jianu et al. [23]. A negative control sample was prepared under the same conditions without adding any additives. All analyses were performed in triplicate.

2.4. 1,1-Diphenyl-2-picrylhydrazyl Radical (DPPH) Free Radical Scavenging Activity

The radical-scavenging activity of the RGEO with DPPH was established on the scavenging capacity of the stable DPPH· free radical following the Brand-Williams method [24]. Shortly, all samples, RGEO and reference positive controls (δ-tocopherol, BHA, BHT) were diluted in methanol to obtain concentrations between 1.5 and 0.093 mg/mL. Samples were pipetted in triplicate into plates with 96 wells and left to incubate at room temperature in the dark for 30 min. Their absorbances were read at 515 nm against methanol as a negative control at a Tecan i-control 1.10.4.0 Infinite 200Pro spectrophotometer. The obtained results were expressed as a DPPH free radical percentage (I%) and calculated based on the equation: $\text{I}\% = (\text{A}_{\text{methanol}} - \text{A}_{\text{sample}}/\text{A}_{\text{methanol}}) \times 100$; $\text{A}_{\text{methanol}}$ is methanol absorbance, and A_{sample} is the tested sample absorbance. IC_{50} index was calculated with the software BioDataFit 1.02 (Chang Bioscience Inc., Fremont, CA, USA).

2.5. β-Carotene Bleaching Test

The experiment measured the coupled autoxidation of β-carotene and linoleic acid as previously described by Jianu et al. [23]. Briefly, β-carotene (0.5 mg) was added to chloroform (1 mL), linoleic acid (25 μL), and Tween 40 (200 mg). The mixture was evaporated at 45 °C for 5 min under vacuum to remove chloroform. The residue was diluted slowly with distilled water saturated with oxygen (100 mL) and vigorously shaken to form an emulsion. The emulsion (2.5 mL) was transferred to the test tubes containing 350 μL of RGEO methanolic solution (2 g/L concentration). BHT in methanol was used as a positive control. The test tubes were gently shaken and incubated for 48 h (room temperature) before their absorbances readings at 490 nm. All experiments were performed in triplicate.

2.6. Determination of Antimicrobial Activity

2.6.1. Bacterial Strains

For determining the RGEO antimicrobial activity, the following microbial reference strains were used: Gram-positive cocci (*Enterococcus faecalis* ATCC 51299, *Staphylococcus aureus* ATCC 25923, *Streptococcus pyogenes* ATCC 19615, and *Streptococcus mutans* ATCC 35668), Gram-negative bacilli (*Escherichia coli* ATCC 25922, *Klebsiella pneumoniae* ATCC 700603, *Salmonella enterica* serotype *Typhimurium* ATCC 14028, *Shigella flexneri* serotype 2b ATCC 12022, and *Pseudomonas aeruginosa* ATCC 27853), and two strains of *Candida* species (*Candida albicans* ATCC 10231 and *Candida parapsilosis* ATCC 22019). The methods used to test RGEO antimicrobial activity were performed according to the European Committee on Antimicrobial Susceptibility Testing (EUCAST) [25] and with minimal adjustments based on our previous studies [20,23,26].

2.6.2. Antimicrobial Screening

The disk diffusion method was used for the initial testing of RGEO antimicrobial activity. The microbial suspension was prepared to 0.5 McFarland using a standard saline solution for each strain and inoculated on Mueller-Hinton agar (bioMérieux, Marcy-l'Etoile, France). Afterward, on these plates, a disk (BioMaxima, Lublin, Poland) containing 10 μL of RGEO to be tested and disks containing 5 μg levofloxacin and 25 μg fluconazole for positive control were placed on the surface. The inhibition zones were measured in millimeters after a 24-hour incubation at 35–37 °C for bacteria species and at 28 °C for *Candida* species.

2.6.3. Minimum Inhibitory Concentration

Using the serial dilution method, the microbial suspension was adjusted to 5×10^5 CFU/mL (colony forming units). Serial dilutions of RGEO in DMSO were prepared, ranging from 400 to 12.5 mg/mL concentrations. The following: 0.5 mL microbial suspension, 0.1 mL of each RGEO dilution, and 0.4 mL Mueller Hinton broth, were transferred in six test tubes, obtaining a final inoculum of 0.5×10^5 CFU/mL and a final RGEO dilution from 40 to 1.25 mg/mL. After 24 h of incubation at 37 and at 30 °C, respectively, the test tube containing the lowest RGEO concentration, and no visible growth was considered MIC interpretation.

2.6.4. Minimum Bactericidal Concentration and Minimum Fungicidal Concentration

From the tubes with MIC, 1 μL was inoculated on Columbia agar and 5% sheep blood for bacterial strains and Sabouraud for Candida strains (bioMérieux, Marcy-l'Etoile, France). The inoculated plates were incubated for appropriately 24 h, and the lowest concentration with no visible growth was considered for MBC or MFC.

2.7. In Silico Molecular Docking

Molecular docking analysis was achieved using a previously described method [27]. All protein target structures were retrieved from the RCSB Protein Data Bank [28] (Table 1). These structures were optimized as suitable docking targets, using Autodock Tools v1.5.6 (The Scripps Research Institute, La Jolla, CA, USA). The protein structure file was prepared by removing water molecules, unlinked atoms/protein chains, and the native ligands after which the potential of the protein target was assigned with Kollman charges, a feature imbedded in the software. The target structures were saved as pdbqt files. The structures of the 37 RGEO compounds were generated based on their available SMILE strings (or isomeric SMILE strings in case of enantiomers), using BIOVIA Draw (Dassault Systems BIOVIA). The 2D structures were converted into 3D structures using PyRx's Open Babel module by using 500 steps of a steepest descent geometry optimization with the MMFF94 forcefield. The lowest energy conformer generation does not alter the stereochemistry of the input structures. Molecular docking was achieved with the PyRx v0.8 virtual screening software (The Scripps Research Institute, La Jolla, CA, USA) using Vina's encoded scoring function [29]. This is a custom scoring function which combines, as their developers describe, empirical information from both the conformational preferences of the receptor-ligand complexes and experimental affinity measurements [29]. The docking software was set to generate/dock 10 conformers of each input molecular structure. The docking protocol was validated by re-docking the native ligands into their original protein binding sites. The root means square deviation (RMSD) between the predicted and experimental docking pose of the native ligand was calculated. Molecular docking was performed only for cases with aforementioned RMSD values not exceeding a 2 Å threshold. The docking grid box coordinates and size were selected to best fit the active binding site (Table 1). Docking scores were recorded as ΔG binding energy values (kcal/mol). Protein-ligand binding interactions were analysed using Accelrys Discovery Studio 4.1 (Dassault Systems BIOVIA, San Diego, CA, USA).

Table 1. Molecular docking parameters for each protein target.

Protein	PDB ID/ Protein Structure Resolution	Grid Box Center Coordinates	Grid Box Size	Native Ligand	References
IARS	1JZQ 3.00 Å	center_x = −27.683 center_y = 7.940 center_z = −28.726	size_x = 15.314 size_y = 10.229 size_z = 10.025	N-[isoleucinyl]-n′-[adenosyl]-diaminosufone	[30]
DNA gyrase	1KZN 2.30 Å	center_x = 17.221 center_y = 31.155 center_z = 36.750	size_x = 11.994 size_y = 13.471 size_z = 14.454	Clorobiocin	[31]
Lipoxygenase	1N8Q 2.10 Å	center_x = 21.088 center_y = 1.163 center_z = 19.056	size_x = 10.291 size_y = 6.561 size_z = 8.855	Protocatechuic acid	[32]
CYP2C9	1OG5 2.55 Å	center_x = −19.429 center_y = 87.688 center_z = 39.060	size_x = 10.873 size_y = 15.624 size_z = 9.745	S-warfarin	[33]

Table 1. *Cont.*

Protein	PDB ID/ Protein Structure Resolution	Grid Box Center Coordinates	Grid Box Size	Native Ligand	References
NADPH-oxidase	2CDU 1.80 Å	center_x = 18.043 center_y = −6.494 center_z = 0.261	size_x = 13.845 size_y = 13.580 size_z = 19.019	ADP 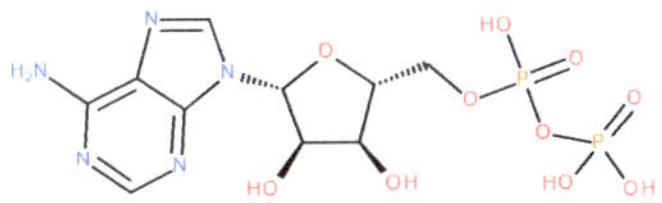	[34]
DDl1	2I80 2.19 Å	center_x = 34.136 center_y = 4.049 center_z = 25.365	size_x = 8.407 size_y = 10.984 size_z = 10.984	3-chloro-2,2-dimethyl-N-[4-(trifluoromethyl)phenyl]propanamide 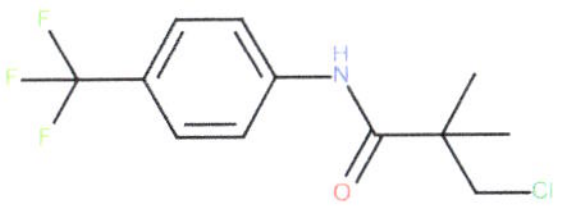	[35]
DHPS	2VEG 2.40 Å	center_x = 29.308 center_y = 47.975 center_z = −0.104	size_x = 11.994 size_y = 10.138 size_z = 8.233	Pterin-6-yl-methyl-monophosphate 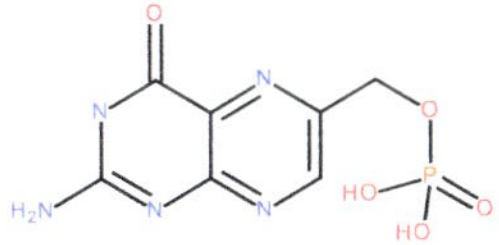	[36]
Xanthine oxidase	3NRZ 1.80 Å	center_x = 37.638 center_y = 19.857 center_z = 17.684	size_x = 6.767 size_y = 10.461 size_z = 8.873	Hypoxanthine 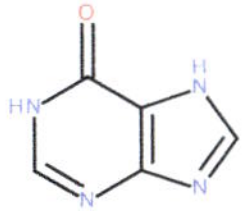	[37]
Type IV topoisomerase	3RAE 2.90 Å	center_x = −33.590 center_y = 67.448 center_z = −25.612	size_x = 14.217 size_y = 9.518 size_z = 6.868	Levofloxacin 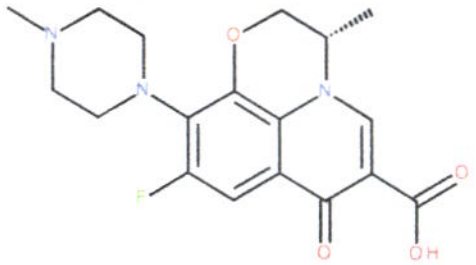	[38]

Table 1. *Cont.*

Protein	PDB ID/ Protein Structure Resolution	Grid Box Center Coordinates	Grid Box Size	Native Ligand	References
DHFR	3SRW 1.70 Å	center_x = −4.932 center_y = −31.078 center_z = 6.811	size_x = 11.512 size_y = 12.532 size_z = 9.986	7-(2-ethoxynaphthalen-1-yl)-6-methylquinazoline-2,4-diamine 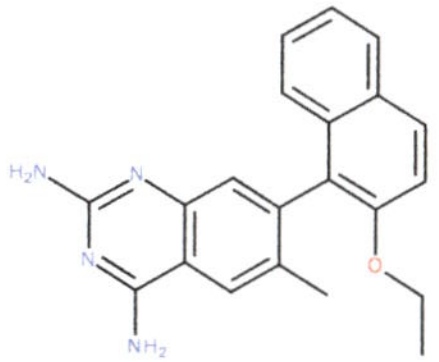	[39]
DNA gyrase subunit B	3TTZ 1.63 Å	center_x = 16.304 center_y = −18.973 center_z = 5.866	size_x = 18.223 size_y = 11.865 size_z = 9.461	2-[(3S,4R)-4-{[(3,4-dichloro-5-methyl-1H-pyrrol-2-yl)carbonyl]amino}-3-fluoropiperidin-1-yl]-1,3-thiazole-5-carboxylic acid 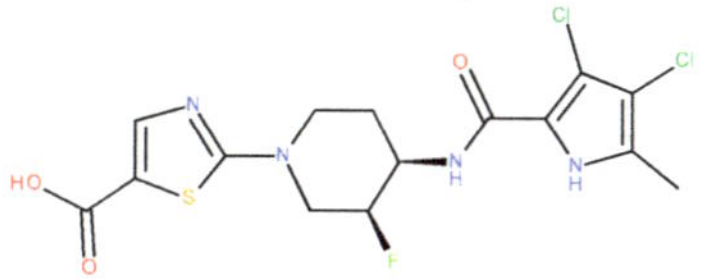	[40]
PBP1a	3UDI 2.60 Å	center_x = 33.807 center_y = −0.788 center_z = 12.228	size_x = 10.343 size_y = 8.297 size_z = 10.636	Penicillin G-open form 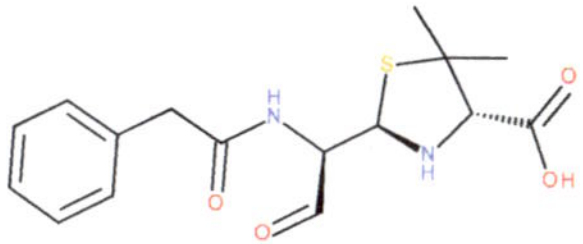	[41]

2.8. Statistical Analysis

The main approach for statistical testing the antioxidant property of RGEO was the ANOVA method, with samples (synthetic antioxidants and RGEO) and incubation period as main effects, followed by a post-hoc analysis. The overall ANOVA analysis shows that the main effects and interaction effects are highly significant ($p < 0.001$) for PV and TBA values. Because the number of observations of each sample per incubation period is low (nine values), the normality assumption was tested on the ANOVA residuals using the Shapiro–Wilk test. The null hypothesis was not rejected in the case of PV ($p = 0.182$) but was rejected in the case of TBA ($p < 0.001$). Consequently, the post-hoc analysis was performed using the Tukey parametric test in the case of PVs. Instead, the post-hoc analysis was performed using Dun's non-parametric test with Bonferroni correction for the TBA values. To take into account the interaction effect, all the pairwise comparisons were performed separately, for each incubation period at a time; also worth mentioning is that the groups have homogenous variances at this level of analysis according to Leven's test. In the case of the scavenging effect on the DPPH radical assay, we faced the situation of non-normality (Shapiro–Wilk, $p = 0.005$) of ANOVA residuals, non-homogeneity of variances across groups (Levene, $p < 0.001$), and a low number of observations per group (nine measurements). Therefore, the data have been normalized by using the natural logarithm transformation. In addition, the Games-Howell test was used in post-hoc analysis to address the lack of variance homogeneity. Finally, the ANOVA approach followed by Tukey's test in the post-hoc analysis was applied to assess the antimicrobial properties. The ANOVA residuals have close to normality distribution (Shapiro–Wilk, $p = 0.844$), and the groups (three inhibition zone measurements per group) are homogenous from a variance point of view (Levens's, $p = 0.28$). Data analysis was completed using JASP (Version 0.15), and p-values < 0.05 were considered as significant.

3. Results

3.1. Chemical Composition of RGEO

Steam distillation of the fresh plant material of *R. graveolens* gave a yellowish oil with an intense and penetrating odour with a yield of 0.29% (v/w). The extraction yields obtained for *R. graveolens* are comparable to those reported in the literature [17,42–45]. As previous reported by Formisano et al. [46], the total EO content of plants was affected by genetic background, environmental conditions, and soil composition.

GC-MS analysis of the RGEO identifies thirty-seven compounds (Figure 1 and Table 2), representing 98.68% of the obtained oil. The main detected compounds are 2-Undecanone and 2-Nonanone at 76.19% and 7.83%, respectively, followed by 2-Undecanol at 1.85% and 2-Tridecanone at 1.42%. The abundance of 2-Undecanone in RGEO is in accord with the previous studies conducted on RGEOs from Egypt [47], Algeria [45], Iran [48], and Saudi Arabia [49]. However, the proportions and nature of the identified chemical compounds of the analysed EOs are not always the same compared with the previous studies. These differences may be due to genetic, distinct environmental and climatic conditions, geographic origins, and plant populations [49,50].

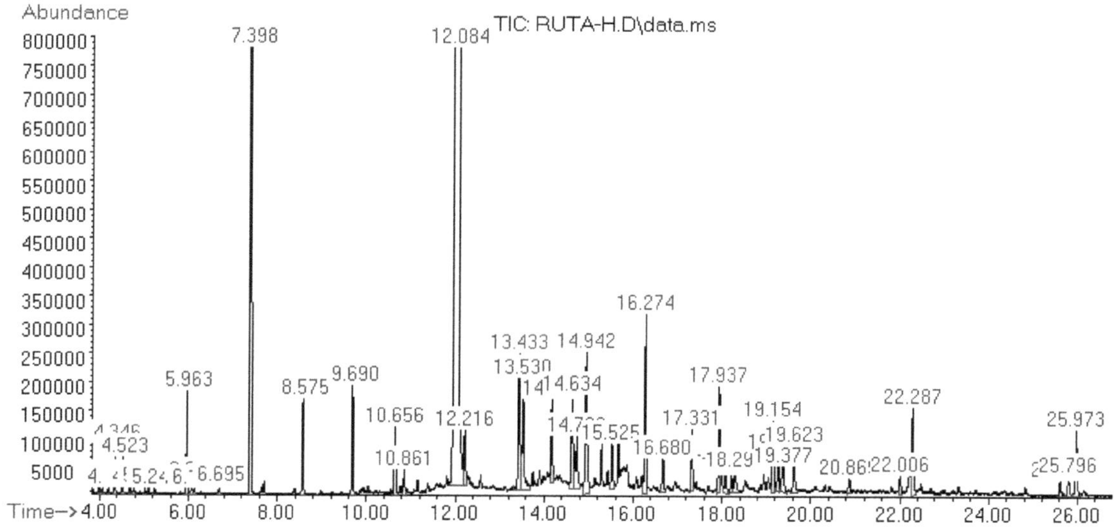

Figure 1. Gas chromatogram of RGEO cultivated in western Romania.

Table 2. Chemical composition of RGEO cultivated in western Romania.

No	Compounds	KI [a]	%
1.	3-Octanone	908	0.06
2.	beta-Thujene	912	tr.
3.	4-Carene, (1S,3S,6R)-(−)	919	tr.
4.	Hydroperoxide, 1-ethylbutyl	925	0.24
5.	Hydroperoxide, 1-methylpentyl	934	0.19
6.	(2E)-2-Hexenyl benzoate	942	tr.
7.	1-Cyclohexyl-2-propen-1-ol	947	tr.
8.	2-Bornene	959	tr.
9.	para-Cymene	1005	0.58
10.	beta-Terpinyl acetate	1011	0.08
11.	Eucalyptol	1014	tr.
12.	4-Carene, (1S,3R,6R)-(−)	1042	0.06
13.	2-Nonanone	1076	7.83
14.	2-Decanone	1190	0.75
15.	Cyclopropanecarboxylic acid, nonyl ester	1238	0.49
16.	(S)-(+)-Carvone	1248	0.18
17.	2-Undecanone	1308	76.19
18.	2-Undecanol	1315	1.85
19.	1-methyl-cycloundecanol	1380	1.13
20.	2-Dodecanone	1412	0.63
21.	beta-Caryophyllene	1439	0.55
22.	2-Acetoxytetradecane	1450	1.34
23.	Germacrene-D	1478	0.42
24.	2-Tridecanone	1515	1.42
25.	Hexa-hydro-farnesol	1536	0.28
26.	Elemol	1568	0.48
27.	9-Methyl-10-methylenetricyclo [4.2.1.1(2,5)]decan-9-ol	1598	0.81
28.	2,5-Octadecadiynoic acid, methyl ester	1605	0.19
29.	4-(3,4-Methylenedioxyphenyl)-2-butanone	1612	0.15
30.	Ascaridole epoxide	1615	0.24
31.	Valeric acid, 2-tridecyl ester	1658	0.59

Table 2. *Cont.*

No	Compounds	KI [a]	%
32.	Geranyl isovalerate	1663	0.32
33.	alpha-Eudesmol	1669	0.38
34.	1,2,3,3a,4,9,10,10a-Octahydrobenzo[f]azulene	1681	0.40
35.	Corymbolone	1743	0.13
36.	Methoxsalen	1977	0.16
37.	Bergaptene	1995	0.56
		Total:	98.68

[a] Retention indices (RIs) calculated upon the calibration curve of alkane C_8–C_{20} standard injected and analyzed in the same conditions as RGEO.; tr. (trace) < 0.05.

3.2. Antioxidant Activity

The formation of primary lipid oxidation products throughout 24 days of storage of the samples was measured using the amount of PV. The effect of RGEO, BHA, and BHT on PV changes in cold-pressed sunflower oil lipids has been shown in Table 3. The PV of samples treated with RGEO, after 8 days and 16 days of incubation, were lower than the values of the control sample and samples treated with BHA and BHT, with a high significance level ($p < 0.001$). After 12 days of incubation, the PV of the sample treated with RGEO was lower than the values of the control sample and sample containing BHA but higher than the values of the sample containing BHT with high significance in all cases ($p < 0.01$). Finally, after 20 days and 24 days of incubation, the PV of the sample treated with RGEO was significantly lower than the control sample values ($p < 0.001$) and significantly higher than the samples treated with BHA and BHT ($p < 0.001$).

Table 3. The antioxidant effects of RGEO, BHA, and BHT in terms of peroxide values (meq of oxygen kg^{-1}).

Storage Time (Days)	PV (meq of Oxygen kg^{-1})			
	Control Sample	BHT	BHA	RGEO
0 days	1.97 ± 0.05 [bc]	1.88 ± 0.04)[a]	1.93 ± 0.06 [ab]	2 ± 0.04 [c]
4 days	3.1 ± 0.06 [a]	2.98 ± 0.06 [c]	3.21 ± 0.06 [b]	3.13 ± 0.09 [ab]
8 days	5.67 ± 0.07 [a]	4.28 ± 0.04 [b]	5.9 ± 0.06 [c]	4.01 ± 0.05 [d]
12 days	7.7 ± 0.05 [a]	5.91 ± 0.06 [b]	7.06 ± 0.07 [c]	6.56 ± 0.08 [d]
16 days	9.73 ± 0.07 [a]	8.66 ± 0.08 [b]	8.68 ± 0.04 [b]	8.19 ± 0.05 [c]
20 days	12.12 ± 0.05 [a]	9.58 ± 0.03 [b]	9.82 ± 0.06 [c]	9.95 ± 0.06 [d]
24 days	15.91 ± 0.07 [a]	10.03 ± 0.05 [b]	10.11 ± 0.08 [b]	11.49 ± 0.08 [c]

Values with different superscripts are significantly different ($p < 0.05$) according to Tukey test; each value is the Mean ± SD.

TBA value has been extensively applied to evaluate the degree of lipid oxidation. TBA reactive substances are reckoning the second stage auto-oxidation, during which peroxides are oxidized to aldehyde and ketone [51]. The changes in TBA value of different treatment samples during 24 days are shown in Table 4. Generally, the values of samples treated with RGEO were closer to BHA values in the case of TBA measurements. From day 0 through day 20, the TBA values of samples treated with RGEO were higher than those of samples treated with BHA, but the difference was not significantly different, with only one exception, day 4. However, after 24 days of incubation, the values of TBA were lower for samples treated with RGEO but not significantly different according to Dunn's test ($p = 0.32$). Regarding the TBA values of samples treated with BHT, samples treated with RGEO were significantly higher after 4 through 20 days of incubation.

Table 4. The antioxidant effects of RGEO, BHA, and BHT in terms of thiobarbituric acid value (TBA) (µg malondialdehyde g^{-1}).

	TBA (µg Malondialdehyde g^{-1})			
Storage Time (Days)	**Control Sample**	**BHT**	**BHA**	**RGEO**
0 days	2.62 ± 0.04 [a]	2.5 ± 0.12 [b]	2.55 ± 0.08 [ab]	2.56 ± 0.06 [ab]
4 days	3.08 ± 0.04 [ab]	2.85 ± 0.04 [c]	2.88 ± 0.05 [ac]	4.4 ± 0.06 [b]
8 days	7.51 ± 0.07 [a]	3.46 ± 0.05 [b]	4.55 ± 0.11 [bc]	6.54 ± 0.11 [ac]
12 days	10.97 ± 0.06 [a]	5.18 ± 0.08 [b]	6.66 ± 0.15 [bc]	9.49 ± 0.1 [ac]
16 days	15.77 ± 0.1 [a]	5.9 ± 0.05 [b]	8.06 ± 0.07 [bc]	10.38 ± 0.11 [ac]
20 days	19.48 ± 0.2 [a]	6.67 ± 0.22 [b]	11.43 ± 0.08 [bc]	12.41 ± 0.16 [ac]
24 days	28.93 ± 0.04 [a]	7.3 ± 0.13 [b]	14.55 ± 0.08 [ac]	14.35 ± 0.12 [bc]

Values with different superscripts are significantly different ($p < 0.05$) according to Dunn's test; each value is the Mean $\pm$ SD.

Antioxidants interact with 1,1-diphenyl-2-picrylhydrazyl radical, a stable free radical, and transform it into 1,1-diphenyl-2-picrylhydrazine. The degree of discoloration demonstrates the radical scavenging potential or the hydrogen-donating ability of the compounds [52]. RGEO was able to to reduce the stable free radical DPPH with an IC_{50} value of 0.25 ± 0.09 mg/mL (Table 5). Even if the effect of the radical scavenging activity of RGEO is comparable to that of the delta-tocopherol (IC_{50}: 0.16 ± 0.02 mg/mL), it is not statistically significant ($p = 0.133$) according to the Games-Howell test. In contrast, BHA (IC_{50}: 0.09 ± 0.01 mg/mL) and BHT (IC_{50}: 0.02 ± 0.02 mg/mL) exhibited significantly ($p < 0.05$) better antioxidant activity than RGEO (Table 5). Recently, Benoli et al. (2020) reported DPPH scavenging abilities for Moroccan oil of R. montana with an IC_{50} value of 0.244 mg/mL [53]. In contrast, Mohammedi et al. (2018) found IC_{50} values ranging from 0.0496 to 0.0634 mg/mL for R. montana oils collected from different regions in Algeria [54]. Similar results were reported by Jaradat et al. (2017) that found IC_{50} values ranging from 0.0069 to 0.0199 mg/mL for Palestinian R. chalepensis volatile oils [55], and by Althaher et al. (2021) that reported an IC_{50} value of 0.035 mg/mL for Jordanian R. chalepensis oil [56].

Table 5. Antioxidant activities of the RGEO by DPPH and β-carotene–linoleic acid bleaching test.

Parameter	RGEO	Delta-Tocopherol	BHA	BHT
DPPH, IC_{50} (mg/mL)	0.25 ± 0.09 [a]	0.16 ± 0.02 [a]	0.09 ± 0.01 [b]	0.02 ± 0.02 [c]
β-carotene bleaching (RAA) (%)	77.42 ± 0.07	Nd	Nd	100

Values with different superscripts are significantly different ($p < 0.05$) according to Games-Howell test; each value is the Mean $\pm$ SD; Nd—not detected.

The β-Carotene bleaching test is based on the discoloration of β-carotene determined to its reaction with radicals produced by linoleic acid oxidation in an emulsion. The antioxidants' presence can decrease the rate of β-carotene bleaching [57,58]. The relative antioxidant activity percentage (RAA%) of RGEO was calculated with the formula RAA $= A_{RGEO}/A_{BHT}$, where A_{RGEO} is the absorption of RGEO, and A_{BHT} is the absorption of BHT (positive control used). Compared with BHT, R. graveolens oil bleached β-carotene by 77.42 ± 0.07% (Table 5). Similar results were reported by Loizzo et al. (2017) for leaf extracts obtained from *R. chalepensis* [59]. However, no previous research were available in the literature concerning the in vitro and in vivo antioxidant activity of RGEO to support us to compare the results directly.

3.3. Antimicrobial Activity

The in vitro antimicrobial activity of RGEO against nine bacteria and fungal strains was evaluated qualitatively and quantitatively by the presence or absence of inhibition zones, MIC, MBC, and MFC values. The diameters of the inhibition zone of RGEO, which include the diameter (6 mm) of the paper disk against the microorganisms tested, are shown in Table 6. The diameters of the inhibition zone induced by RGEO against the tested microorganism strains ranged between 15.21 ± 0.14 mm and 20.61 ± 0.21 mm, suggesting that the oil exerts low to moderate antimicrobial effects. The results revealed that S. pyogenes, S. aureus, and S. mutans were the most susceptible tested strain to the RGEO action, followed by *C. albicans > C. parapsilosis > E. faecalis > P. aeruginosa > E. coli > K. pneumoniae > S. enterica > S. flexneri*. Our results agree with previous studies [6,45,49], which reported that RGEO exhibited antimicrobial activity against *S. aureus, E. faecalis, E. coli, K. pneumoniae*, and *C. albicans*. The recorded MICs, MBCs, and MFCs for the tested strains were 1.25, 2.5, and 5 mg/mL, respectively. According to Aligiannis et al. [60], a strong MIC EOs can hold up to 0.5 mg/mL, moderate for MIC 0.6–1.5 mg/mL, and low for MIC above 1.5 mg/mL. The RGEO exhibited a moderate MIC for S. mutans and S. pyogenes and showed low activity against the rest of the analyzed bacteria. Overall, RGEO showed low efficiency in inhibiting the Gram-negative strains compared to Gram-positive strains, following previous studies [6,27,49,61,62]. These differences in susceptibility could be associated with different rates of penetration of EO constituents into the cell wall and cell membrane structures. Therefore, the ability of EO to disrupt the permeability barrier of cell membrane structures and the accompanying loss of chemiosmotic control are the most likely reasons for its lethal action [63].

Table 6. Antimicrobial of the RGEO by disk diffusion, minimum inhibitory concentration (MIC), minimum bactericidal concentration (MBC), and minimum fungicidal concentration (MFC).

Bacterial and Yeast Strains	Disk Diffusion (mm)	MIC Value (mg/mL)	MBC Value (mg/mL)	MFC Value (mg/mL)
Streptococcus mutans	19.77 ± 0.26 [b]	1.25	1.25	-
Streptococcus pyogenes	20.61 ± 0.21 [a]	1.25	1.25	-
Staphylococcus aureus	19.89 ± 0.14 [b]	2.5	2.5	-
Enterococcus faecalis	16.56 ± 0.17 [d]	2.5	2.5	-
Escherichia coli	15.76 ± 0.1 [fe]	5.0	5.0	-
Klebsiella pneumoniae	15.71 ± 0.13 [fe]	5.0	5.0	-
Salmonella enterica	15.65 ± 0.23 [fe]	5.0	5.0	-
Shigella flexneri	15.21 ± 0.14 [f]	5.0	5.0	-
Pseudomonas aeruginosa	15.95 ± 0.08 [de]	5.0	5.0	-
Candida albicans	18.66 ± 0.26 [c]	2.5	-	2.5
Candida parapsilosis	18.42 ± 0.39 [c]	2.5	-	2.5

Values with different superscript are significantly different ($p < 0.05$) according to Tukey test; each value is the Mean $\pm$ SD.

3.4. In Silico Prediction of Mechanism by Molecular Docking Analysis

Ligand-based molecular docking is a computational technique that can be used, among other methods, as a starting point for understanding the targeted mechanism of action of a given molecular structure. In the form of free-binding energy values, the obtained results may indicate an increased/decreased affinity of the analyzed molecule towards the selected target compared to the native ligand (a known inhibitor), given that the binding energy decreases when the compounds' affinity increases [64–66]. For our current study, we used a molecular docking-based protocol to identify possible protein targets for the 37 RGEO components, whose inhibition could be correlated with their in vitro antimicrobial activity. The same method was also employed to provide an insight regarding a potential in vitro protein targeted based antioxidant activity, apart from the chemical structure related antioxidant activity described above. Protein targets, usually associated with bactericidal/bacteriostatic effects, such as dihydropteroate synthase (DHPS), dihydrofolate reductase (DHFR), D-alanine: D-alanine ligase (Ddl), penicillin binding protein 1a (PBP1a), DNA gyrase, type IV topoisomerase, and isoleucyl-tRNA synthetase (IARS), were used in the present work [10]. Furthermore, molecular docking screening was also employed to assess the RGEO components inhibitory potential towards protein targets that play active roles in intracellular antioxidant mechanisms. To achieve this goal CYP2C9, lipoxygenase xanthine oxidase, and NADPH-oxidase [67] were selected as protein targets.

Docking scores recorded as ΔG values (kcal/mol) corresponding to the 37 docked compounds and the native ligands of each protein are presented in Table 7. However, we intended to spot a trend related to a possible protein-targeted cumulative mechanism of action associated with the set of 37 RGEO compounds in the sense that if a majority of the RGEO components score better or comparable docking results with the native ligand of a target protein, then that cumulative effect may lead to the observed biological activity. To better visualize this tendency, firstly, the docking scores corresponding to each protein target column were reordered in descending order, the lowest ΔG value representing the highest affinity for that specific target. Subsequently, we generated a heatmap based on the rearranged table. Each table column was colored with a three-color scheme gradient, ranging from red for the ΔG value scored by the native ligand (used as control), through white for the midpoint interval, and to blue for the highest value (the structure with the lowest affinity), respectively. Thus, the columns of the target proteins where most of the compounds obtained good docking scores compared to the native ligands will be colored predominantly red (Figure 2).

Table 7. Docking results (binding energy, ΔG kcal/mol) for RGEO 37 compounds.

	1N8Q	1OG5	2CDU	3NRZ	1JZQ	1KZN	2VEG	3RAE	3SRW	3TTZ	3UDI	2I80
					Free Binding Energy ΔG (kcal/mol)							
NL	−5.8	−9.8	−9.3	−6.7	−8.8	−9.3	−6.9	−4.3	−9.9	−8.5	−7.4	−8.4
1	−4.8	−4.7	−4.7	−5.8	−4.3	−4.4	−4	−2.3	−4.5	−4.7	−3.9	−5.4
2	−5.8	−5.5	−5.2	−4.5	−4.8	−4.7	−3.8	−2.1	−5.7	−4.9	−4.4	−5.8
3	−7.1	−5.6	−5.8	−4.3	−5.2	−5.2	−4	−2.4	−6.2	−5.5	−4.6	−6.3
4	−4.7	−4.2	−4.3	−5.6	−3.9	−4.4	−4.1	−2.6	−4.3	−4.6	−4	−4.9
5	−4.8	−4.3	−4.3	−5.5	−4	−4.5	−4	−2.7	−4.6	−4.8	−4.1	−4.9
6	−5.2	−6.8	−6.3	−6.8	−5.8	−5.6	−5.3	−2.9	−6.2	−6.2	−5.3	−6.9
7	−5.8	−5.6	−5.5	−6.7	−5.1	−5	−4.7	−2.9	−5.7	−5.6	−4.6	−6
8	−3.3	−5.6	−5.7	1.7	−4.6	−4.2	−3.7	−2.1	−5.5	−4.7	−4.5	−6.2
9	−6.1	−6.2	−5.7	−6.9	−5.1	−5.3	−4.4	−2.3	−5.6	−5.8	−4.8	−5.9
10	−3.7	−6.8	−6.2	−6.5	−5.9	−5.7	−5.2	−3.4	−6.3	−6.3	−5.7	−6.8
11	−3.6	−5.6	−6	2.8	−4.8	−4.6	−3.7	−2.6	−5.9	−5	−4.8	−6
12	−7.1	−5.7	−5.8	−4.3	−5.1	−5.2	−4	−2.4	−6.2	−5.5	−4.6	−6.3
13	−4.7	−5	−4.9	−5.9	−4.6	−4.5	−3.7	−2.4	−4.8	−5.1	−3.9	−5.8
14	−4.8	−5.4	−4.8	−6.2	−4.7	−4.6	−4.2	−2.3	−5	−5.4	−4	−5.8
15	−4.3	−5.7	−5.3	−5.5	−4.9	−4.6	−4.8	−2.6	−5.4	−5.3	−4.6	−6.6
16	−5.3	−6.4	−6.2	−7.3	−5.7	−5.5	−4.8	−3.2	−5.9	−6	−5.2	−6.2
17	−4.7	−5.3	−5.2	−6	−4.9	−4.4	−4.1	−2.4	−5	−5.1	−4.2	−6
18	−5.3	−5.4	−5	−5.7	−5	−4.5	−4.3	−2.2	−5.2	−5.1	−4.3	−6.1

Table 7. *Cont.*

	1N8Q	1OG5	2CDU	3NRZ	1JZQ	1KZN	2VEG	3RAE	3SRW	3TTZ	3UDI	2I80
						Free Binding Energy ΔG (kcal/mol)						
19	−2.5	−6.2	−6.3	−3	−5.8	−5.3	−4.8	−3.4	−7.1	−6.5	−5.5	−3.4
20	−4.8	−5.5	−4.8	−5.7	−4.8	−4.7	−4.5	−2.4	−5.2	−5	−4.2	−6.4
21	−0.5	−7.3	−6.5	0.4	−6.3	−6.1	−4.9	−3	−7.6	−6.5	−6.2	−4.3
22	−2	−5.7	−5.8	−2.8	−5.4	−4.8	−4.2	−2.2	−5.9	−5.7	−5.1	−6.2
23	−1.3	−7.1	−6.5	−1.2	−6.8	−6.4	−5.3	−3	−7.7	−6.6	−6.2	−4.4
24	−4.7	−5.6	−5.1	−5.3	−4.9	−4.7	−4.4	−2.1	−5.4	−5.5	−4.5	−6.4
25	−2.1	−6.4	−6.3	−4.3	−5.7	−5.4	−5.1	−2.7	−6.1	−6.4	−5.6	−6.3
26	−2	−6.7	−6.6	−0.8	−6	−6.1	−4.6	−3	−7	−6.7	−5.9	−6.3
27	−0.4	−6.2	−6.7	2.5	−5.3	−5.1	−4.1	−3.2	−6.6	−5.6	−5.8	−5.3
28	0.7	−6	−5.6	−1.1	−5.1	−5.2	−4.8	−2.4	−6.4	−6	−5.1	−5.4
29	−2.9	−6.4	−6.5	−8	−5.8	−5.7	−5.2	−3.3	−6.6	−6.6	−5.8	−7.1
30	−4.7	−5.9	−6.2	−1.2	−5.9	−5.1	−5.1	−3.8	−6.2	−5.6	−5.9	−5.7
31	0.6	−5.5	−5.3	0.3	−4.8	−4.6	−4.5	−2.4	−5.6	−5.5	−4.9	−5.9
32	−2.3	−6.4	−5.8	−3.6	−5.6	−5.5	−5	−2.9	−6.3	−5.9	−5.8	−6.1
33	−2.7	−7.2	−7.2	−2.3	−6.7	−7.6	−5.4	−3.3	−7.8	−7	−6.6	−6.4
34	−0.8	−8.5	−7	−6.7	−7	−6.9	−5.7	−3.1	−7.7	−7.3	−6	−5.7
35	2.9	−7	−6.2	6.1	−6.2	−5.7	−4.6	−3.6	−7.8	−6	−5.4	−2.9
36	−3.7	−7	−7.4	−6.5	−6.8	−6.7	−6.4	−4.3	−7.3	−7.2	−7	−5.4
37	−3.7	−7.1	−7.1	−6.2	−6.3	−6.9	−6.2	−4.3	−7.4	−7.2	−6.4	−6.2

NL—native ligand; highlighted values represent cases were ΔG values of the respective compounds are lower than ΔG of the NL.

Concerning the set of proteins related to RGEO's antibacterial activity, our results show an increased affinity of the majority of docked molecules towards the DDl protein (2I80). DDl is an essential key enzyme involved in bacterial wall biosynthesis and an important drug target for developing new antibiotic agents. This enzyme is responsible for the formation of the dipeptide D-alanine: D-alanine, in a two-step reaction, sequentially by using D-alanine and ATP as substrates for the first reaction step and another D-alanine to complete the reaction [35]. The analyzed compounds were docked in an allosteric pocket adjacent to the D-Ala and ATP binding site. Of the docked compounds, various structures showed a good affinity towards DDl compared to the native ligand, in the range of 2.1 kcal/mol. These structures include monoterpenoids (β-Terpinyl acetate, −6.8 kcal/mol; 4-Carene, −6.3 kcal/mol), sesquiterpenes (Hexa-hydro-farnesol, −6.3 kcal/mol; Elemol −6.3 kcal/mol, α-Eudesmol, −6.4 kcal/mol), esters (Cyclopropanecarboxylic acid, nonyl ester, −6.6 kcal/mol; (2E)-2-Hexenyl benzoate, −6.9 kcal/mol), and ketones (2-Tridecanone, −6.4 kcal/mol; 4-(3,4-Methylenedioxyphenyl)-2-butanone, −7.1 kcal/mol). These findings can be correlated with previous studies that have clearly shown that monoterpenes or terpene-rich EOs are bactericidal and induce bacterial wall disruption, causing the loss of essential nutrients [68,69]. Additionally, based on the present computational data, the antibacterial effect of RGEO may be attributed more to the lower occuring components.

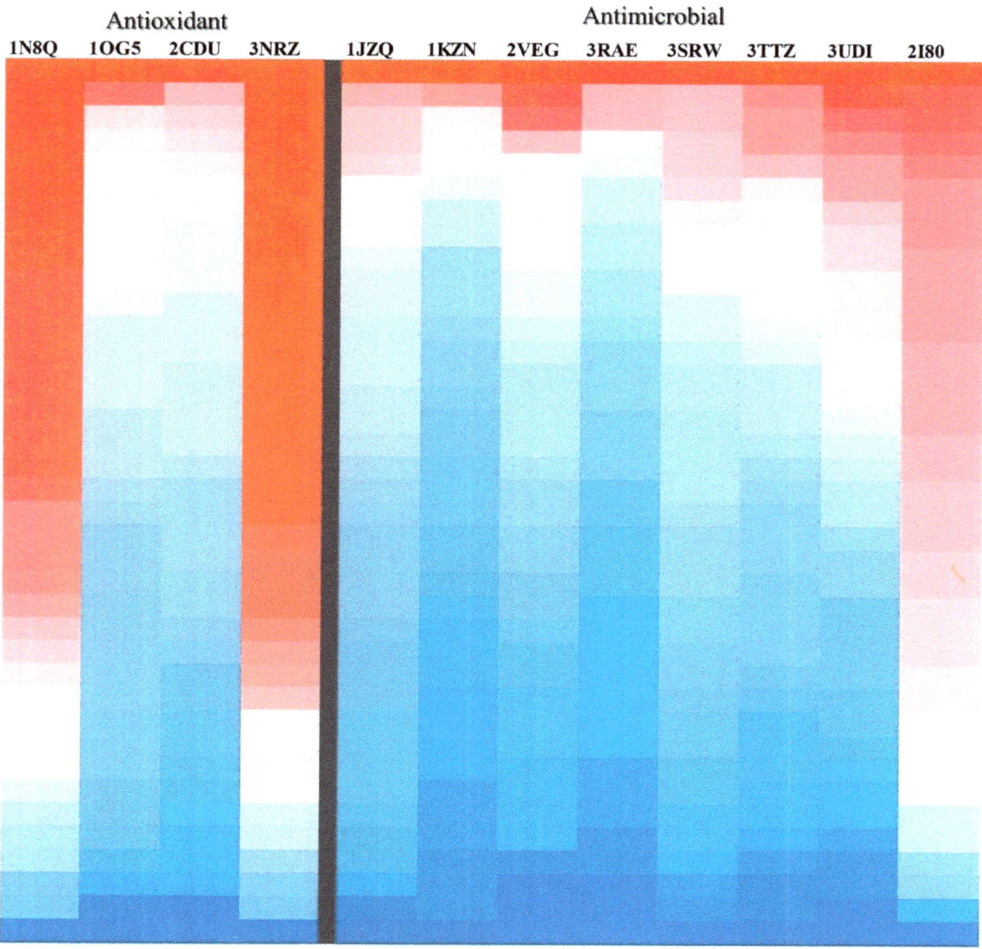

Figure 2. Three-colored heat map (red-white-blue gradient) obtained after coloring each reordered column of the docking scores table in descending order, from red for the ΔG value scored by the native ligand (control), through white for the midpoint interval, and to blue for the highest value (the structure with the lowest affinity).

Compound 29 (4-(3,4-Methylenedioxyphenyl)-2-butanone) was recorded as the highest-scoring structure towards DDl. Binding analysis revealed a good accommodation of the structure in the protein binding pocket (Figure 3). The compound forms four hydrogen bonds (HBs) (Glu16, Leu95, Thr23, and Gly118), two hydrophobic interactions (Phe313, Leu94), and one S-Pi interaction with Met310. This binding pattern is highly similar to that of the native-ligand, which also forms three of the four HB mentioned above (Figure 3).

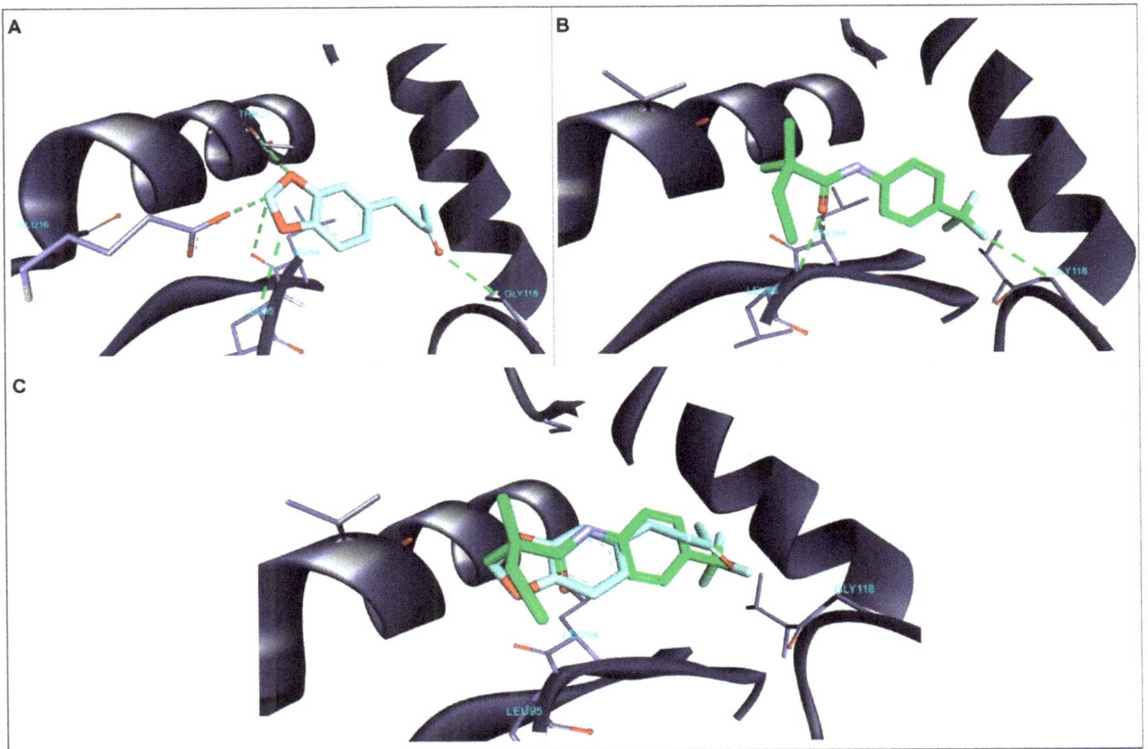

Figure 3. Structure of DDl (2I80) in presence of 4-(3,4-Methylenedioxyphenyl)-2-butanone (29). (**A**) the native ligand, 3-chloro-2,2-dimethyl-N-[4-(trifluoromethyl)phenyl]propenamide and (**B**) compound 29 and the native ligand's structures superimposed; (**C**) HB interactions are depicted as green dotted lines; interacting amino acids are shown as violet sticks.

Terpenes, represent a varied structural group of naturally occurring compounds, with a wide range of pharmacological proprieties. Given their well-documented antioxidant potency, terpenes were shown to induce significant protection against oxidative stress environments in the case of different types of diseases, such as neurodegenerative liver, cardiovascular and renal diseases, cancer, diabetes, and aging [70]. The docking data for the second subset of protein targets, corelated to the antioxidant activity, showed a tendency for the majority of compounds to potentially inhibit xanthine oxidase (3NRZ) and lipoxygenase (1N8Q), the results being very close.

Xanthine oxidase (XO) is the enzyme responsible for the metabolisation of hypoxanthine to xanthine and further to uric acid. The inhibition of XO was shown to reduce vascular oxidative stress and circulating levels uric acid [71]. Docking results show that four of the assessed compounds recorded a superior affinity compared to that of the native ligand, hypoxanthine (−6.7 kcal/mol). These compounds include 4-(3,4-Methylenedioxyphenyl)-2-butanone (29), (2E)-2-Hexenyl benzoate (6), (S)-(+)-Carvone (16), and 2-Bornene, with compound 29 showing the highest calculated affinity for XO. Binding analysis reveals the formation of 3HB (Val1011, Thr1010, Ala1079), two of which are also observed in the case of the native ligand hypoxanthine and several other hydrophobic interactions, which stabilize the molecule in a tight conformation (Figure 4). Previous reports showed that rich monoterpene EOs exerted a significant antioxidant effect, assessed by a HPLC-based assay that quantifies the activity of xanthine oxidase [72].

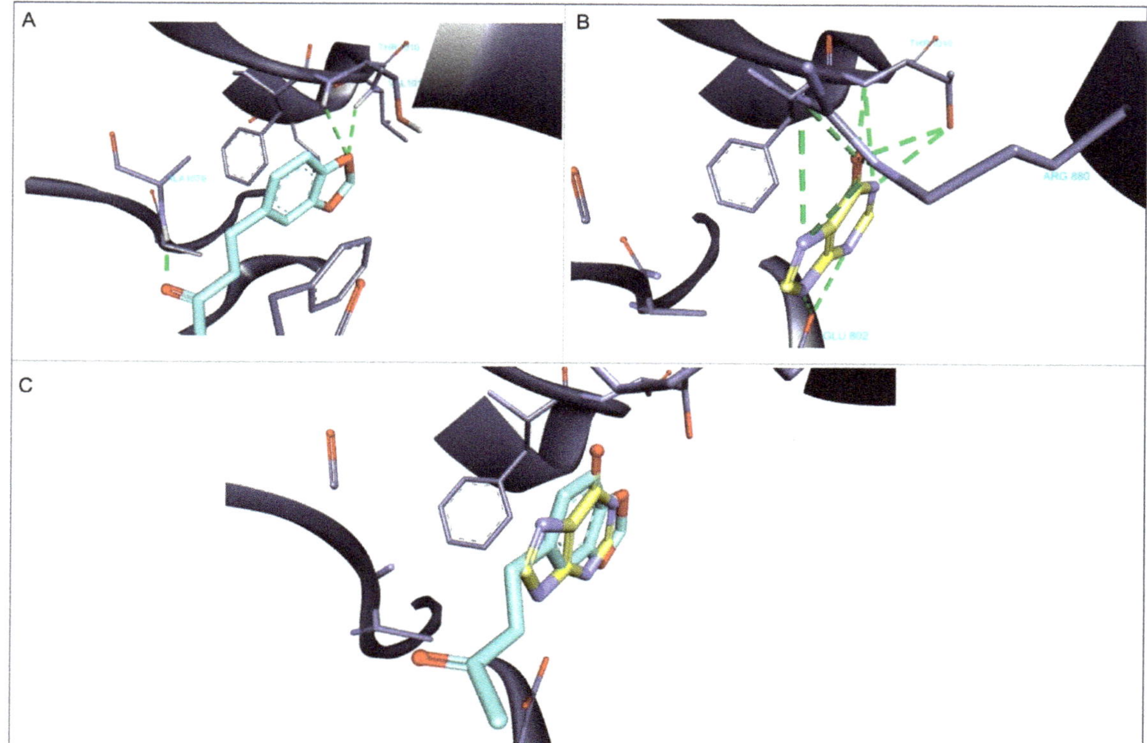

Figure 4. Structure of XO (3NRZ) in complex with 4-(3,4-Methylenedioxyphenyl)-2-butanone (29). (**A**) the native ligand, hypoxanthine; (**B**) compound 29 and the native ligand's structures superimposed; and (**C**) HB interactions are depicted as green dotted lines; interacting amino acids are shown as violet sticks.

In the case of lipoxygenase (LOX) (1N8Q), three compounds scored higher than the native ligand. These structures include the two stereoisomers of 4-Carene (3,12) and p-Cymene (9). The most active compound, 4-Carene interacts with the active site of LOX through multiple hydrophobic interactions, as presented in Figure 5. LOX is, among others, a polyunsaturated fatty acid (PUFA) metabolizing enzyme. PUFA metabolites profoundly affect inflammatory diseases and cancer progression [32]. Therefore, antioxidant compounds that act as LOX inhibitors may reduce these problems. These findings are in line with a previous study that showed the inhibitory LOX activity of terpene-containing orange juice extracts. The study also showed that some extracts elicited LOX inhibitory activity comparable to the known inhibitor quercitin [73].

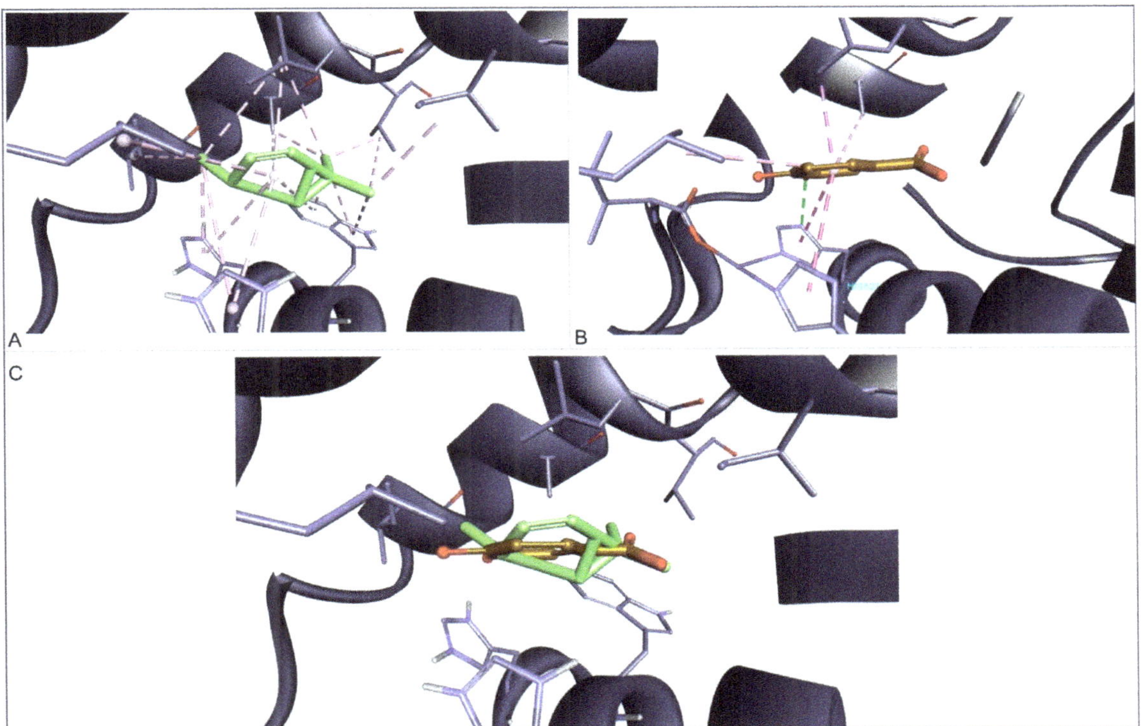

Figure 5. Structure of LOX (1N8Q) in presence of 4-Carene (3). (**A**) the native ligand, protocatechuic acid; (**B**) compound 3 and the native ligand's structures superimposed; and (**C**) hydrophobic interactions are depicted as purple dotted lines and HB as green dotted lines; interacting amino acids are shown as violet sticks.

4. Conclusions

The current research reveals that the volatile oil extracted from the aerial parts of *R. graveolens* L. is rich in ketone compounds, mainly 2-Undecanone and 2-Nonanone. The oil exhibits broad-spectrum antifungal and antibacterial effect along with moderate antioxidants properties revealed by DPPH and β-carotene/linoleic acid bleaching assays. However, the oil inhibited the formation of primary oxidation products is significantly stronger ($p < 0.001$) than BHA between the 8th and 16th days of the incubation period. Furthermore, molecular docking analysis showed that the RGEO could exert its antimicrobial activity by inhibiting the DDl enzyme. The compound with the highest affinity (29) binds in the active site through HB interactions (Glu16, Leu95, Thr23, and Gly118), sharing a high similarity with the native ligand. RGEO compounds may also induce an in vitro antioxidant effect through cumulative XO and LOX inhibition. The highest in silico active compounds showed increased affinity for XO inhibition (compound 29 through HB formation with Val1011, Thr1010, Ala1079) and LOX inhibition (compound 3) mainly through a high number of hydrophobic interactions. Consequently, the analyzed oil could be a new source of natural preservatives and antioxidants in various food and pharmaceutical industry applications.

Author Contributions: Conceptualization and methodology, C.J., C.Ș., and D.S.; investigation, D.M., C.M., G.B., A.T.L.-G., L.-C.R., A.M., M.M., and I.C.; statistical analysis, I.G.; writing—original draft preparation, C.J., D.I.H., and C.M.; writing—review and editing, C.J., D.I.H., and C.M. All authors have read and agreed to the published version of the manuscript.

Funding: This research received no external funding.

Institutional Review Board Statement: Not applicable.

Informed Consent Statement: Not applicable.

Data Availability Statement: The data published in this research are available on request from the first author and corresponding authors.

Acknowledgments: The authors would like to thank the entire team of the Interdisciplinary Research Platform belonging to Banat University of Agricultural Sciences and Veterinary Medicine "King Michael I of Romania" from Timisoara for their support during our study. This paper is published from the own funds of the Banat University of Agricultural Sciences and Veterinary Medicine "King Michael I of Romania" from Timisoara.

Conflicts of Interest: The authors declare no conflict of interest.

References

1. Bautista, D.A. Spoilage Problems. Problems Caused by Bacteria. In *Encyclopedia of Food Microbiology*, 2nd ed.; Batt, C.A., Tortorello, M.L., Eds.; Academic Press: Oxford, UK, 2014; pp. 465–470.
2. Petruzzi, L.; Corbo, M.R.; Sinigaglia, M.; Bevilacqua, A. Microbial Spoilage of Foods: Fundamentals. In *The Microbiological Quality of Food*; Bevilacqua, A., Corbo, M.R., Sinigaglia, M., Eds.; Woodhead Publishing: Kidlington, UK, 2017; Chapter 1; pp. 1–21.
3. Roberta, A.M.; Alejandra, P.G. Quorum Sensing as a Mechanism of Microbial Control and Food Safety. In *Microbial Contamination and Food Degradation*; Holban, A.M., Grumezescu, A.M., Eds.; Academic Press: London, UK, 2018; Chapter 4; pp. 85–107.
4. Gustavsson, J.; Cederberg, C.; Sonesson, U. *Global Food Losses and Food Waste*; Swedish Institute for Food and Biotechnology SIK: Gothenburg, Sweden, 2011.
5. Hossain, A.; Shah, M.D.; Sang, S.V.; Sakari, M. Chemical composition and antibacterial properties of the essential oils and crude extracts of *Merremia borneensis*. *J. King Saud Univ. Sci.* **2012**, *24*, 243–249. [CrossRef]
6. Orlanda, J.F.F.; Nascimento, A. Chemical composition and antibacterial activity of *Ruta graveolens* L. (Rutaceae) volatile oils, from São Luís, Maranhão, Brazil. *S. Afr. J. Bot.* **2015**, *99*, 103–106. [CrossRef]
7. Xu, X.; Liu, A.; Hu, S.; Ares, I.; Martínez-Larrañaga, M.-R.; Wang, X.; Martínez, M.; Anadón, A.; Martínez, M.-A. Synthetic phenolic antioxidants: Metabolism, hazards and mechanism of action. *Food Chem.* **2021**, *353*, 129488. [CrossRef]
8. Sahu, S.C.; Green, S. Food Antioxidants: Their Dual Role in Carcinogenesis. In *Oxidants, Antioxidants, and Free Radicals*; CRC Press: Boca Raton, FL, USA, 2017; pp. 327–340.
9. Álvarez-Martínez, F.; Barrajón-Catalán, E.; Herranz-López, M.; Micol, V. Antibacterial plant compounds, extracts and essential oils: An updated review on their effects and putative mechanisms of action. *Phytomedicine* **2021**, *90*, 153626. [CrossRef] [PubMed]
10. Alves, M.J.; Froufe, H.J.; Costa, A.F.; Santos, A.F.; Oliveira, L.G.; Osório, S.R.; Abreu, R.; Pintado, M.; Ferreira, I.C. Docking studies in target proteins involved in antibacterial action mechanisms: Extending the knowledge on standard antibiotics to antimicrobial mushroom compounds. *Molecules* **2014**, *19*, 1672–1684. [CrossRef] [PubMed]
11. Wei, L.; Wang, Y.Z.; Li, Z.Y. Floral ontogeny of Ruteae (Rutaceae) and its systematic implications. *Plant Biol.* **2012**, *14*, 190–197. [CrossRef]
12. Wei, L.; Xiang, X.-G.; Wang, Y.-Z.; Li, Z.-Y. Phylogenetic relationships and evolution of the androecia in Ruteae (Rutaceae). *PLoS ONE* **2015**, *10*, e0137190. [CrossRef]
13. Coimbra, A.T.; Ferreira, S.; Duarte, A.P. Genus Ruta: A natural source of high value products with biological and pharmacological properties. *J. Ethnopharmacol.* **2020**, *260*, 113076. [CrossRef]
14. Attia, E.Z.; Abd El-Baky, R.M.; Desoukey, S.Y.; El Hakeem Mohamed, M.A.; Bishr, M.M.; Kamel, M.S. Chemical composition and antimicrobial activities of essential oils of Ruta graveolens plants treated with salicylic acid under drought stress conditions. *Future J. Pharm. Sci.* **2018**, *4*, 254–264. [CrossRef]
15. Alexan, M.; Bojor, O.; Craciun, F. *The Drug Flora in Romania*; Ceres Publishing House: Bucharest, Romania, 1991; pp. 161–162. (In Romanian)
16. Stashenko, E.E.; Acosta, R.; Martínez, J.R. High-resolution gas-chromatographic analysis of the secondary metabolites obtained by subcritical-fluid extraction from Colombian rue (*Ruta graveolens* L.). *J. Biochem. Biophys. Methods* **2000**, *43*, 379–390. [CrossRef]
17. De Feo, V.; De Simone, F.; Senatore, F. Potential allelochemicals from the essential oil of *Ruta graveolens*. *Phytochemistry* **2002**, *61*, 573–578. [CrossRef]
18. Mejri, J.; Abderrabba, M.; Mejri, M. Chemical composition of the essential oil of *Ruta chalepensis* L.: Influence of drying, hydro-distillation duration and plant parts. *Ind. Crop. Prod.* **2010**, *32*, 671–673. [CrossRef]
19. Ahmad, N.; Faisal, M.; Anis, M.; Aref, I. In vitro callus induction and plant regeneration from leaf explants of *Ruta graveolens* L. *S. Afr. J. Bot.* **2010**, *76*, 597–600. [CrossRef]
20. Muntean, D.; Licker, M.; Alexa, E.; Popescu, I.; Jianu, C.; Buda, V.; Dehelean, C.A.; Ghiulai, R.; Horhat, F.; Horhat, D.; et al. Evaluation of essential oil obtained from *Mentha × piperita* L. against multidrug-resistant strains. *Infect. Drug Resist.* **2019**, *12*, 2905–2914. [CrossRef] [PubMed]

21. Adams, R.P. *Identification of essential oil components by gas chromatography/mass spectrometry*, 4th ed.; Allured Publishing Corporation: Carol Stream, IL, USA, 2007.

22. ISO. *Animal and Vegetable Fats and Oils-Determination of Peroxide Value-Potentiometric End-Point Determination*; ISO 27107:2010; International Organization for Standardization: Geneva, Switzerland, 2010.

23. Jianu, C.; Goleţ, I.; Stoin, D.; Cocan, I.; Lukinich-Gruia, A.T. Antioxidant activity of *Pastinaca sativa* L. ssp. sylvestris [Mill.] Rouy and Camus Essential Oil. *Molecules* **2020**, *25*, 869. [CrossRef]

24. Jianu, C.; Mişcă, C.; Muntean, S.G.; Lukinich-Gruia, A.T. Composition, antioxidant and antimicrobial activity of the essential oil of *Achillea collina* Becker growing wild in Western Romania. *Hem. Ind.* **2015**, *69*, 381–386. [CrossRef]

25. Rodriguez-Tudela, J.L.; Arendrup, M.C.; Barchiesi, F.; Bille, J.; Chryssanthou, E.; Cuenca-Estrella, M.; Dannaoui, E.; Denning, D.W.; Donnelly, J.P.; Dromer, F.; et al. EUCAST Definitive Document EDef 7.1: Method for the determination of broth dilution MICs of antifungal agents for fermentative yeasts: Subcommittee on Antifungal Susceptibility Testing (AFST) of the ESCMID European Committee for Antimicrobial Susceptibility Testing (EUCAST). *Clin. Microbiol. Infect.* **2008**, *14*, 398–405.

26. Jianu, C.; Moleriu, R.; Stoin, D.; Cocan, I.; Bujancă, G.; Pop, G.; Lukinich-Gruia, A.T.; Muntean, D.; Rusu, L.-C.; Horhat, D.I. Antioxidant and Antibacterial Activity of *Nepeta × faassenii* Bergmans ex Stearn Essential Oil. *Appl. Sci.* **2021**, *11*, 442. [CrossRef]

27. Jianu, C.; Stoin, D.; Cocan, I.; David, I.; Pop, G.; Lukinich-Gruia, A.T.; Mioc, M.; Mioc, A.; Şoica, C.; Muntean, D. In silico and in vitro evaluation of the antimicrobial and antioxidant potential of *Mentha×smithiana* R. GRAHAM Essential Oil from Western Romania. *Foods* **2021**, *10*, 815. [CrossRef] [PubMed]

28. Berman, H.M.; Westbrook, J.; Feng, Z.; Gilliland, G.; Bhat, T.N.; Weissig, H.; Shindyalov, I.N.; Bourne, P.E. The protein data bank. *Nucleic Acids Res.* **2000**, *28*, 235–242. [CrossRef]

29. Trott, O.; Olson, A.J. AutoDock Vina: Improving the speed and accuracy of docking with a new scoring function, efficient optimization, and multithreading. *J. Comput. Chem.* **2010**, *31*, 455–461. [CrossRef] [PubMed]

30. Nakama, T.; Nureki, O.; Yokoyama, S. Structural basis for the recognition of isoleucyl-adenylate and an antibiotic, mupirocin, by isoleucyl-tRNA synthetase. *J. Biol. Chem.* **2001**, *276*, 47387–47393. [CrossRef] [PubMed]

31. Lafitte, D.; Lamour, V.; Tsvetkov, P.O.; Makarov, A.A.; Klich, M.; Deprez, P.; Moras, D.; Briand, A.C.; Gilli, R. DNA gyrase interaction with coumarin-based inhibitors: The role of the hydroxybenzoate isopentenyl moiety and the 5'-methyl group of the noviose. *Biochemistry* **2002**, *41*, 7217–7223. [CrossRef]

32. Borbulevych, O.Y.; Jankun, J.; Selman, S.H.; Skrzypczak-Jankun, E. Lipoxygenase interactions with natural flavonoid, quercetin, reveal a complex with protocatechuic acid in its X-ray structure at 2.1 Å resolution. *Proteins Struct. Funct. Bioinform.* **2004**, *54*, 13–19. [CrossRef]

33. Williams, P.A.; Cosme, J.; Ward, A.; Angove, H.C.; Vinković, D.M.; Jhoti, H. Crystal structure of human cytochrome P450 2C9 with bound warfarin. *Nature* **2003**, *424*, 464–468. [CrossRef]

34. Lountos, G.T.; Jiang, R.; Wellborn, W.B.; Thaler, T.L.; Bommarius, A.S.; Orville, A.M. The crystal structure of NAD(P)H oxidase from *Lactobacillus sanfranciscensis*: Insights into the Conversion of O2 into Two Water Molecules by the Flavoenzyme. *Biochemistry* **2006**, *45*, 9648–9659. [CrossRef]

35. Liu, S.; Chang, J.S.; Herberg, J.T.; Horng, M.-M.; Tomich, P.K.; Lin, A.H.; Marotti, K.R. Allosteric inhibition of Staphylococcus aureusd-alanine: D-alanine ligase revealed by crystallographic studies. *Proc. Nat. Acad. Sci. USA* **2006**, *103*, 15178–15183. [CrossRef]

36. Levy, C.; Minnis, D.; Derrick, J.P. Dihydropteroate synthase from Streptococcus pneumoniae: Structure, ligand recognition and mechanism of sulfonamide resistance. *Biochem. J.* **2008**, *412*, 379–388. [CrossRef]

37. Cao, H.; Pauff, J.M.; Hille, R. Substrate orientation and catalytic specificity in the action of xanthine oxidase: The sequential hydroxylation of hypoxanthine to uric acid. *J. Biol. Chem.* **2010**, *285*, 28044–28053. [CrossRef] [PubMed]

38. Veselkov, D.A.; Laponogov, I.; Pan, X.-S.; Selvarajah, J.; Skamrova, G.B.; Branstrom, A.; Narasimhan, J.; Prasad, J.V.V.; Fisher, L.M.; Sanderson, M.R. Structure of a quinolone-stabilized cleavage complex of topoisomerase IV from Klebsiella pneumoniae and comparison with a related Streptococcus pneumoniae complex. *Acta Crystallogr. Sect. D Struct. Biol.* **2016**, *72*, 488–496. [CrossRef] [PubMed]

39. Li, X.; Hilgers, M.; Cunningham, M.; Chen, Z.; Trzoss, M.; Zhang, J.; Kohnen, L.; Lam, T.; Creighton, C.; Kedar, G.; et al. Structure-based design of new DHFR-based antibacterial agents: 7-aryl-2,4-diaminoquinazolines. *Bioorganic Med. Chem. Lett.* **2011**, *21*, 5171–5176. [CrossRef]

40. Sherer, B.A.; Hull, K.; Green, O.; Basarab, G.; Hauck, S.; Hill, P.; Loch, J.T., III; Mullen, G.; Bist, S.; Bryant, J. Pyrrolamide DNA gyrase inhibitors: Optimization of antibacterial activity and efficacy. *Bioorgan. Med. Chem. Lett.* **2011**, *21*, 7416–7420. [CrossRef]

41. Han, S.; Caspers, N.; Zaniewski, R.P.; Lacey, B.M.; Tomaras, A.P.; Feng, X.; Geoghegan, K.F.; Shanmugasundaram, V. Distinctive Attributes of β-Lactam Target Proteins in Acinetobacter baumannii Relevant to Development of New Antibiotics. *J. Am. Chem. Soc.* **2011**, *133*, 20536–20545. [CrossRef] [PubMed]

42. De Pooter, H.L.; Nicolai, B.; De Laet, J.; De Buyck, L.F.; Schamp, N.M.; Goetghebeur, P. The essential oils of fiveNepeta Species. A preliminary evaluation of their use in chemotaxonomy by cluster analysis. *Flavour Fragr. J.* **1988**, *3*, 155–159. [CrossRef]

43. Radulovic, N.; Blagojevic, P.D.; Rabbitt, K.; de Sousa Menezes, F. Essential oil of *Nepeta x faassenii* Bergmans ex Stearn (*N. mussinii* Spreng. × N. nepetella L.): A comparison study. *Nat. Prod. Commun.* **2011**, *6*, 1015–1022. [CrossRef]

44. Yaacob, K.B.; Abdullah, C.M.; Joulain, D. Essential Oil of *Ruta graveolens* L. *J. Essent. Oil Res.* **1989**, *1*, 203–207. [CrossRef]

45. Haddouchi, F.; Chaouche, T.M.; Zaouali, Y.; Ksouri, R.; Attou, A.; Benmansour, A. Chemical composition and antimicrobial activity of the essential oils from four Ruta species growing in Algeria. *Food Chem.* **2013**, *141*, 253–258. [CrossRef] [PubMed]
46. Formisano, C.; Delfine, S.; Oliviero, F.; Tenore, G.C.; Rigano, D.; Senatore, F. Correlation among environmental factors, chemical composition and antioxidative properties of essential oil and extracts of chamomile (*Matricaria chamomilla* L.) collected in Molise (South-central Italy). *Ind. Crop. Prod.* **2015**, *63*, 256–263. [CrossRef]
47. El-Sherbeny, S.; Hussein, M.; Khalil, M. Improving the production of *Ruta graveolens* L. plants cultivated under different compost levels and various sowing distance. *Am.-Euras. J. Agric. Environ. Sci.* **2007**, *23*, 271–281.
48. Soleimani, M.; Azar, P.A.; Saber-Tehranil, M.; Rustaiyan, A. Volatile composition of *Ruta graveolens* L. of North of Iran. *World Appl. Sci. J.* **2009**, *7*, 124–126.
49. Reddy, D.N.; Al-Rajab, A.J. Chemical composition, antibacterial and antifungal activities of Ruta graveolens L. volatile oils. *Cogent Chem.* **2016**, *2*, 1220055. [CrossRef]
50. Narimani, R.; Moghaddam, M.; Ghasemi Pirbalouti, A.; Mojarab, S. Essential oil composition of seven populations belonging to two Nepeta species from Northwestern Iran. *Int. J. Food Prop.* **2017**, *20* (Suppl. S2), 2272–2279.
51. Li, T.; Hu, W.; Li, J.; Zhang, X.; Zhu, J.; Li, X. Coating effects of tea polyphenol and rosemary extract combined with chitosan on the storage quality of large yellow croaker (*Pseudosciaena crocea*). *Food Control* **2012**, *25*, 101–106. [CrossRef]
52. Singh, G.; Kapoor, I.; Singh, P.; de Heluani, C.S.; de Lampasona, M.P.; Catalan, C.A.N. Chemistry, antioxidant and antimicrobial investigations on essential oil and oleoresins of Zingiber officinale. *Food Chem. Toxicol.* **2008**, *46*, 3295–3302. [CrossRef]
53. Benali, T.; Habbadi, K.; Khabbach, A.; Marmouzi, I.; Zengin, G.; Bouyahya, A.; Chamkhi, I.; Chtibi, H.; Aanniz, T.; Achbani, E.H.; et al. GC–MS analysis, antioxidant and antimicrobial activities of *Achillea Odorata* Subsp. Pectinata and Ruta Montana essential oils and their potential use as food preservatives. *Foods* **2020**, *9*, 668. [CrossRef]
54. Mohammedi, H.; Mecherara-Idjeri, S.; Hassani, A. Variability in essential oil composition, antioxidant and antimicrobial activities of Ruta montana L. collected from different geographical regions in Algeria. *J. Essent. Oil Res.* **2019**, *32*, 88–101. [CrossRef]
55. Jaradat, N.; Adwan, L.; K'Aibni, S.; Zaid, A.N.; Shtaya, M.; Shraim, N.; Assali, M. Variability of Chemical Compositions and Antimicrobial and Antioxidant Activities of Ruta chalepensis Leaf Essential Oils from Three Palestinian Regions. *BioMed Res. Int.* **2017**, *2017*, 2672689. [CrossRef]
56. Althaher, A.R.; Oran, S.A.; Bustanji, Y.K. Phytochemical analysis, in vitro assessment of antioxidant properties and cytotoxic potential of *Ruta chalepensis* L. essential oil. *J. Essent. Oil Bear. Plants* **2020**, *23*, 1409–1421. [CrossRef]
57. Oke, F.; Aslim, B.; Ozturk, S.; Altundag, S. Essential oil composition, antimicrobial and antioxidant activities of *Satureja cuneifolia* Ten. *Food Chem.* **2009**, *112*, 874–879. [CrossRef]
58. Kulisic, T.; Radonic, A.; Katalinic, V.; Milos, M. Use of different methods for testing antioxidative activity of oregano essential oil. *Food Chem.* **2004**, *85*, 633–640. [CrossRef]
59. Loizzo, M.; Falco, T.; Bonesi, M.; Sicari, V.; Tundis, R.; Bruno, M. *Ruta chalepensis* L. (Rutaceae) leaf extract: Chemical composition, antioxidant and hypoglicaemic activities. *Nat. Prod. Res.* **2018**, *32*, 521–528. [CrossRef]
60. Aligiannis, N.; Kalpoutzakis, E.; Mitaku, S.; Chinou, I.B. Composition and antimicrobial activity of the essential oils of two Origanum species. *J. Agric. Food Chem.* **2001**, *49*, 4168–4170. [CrossRef]
61. Angienda, P.O.; Onyango, D.M.; Hill, D.J. Potential application of plant essential oils at sub-lethal concentrations under extrinsic conditions that enhance their antimicrobial effectiveness against pathogenic bacteria. *Afr. J. Microbiol. Res.* **2010**, *4*, 1678–1684.
62. Gilles, M.; Zhao, J.; An, M.; Agboola, S. Chemical composition and antimicrobial properties of essential oils of three Australian Eucalyptus species. *Food Chem.* **2010**, *119*, 731–737. [CrossRef]
63. Cox, S.; Mann, C.; Markham, J.; Bell, H.C.; Gustafson, J.; Warmington, J.; Wyllie, S.G. The mode of antimicrobial action of the essential oil of Melaleuca alternifolia (tea tree oil). *J. Appl. Microbiol.* **2000**, *88*, 170–175. [CrossRef] [PubMed]
64. Franchini, S.; Prandi, A.; Baraldi, A.; Sorbi, C.; Tait, A.; Buccioni, M.; Marucci, G.; Cilia, A.; Pirona, L.; Fossa, P. 1,3-Dioxolane-based ligands incorporating a lactam or imide moiety: Structure–affinity/activity relationship at α1-adrenoceptor subtypes and at 5-HT1A receptors. *Eur. J. Med. Chem.* **2010**, *45*, 3740–3751. [CrossRef]
65. Fossa, P.; Cichero, E. In silico evaluation of human small heat shock protein HSP27: Homology modeling, mutation analyses and docking studies. *Bioorgan. Med. Chem.* **2015**, *23*, 3215–3220. [CrossRef] [PubMed]
66. Tonelli, M.; Espinoza, S.; Gainetdinov, R.; Cichero, E. Novel biguanide-based derivatives scouted as TAAR1 agonists: Synthesis, biological evaluation, ADME prediction and molecular docking studies. *Eur. J. Med. Chem.* **2017**, *127*, 781–792. [CrossRef]
67. Goncharov, N.V.; Avdonin, P.V.; Nadeev, A.D.; Zharkikh, I.L.; Jenkins, R.O. Reactive oxygen species in pathogenesis of atherosclerosis. *Curr. Pharm. Des.* **2015**, *21*, 1134–1146. [CrossRef]
68. Trombetta, D.; Castelli, F.; Sarpietro, M.G.; Venuti, V.; Cristani, M.; Daniele, C.; Saija, A.; Mazzanti, G.; Bisignano, G. Mechanisms of antibacterial action of three monoterpenes. *Antimicrob. Agents Chemother.* **2005**, *49*, 2474–2478. [CrossRef] [PubMed]
69. Zengin, H.; Baysal, A.H. Antibacterial and Antioxidant activity of essential oil terpenes against pathogenic and spoilage-forming bacteria and cell structure-activity relationships evaluated by SEM microscopy. *Molecules* **2014**, *19*, 17773–17798. [CrossRef] [PubMed]
70. Gonzalez-Burgos, E.; Gómez-Serranillos, M. Terpene compounds in nature: A review of their potential antioxidant activity. *Curr. Med. Chem.* **2012**, *19*, 5319–5341. [CrossRef]
71. Kostić, D.A.; Dimitrijević, D.S.; Stojanović, G.; Palić, I.R.; Đorđević, A.; Ickovski, J. Xanthine oxidase: Isolation, assays of activity, and inhibition. *J. Chem.* **2015**, *2015*, 294858. [CrossRef]

72. Salles Trevisan, M.T.; Vasconcelos Silva, M.G.; Pfundstein, B.; Spiegelhalder, B.; Owen, R.W. Characterization of the volatile pattern and antioxidant capacity of essential oils from different species of the genus *Ocimum*. *J. Agric. Food Chem.* **2006**, *54*, 4378–4382. [CrossRef] [PubMed]
73. Sánchez-Martínez, J.D.; Bueno, M.; Alvarez-Rivera, G.; Tudela, J.; Ibañez, E.; Cifuentes, A. In vitro neuroprotective potential of terpenes from industrial orange juice by-products. *Food Funct.* **2021**, *12*, 302–314. [CrossRef]

Article

Anti-Yeast Synergistic Effects and Mode of Action of Australian Native Plant Essential Oils

Fahad Alderees [1], Ram Mereddy [2], Stephen Were [2], Michael E. Netzel [1] and Yasmina Sultanbawa [1,*]

[1] ARC Industrial Transformation Training Centre for Uniquely Australian Foods, Queensland Alliance for Agriculture and Food Innovation, The University of Queensland, Indooroopilly, Brisbane, QLD 4068, Australia; alderees@gmail.com (F.A.); m.netzel@uq.edu.au (M.E.N.)

[2] Department of Agriculture and Fisheries, Health and Food Sciences Precinct, Coopers Plains, Brisbane, QLD 4108, Australia; ram.mereddy@daf.qld.gov.au (R.M.); stephen.were@daf.qld.gov.au (S.W.)

* Correspondence: y.sultanbawa@uq.edu.au

Abstract: Yeasts are the most common group of microorganisms responsible for spoilage of soft drinks and fruit juices due to their ability to withstand juice acidity and pasteurization temperatures and resist the action of weak-acid preservatives. Food industries are interested in the application of natural antimicrobial compounds as an alternative solution to the spoilage problem. This study attempts to investigate the effectiveness of three Australian native plant essential oils (EOs) Tasmanian pepper leaf (TPL), lemon myrtle (LM) and anise myrtle (AM) against weak-acid resistant yeasts, to identify their major bioactive compounds and to elucidate their anti-yeast mode of action. The minimum inhibitory concentration (MIC), minimum fungicidal concentration (MFC) and minimum bactericidal concentration (MBC) were assessed for EOs against weak-acid resistant yeasts (*Candida albicans, Candida krusei, Dekkera anomala, Dekkera bruxellensis, Rhodotorula mucilaginosa, Saccharomyces cerevisiae, Schizosaccharomyces pombe, Zygosaccharomyces bailii* and *Zygosaccharomyces rouxii*) and bacteria (*Staphylococcus aureus* and *Escherichia coli*). The EOs showed anti-yeast and antibacterial activity at concentrations ranging from 0.03–0.07 mg/mL and 0.22–0.42 mg/mL for TPL and 0.07–0.31 mg/mL and 0.83–1.67 mg/mL for LM, respectively. The EOs main bioactive compounds were identified as polygodial in TPL, citral (neral and geranial) in LM and anethole in AM. No changes in the MICs of the EOs were observed in the sorbitol osmotic protection assay but were found to be increased in the ergosterol binding assay after the addition of exogenous ergosterol. Damaging of the yeast cell membrane, channel formation, cell organelles and ion leakage could be identified as the mode of action of TPL and LM EOs. The studied Australian native plant EOs showed potential as natural antimicrobials that could be used in the beverage and food industry against the spoilage causing yeasts.

Keywords: Australian native plants; essential oils; weak-acid resistant yeast; natural antimicrobials; anti-yeast activity

Citation: Alderees, F.; Mereddy, R.; Were, S.; Netzel, M.E.; Sultanbawa, Y. Anti-Yeast Synergistic Effects and Mode of Action of Australian Native Plant Essential Oils. *Appl. Sci.* **2021**, *11*, 10670. https://doi.org/10.3390/app112210670

Academic Editors: Ionel Calin Jianu and Georgeta Pop

Received: 12 October 2021
Accepted: 6 November 2021
Published: 12 November 2021

Publisher's Note: MDPI stays neutral with regard to jurisdictional claims in published maps and institutional affiliations.

1. Introduction

Plant essential oils (EOs) have been utilized for many centuries in traditional medicine, as spices, coloring, perfume and aromatherapy [1,2]. Due to their beneficial properties, they have been incorporated in recent years into pharmaceuticals, food preservation, agriculture, disinfection and sanitation products, and cosmetics [3]. There are about 3000 known EOs; however, only 10% have been commercially utilized in different industries [4,5]. Plant EOs are unique and worth deeper investigation due to their multiple bioactive properties such as being antibacterial, antiviral, antifungal, anti-toxigenic, anti-carcinogenic, antioxidant, anti-parasitic, insecticidal, sedative, anti-inflammatory, antiseptic, spasmolytic and anesthetic [1,3,4]. There is a huge opportunity to study and exploit the biological properties of these EOs for novel applications in the food (e.g., as natural preservatives) and pharmaceutical industries.

In recent years, EOs have been studied for their antimicrobial properties against bacteria, fungi and viruses. However, little is known about their efficacy when compared to synthetic food preservatives, their activity against weak-acid resistant yeasts and mode of action [6–10].

Yeasts are well-known for their positive contributions in the production and development of various foods and beverages. However, some species of yeasts cause negative effects by spoilage [11]. These yeasts can grow in an environment that has a low water activity, low pH, high concentration of sugar/salt, cold temperatures and are able to resist and survive within the legally permitted levels of synthetic preservatives, which creates a real challenge to the food and beverage industries [6,12,13]. The scientific literature shows numerous studies referring to the emergence of spoilage yeast resistant to weak-acid preservatives such as acetic acid, benzoic acid, sulfur dioxide and sorbic acid [6,12,13]. Almost every food industry that manufactures acid beverages was exposed to one or more occasions of great economic losses due to yeasts that are resistant to weak-acid preservatives [14]. A great example of the seriousness of this challenge is when yeasts spoiled about 300,000 bottles of orange-based carbonated beverages from one of the manufacturers in Europe [6]. The observed increased resistance is a "wake up call" to look for alternative solutions, and the food industry is interested in exploring the utilization of natural antimicrobial compounds as an alternative pathway to control this spoilage problem [15]. In addition, the pharmaceutical industry is also focusing on incorporating natural products in medications, where about 28% of all the newly approved drugs were derived from either natural or derivatives of natural compounds [16,17].

The rich Australian indigenous flora contains a variety of plants that are abundant in EOs and phytochemicals with promising biological activities. Three commercially produced plants, Tasmanian pepper leaf (TPL) (*Tasmannia lanceolata*), lemon myrtle (LM) (*Backhousia citriodora*) and anise myrtle (AM) (*Syzygium anisatum*) are known to possess antimicrobial activities. TPL belongs to the Winteraceae family and is located in forested regions in Tasmania, Victoria and north to the Blue Mountains [18]. Polygodial, sesquiterpene dialdehyde, is reported to be the major bioactive compound (37% to 64%) in TPL EO contributing to its antibacterial and antifungal activities [18–22]. TPL is utilized commercially in food products as a flavoring, seasoning, coloring and preservative agent [19,23,24]. LM is a member of the Myrtaceae family and grows in the subtropical rainforests of central and southeast regions of Queensland, Australia spreading along the coastal regions from Brisbane to Cairns [25]. LM EO was first steam distilled in 1890 and is used as a flavoring agent in food and beverage systems, cosmetic products, fragrances and aromatherapy [26–29]. The antimicrobial properties of LM are due to its predominant compound (82% to 91%) citral, a monoterpene aldehyde [30]. Anise myrtle is a rare Australian native plant found in the moist rainforest areas located in the Bellingen Valley of northeast New South Wales and some nearby parts of Queensland [31]. Its leaves are used as an herb either fresh or as ground powder to provide aniseed flavors in Australian cuisines, and its EO is utilized in cosmetic products, alcoholic beverages and in the pharmaceutical industry [32]. EO of AM consists of anethole (94.97%) and its isomer methyl chavicol (4.43%), known as estragole, which contribute to the antimicrobial property [33].

There is a knowledge gap regarding the antimicrobial activity of Australian native essential oils against weak-acid resistant yeasts that are commonly reported to cause spoilage in fruit juices. Therefore, the aims of this study are to evaluate the antimicrobial activities of TPL, LM and AM EOs against nine weak-acid resistant yeasts, *Candida albicans, Candida krusei, Dekkera anomala, Dekkera bruxellensis, Rhodotorula mucilaginosa, Saccharomyces cerevisiae, Schizosaccharomyces pombe, Zygosaccharomyces bailii* and *Zygosaccharomyces rouxii* and two food related bacteria *Staphylococcus aureus* and *Escherichia coli*, to determine the EOs minimum inhibitory concentration (MIC), minimum fungicidal concentration (MFC), and minimum bactericidal concentration (MBC), and to investigate possible synergistic effects between the three EOs and to elucidate the potential anti-yeast mode of action of these EOs.

2. Materials and Methods

2.1. Essential Oils and Other Chemicals

EOs of LM and AM were supplied by Australian Rainforest Products Pty Ltd. (Lismore, NSW, Australia) and EO of TPL was supplied by Essential Oils of Tasmania Pty Ltd. (Margate, TAS, Australia). Oils were kept in their original bottles protected from light and stored at 4 °C until further use. In all experiments, EOs were applied as emulsions in sterile water with Tween-80 at a ratio of 10:1 (*v/v*). Nutrient broth (CM0001, Oxoid, Basingstoke, UK), Sabouraud dextrose broth (CM0147, Oxoid, Basingstoke, UK), Tween-80 (9005656, Sigma-Aldrich, St. Louis, MO, USA), ergosterol (57874, Sigma-Aldrich, St. Louis, MO, USA), sorbitol (50704, Sigma-Aldrich, St. Louis, MO, USA), Cesium chloride (15507023, Thermo Fisher Scientific, Scoresby, VIC, Australia), Sodium Chloride (Thermo Fisher Scientific, Scoresby, VIC, Australia), DAPI (4′,6-diamidino-2-phenylindole) (62247, Thermo Fisher Scientific, Scoresby, VIC, Australia), paraformaldehyde (R37814, Thermo Fisher Scientific, Scoresby, VIC, Australia), poly-L-lysine (25988630, Sigma-Aldrich, St. Louis, MO, USA), glutaraldehyde (ProSciTech, Townsville, QLD, Australia), osmium tetroxide (ProSciTech, Townsville, QLD, Australia), phosphate buffered saline (Sigma P4417, Sigma-Aldrich, Castle Hill, NSW, Australia).

2.2. Microorganisms

Antimicrobial testing was performed against two bacteria, Gram-positive *Staphylococcus aureus* (ATCC 9144) and Gram-negative *Escherichia coli* (ATCC 11775), and nine yeasts, *Candida albicans* (ATCC 10231), *Candida krusei* (ATCC 6258), *Dekkera anomala* (ATCC 58985), *Dekkera bruxellensis* (ATCC 56866), *Rhodotorula mucilaginosa* (ATCC 20129), *Saccharomyces cerevisiae* (ATCC 38555), *Schizosaccharomyces pombe* (ATCC 26189), *Zygosaccharomyces bailii* (ATCC 38923) and *Zygosaccharomyces rouxii* (ATCC 10682). Microorganisms were purchased from In Vitro Technologies (Melbourne, VIC, Australia).

2.3. Determination of Antimicrobial Activity

The minimum inhibitory concentration (MIC), defined as the lowest oil concentration to inhibit visible microbial growth, was performed by a broth serial microdilution method using 96-well plates according to Ahmad and Viljoen [34] with a few modifications. Cultures were grown for 24 h at 35 °C in a nutrient broth (CM0001, Oxoid, Basingstoke, UK) for bacteria and at 25 °C in a Sabouraud dextrose broth (CM0147, Oxoid) for yeasts, except for *C. albicans* and *S. pombe* which were grown at 30 °C. Cultures were measured at OD540 = 0.5 McFarland, diluted (using their designated double strength broth) into 1×10^5 colony forming units (cfu) per ml and dispensed (100 µL) onto the 96-well plates. A stock solution of EO was diluted serially two-fold in 25 mL centrifuge tubes to give solutions from 1 to 0.002% (*v/v*) and 100 µL of each concentration was dispensed in triplicate from highest to lowest on the plate. Growth control (culture + broth + Tween-80), sterility control (broth + tween-80 + EO) and standard drugs as positive control (amphotericin B for yeasts and chloramphenicol for bacteria) were included in every microplate assay test. After incubation for 48 h for yeasts and 24 h for bacteria, the absence of a white growth on the bottom of the well was used as an indication of the MIC. The minimum bactericidal concentration (MBC) and minimum fungicidal concentration (MFC) were determined by inoculating 20 µL from wells with no observed growth onto a new 96-well plate containing 100 µL broth (normal strength) and incubated at the same conditions as described above. Wells with no visible growth and the lowest EO concentration were determined as the MBC or MFC. The whole procedure was performed in triplicate and the MICs, MBCs and MFCs were expressed in mg/mL.

2.4. Ergosterol Binding Assay

The interaction between EOs and ergosterol was evaluated by observing the MIC changes in the absence and presence of ergosterol at concentrations of 25, 50 and 100 µg/mL using a broth serial dilution method on 96-well plates [35]. Ergosterol was dissolved in ethanol and Tween-80 (1:1, v/v) at 40 °C and 50 µL of that solution was dispensed onto the plate. Each well contained 100 µL double strength broth and 100 µL of EO as previously prepared, with a yeast culture concentration of 1–2 $\times$ 10^5. Growth control (culture + broth + ethanol + tween-80) and sterility controls were included on the same 96-well plates.

2.5. Sorbitol Osmotic Protection Assay

The sorbitol assay was performed according to Frost et al. [36] with some modifications using a serial broth microdilution method. The MICs of the EOs were tested without and with the addition of 0.8 M sorbitol in 96-well plates. Sorbitol was dissolved in a double strength broth and 100 µL was dispensed onto each well containing 100 µL EO at different concentrations as previously prepared. Growth control (culture + broth + 0.8 M sorbitol) and sterility controls were included on the same 96-well plates.

2.6. Leakage of Potassium and Magnesium Ions

The concentrations of potassium and magnesium ions were measured before and after exposure of *S. cerevisiae* to LM, TPL and AM EOs using an atomic absorption spectrophotometer (contrAA, Analytik Jena AG, Jena, Germany) following the method of Prashar and co-workers [37] with a few modifications. *S. cerevisiae* cell suspensions were obtained at turbidity of OD540 = 2.0 in 5 mL of 0.1 M NaCl solution and samples were treated with 0.5% (v/v) EO for 0, 2, 4 and 6 h at 25 °C while shaking on an orbital shaker at a speed of 150 rpm. Positive control samples (yeast + NaCl) and negative control samples (oil + NaCl) were exposed to the same conditions as the treated samples and measured at 0, 2, 4 and 6 h. A solution of 0.1% (w/v) cesium chloride was used as an ion suppressant and depending on the concentration of ions, samples were diluted 1:10 or 1:40 with the ion suppressant. Acetylene was used as a fuel and wavelength measurement for potassium and magnesium was set at 766.49 nm and 285.21 nm, respectively.

2.7. Synergy Checkerboard (Interactions between EOs)

The MIC was obtained by testing the EOs individually or in combination to determine if their interaction is of synergistic, additive, indifferent or antagonistic nature, using the checkerboard assay on a 96-well plate according to Zengin and Baysal [38]. The first oil (component A) was dispensed vertically, and the second oil (component B) was dispensed horizontally on the plate from highest to lowest concentration. The Fractional Inhibitory Concentration Index (FICI) is the sum of Fractional Inhibitory Concentration (FIC) of component A (FICA) and FIC of component B (FICB) and was determined based on the following formula: FICA = MIC of component A in combination/MIC of component A alone, FICB = MIC of component B in combination/MIC of component B alone, FICI = (FICA + FICB). FICI was considered synergistic, additive, indifferent or antagonistic when values were $\leq$0.5, 0.5 to $\leq$1, 1 to 4 or >4, respectively.

2.8. Fluorescence and Scanning Electron Microscopy

The morphological changes of *S. cerevisiae* cells treated with EOs were monitored by fluorescence and scanning electron microscopy. An overnight culture suspension was adjusted to 1 $\times$ 10^5 cfu/mL, pelleted by centrifugation at 2500$\times$ g rpm for 5 min and treated with 0.5% oil for 2 h at 25 °C. The control group was treated the same way, except that 0.1 M phosphate buffered saline (Sigma P4417, Sigma-Aldrich, Castle Hill, NSW, Australia) replaced the EOs. After the treatment, *S. cerevisiae* cells were harvested by centrifugation at 2500$\times$ g rpm for 5 min and washed twice with 0.1 M phosphate buffered saline.

2.9. Fluorescence Microscopy and Yeast Cell Staining

Two fluorescent dyes, DAPI (4′,6-diamidino-2-phenylindole) and SYTO 9 (Thermo Fisher Scientific, Scoresby, VIC, Australia) were used to stain the yeast cells for fluorescence microscopy [39]. *S. cerevisiae* cells were fixed in 4% paraformaldehyde for 1 h, harvested by centrifugation at 2500$\times$ g rpm for 5 min and washed twice with 0.1 M phosphate buffered saline. DAPI (50 ng/mL) was mixed with the yeast cell suspension at a 1:1 (*v*/*v*) ratio. SYTO 9 (5 mM) was mixed with the cell suspension at a 1:1000 (*v*/*v*) ratio. Cells were allowed to stain in the dark at 25 °C for 15 min, then 8 µL were transferred to a microscope glass slide and covered with a coverslip. Cells were observed with a Leica DM6000B microscope using a 100× objective (Leica, Wetzlar, Germany). Images were captured with a Leica DFC420C digital camera and Microsystem LAS AF6000 software (Leica).

2.10. Scanning Electron Microscopy

Yeast cells were initially fixed in glutaraldehyde and then in 1% osmium tetroxide. Cells were dehydrated through a series of ethanol washes, transferred to coverslips coated with 1 mg/mL poly-L-lysine, dried in a critical point dryer (Tousimis Research Corp., Rockville, MD, USA) and coated with gold according to the manufacturer's instructions using a sputter coater (Agar Scientific Ltd., Essex, UK). Images were captured with a JCM-5000 NeoScopeTM Table Top scanning electron microscope (Jeol Neoscope/Australasia, Frenchs Forest, NSW, Australia) at an accelerating voltage of 10 kV.

2.11. Gas Chromatography (GC)–Mass Spectrometry Analysis

EOs were analyzed for bioactive compounds as described by Chaliha et al. [40] with slight modifications. A Trace GC Ultra system coupled with a DSQ mass spectrometer (MS) (Thermo Fisher Scientific) was used. The bioactive compounds were separated by a Factor Four capillary column VF-5 ms (30 m × 0.25 mm × 0.25 µm; Varian Inc., Palo Alto, CA, USA) under the following conditions: sample injection volume of 1 µL; temperature program, 1.5 min at 30 °C, increase to 200 °C at 10 °C/min intervals, 1 min at 200 °C, then increase to 220 °C within 1 min, 2 min at 220 °C before returning to the initial setting; helium was used as a carrier gas with a flow rate of 1.5 mL/min. Mass spectra were recorded from 40–390 amu (atomic mass unit) with a scan time of 0.14 s.

2.12. Statistical Analysis

Statistical analysis was performed using GraphPad Prism version 7.00 (GraphPad, San Diego, CA, USA) and figures were generated in Microsoft Excel (Microsoft Corp., Redmond, WA, USA). Differences between treatment groups were evaluated using one-way ANOVA followed by Tukey's HSD test and $p < 0.05$ was considered significant.

3. Results

3.1. Antimicrobial Properties

The MICs, MFCs and MBCs of EOs and standard antibiotic drugs against yeasts and bacteria are presented in Table 1. The EO of AM did not show antimicrobial activity against the tested microorganisms below 2% (20 mg/mL). TPL and LM EOs demonstrated a broad-spectrum of weak-acid resistant anti-yeast and antibacterial activity. The range of anti-yeast and antibacterial MICs for TPL EO was 0.03–0.07 mg/mL and 0.22–0.42 mg/mL, respectively, and 0.07–0.31 mg/mL and 0.83–1.67 mg/mL for LM EO. However, the MICs of chloramphenicol against *S. aureus* and *E. coli* were 20.82 and 26.03 µg/mL, which is considerably lower than the MICs of LM (0.83 and 1.67 mg/mL) and TPL (0.22 and 0.42 mg/mL) EOs. Overall, TPL and LM EOs demonstrated anti-yeast and antibacterial activities, but were less potent than the standard antifungal agent, amphotericin B, and standard antibacterial agent, chloramphenicol. TPL EO had a stronger anti-yeast activity than LM EO, with MICs ranging from 0.03 to 0.07 mg/mL vs. 0.07 to 0.31 mg/mL. Furthermore, the MICs of TPL EO were equal to the MFCs against *C. albicans*, *S. cerevisiae*, *R. mucilaginosa*, *R. glutinis*, *S. pombe*, *D. anomala* and *D. bruxellensis*, but lower than the MFCs against *Z. rouxii*, *Z. bailii* and *C. krusei*. In the experiments with LM EO, the MIC and MFC was equal only for *C. albicans* and lower for all other yeasts (MIC < MFC).

Table 1. Minimum inhibitory concentration, minimum fungicidal concentration and minimum bactericidal concentration of Tasmanian pepper leaf and lemon myrtle essential oils, amphotericin B and chloramphenicol against bacteria and yeasts.

Microorganisms	TPL EO		LM EO		Amphotericin B		Chloramphenicol	
	MIC (mg/mL)	MBC/MFC (mg/mL)	MIC (mg/mL)	MBC/MFC (mg/mL)	MIC (µg/mL)	MFC (µg/mL)	MIC (µg/mL)	MBC (µg/mL)
E. coli	0.42 ± 0.02 [a]	0.42 ± 0.02	1.67 ± 0.07 [a]	3.33 ± 0.14	NT	NT	26.03 ± 1.81 [b]	52.08 ± 3.61
S. aureus	0.22 ± 0.01 [a]	0.30 ± 0.02	0.83 ± 0.04 [a]	0.83 ± 0.04	NT	NT	20.82 ± 1.81 [b]	41.67 ± 3.61
Z. rouxii	0.05 ± 0.002 [a]	0.08 ± 0.01	0.31 ± 0.001 [a]	0.52 ± 0.02	5.20 ± 0.45 [b]	7.80 ± 0.001	NT	NT
Z. bailii	0.07 ± 0.002 [a]	0.09 ± 0.01	0.26 ± 0.01 [a]	0.31 ± 0.001	13.02 ± 0.90 [b]	13.02 ± 0.90	NT	NT
C. albicans	0.03 ± 0.001 [a]	0.03 ± 0.001	0.26 ± 0.01 [a]	0.26 ± 0.01	41.67 ± 3.61 [b]	52.08 ± 3.61	NT	NT
S. cerevisiae	0.05 ± 0.003 [a]	0.05 ± 0.002	0.21 ± 0.01 [a]	0.26 ± 0.01	3.25 ± 0.23 [b]	7.80 ± 0.001	NT	NT
R. mucilaginosa	0.05 ± 0.003 [a]	0.05 ± 0.003	0.21 ± 0.01 [a]	0.42 ± 0.02	3.25 ± 0.23 [b]	6.50 ± 0.45	NT	NT
C. krusei	0.03 ± 0.001 [a]	0.04 ± 0.001	0.21 ± 0.01 [a]	0.26 ± 0.01	3.25 ± 0.23 [b]	3.90 ± 0.45	NT	NT
R. glutinis	0.03 ± 0.001 [a]	0.03 ± 0.001	0.21 ± 0.01 [a]	0.37 ± 0.02	2.60 ± 0.23 [b]	5.21 ± 0.45	NT	NT
S. pombe	0.07 ± 0.008 [a]	0.07 ± 0.008	0.16 ± 0.001 [a]	0.21 ± 0.01	2.60 ± 0.23 [b]	6.50 ± 0.45	NT	NT
D. anomala	0.03 ± 0.001 [a]	0.03 ± 0.001	0.11 ± 0.01 [a]	0.13 ± 0.005	2.93 ± 0.34 [b]	3.90 ± 0.001	NT	NT
D. bruxellensis	0.03 ± 0.001 [a]	0.03 ± 0.001	0.07 ± 0.002 [a]	0.11 ± 0.004	13.02 ± 0.90 [b]	13.02 ± 0.90	NT	NT

Data are means ± SD ($n = 3$). Results for anise myrtle are not shown due to the lack of inhibitory activity below 20 mg/mL. NT: not tested. Minimum inhibitory concentration (MIC) values between treatments sharing different letters (a, b) are significantly ($p < 0.05$) different in each row. Minimum fungicidal concentration (MFC), minimum bactericidal concentration (MBC).

3.2. Effect of Ergosterol on Yeast Cell Membranes

The ergosterol binding assay was performed to examine if the EOs of LM and TPL are binding to the sterol (ergosterol) of the yeast plasma membrane. The affinity of EOs for ergosterol can be determined by comparing the MICs of EOs against yeasts in the absence or presence of exogenous ergosterol. If the EOs bind to ergosterol, they will readily form a complex and prevent the membrane from utilizing or interacting with them. The changes in the MICs of TPL and LM EOs against yeasts after adding ergosterol at different concentrations (0, 25, 50 and 100 µg/mL) are presented in Figures 1 and 2. Increasing the exogenous ergosterol concentrations resulted in increased MICs for both EOs. The calculated correlation coefficient for LM EO ($R^2 = 0.771–0.996$) and TPL EO ($R^2 = 0.771–0.986$) showed a positive correlation between the added ergosterol concentrations and increased MICs, indicating a significant binding activity between the EOs and the added ergosterol (Table 2).

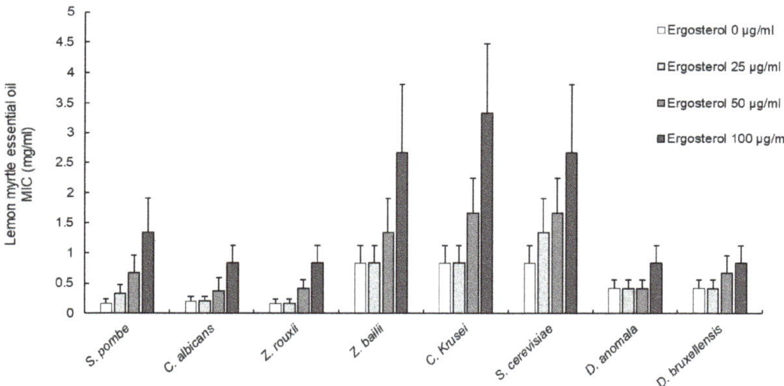

Figure 1. Changes in minimum inhibitory concentrations of lemon myrtle essential oil against yeasts in the absence and presence of exogenous ergosterol at different concentrations (comparative presentation). Data are means ± SD ($n = 3$).

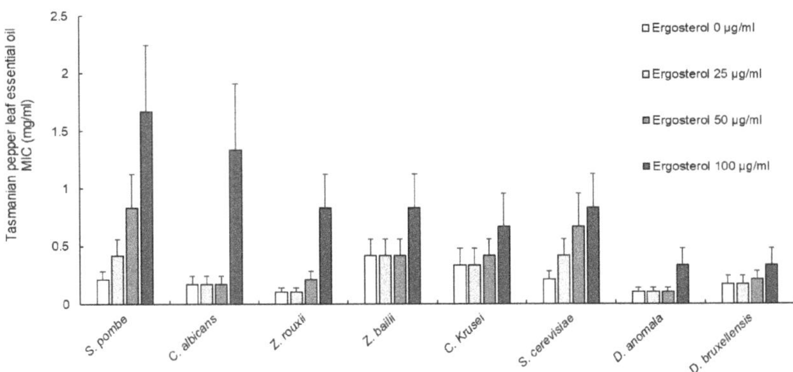

Figure 2. Changes in minimum inhibitory concentrations of Tasmanian pepper leaf essential oil against yeasts in the absence and presence of exogenous ergosterol at different concentrations (comparative presentation). Data are means $\pm$ SD ($n = 3$).

Table 2. Correlations between the added ergosterol (0–100 µg/mL) and corresponding minimum inhibitory concentrations of Tasmanian pepper leaf and lemon myrtle essential oils against yeasts.

		S. pombe	*C. albicans*	*Z. rouxii*	*Z. bailii*	*C. Krusei*	*S. cerevisiae*	*D. anomala*	*D. bruxellensis*
TPL EO and ergosterol	R^2	0.986 **	0.771 ns	0.861 ns	0.771 ns	0.910 *	0.938 *	0.771 ns	0.910 *
LM EO and ergosterol	R^2	0.986 **	0.916 *	0.940 *	0.918 *	0.933 *	0.996 **	0.771 ns	0.914 *

Values marked with an asterisk (*) indicate a correlation coefficient (R^2) significant at p value < 0.05 and with double asterisk (**) at p value < 0.01. Values marked with the letters (ns) indicate a non-significant correlation coefficient (R^2) at p value > 0.05. Correlations determined based on changes in minimum inhibitory concentrations of Tasmanian pepper leaf (TPL) and lemon myrtle (LM) essential oils (EOs) when mixed with different exogenous ergosterol concentrations of 0, 25, 50 and 100 µg/mL.

3.3. Effect of Sorbitol on Yeast Cell Wall

The MICs of LM and TPL EOs against yeasts did not change after the addition of sorbitol to the growth medium, indicating that both EOs did not affect the yeast cell wall (data not shown).

3.4. Quantification of Cellular Potassium and Magnesium Ions

Penetration and disruption of the yeast cell membrane by EOs will destroy the cellular lipid bilayer structure and increase membrane fluidity and permeability, which leads to ion leakage. The leakage of potassium and magnesium ions from *S. cerevisiae* cells after 0, 2, 4 and 6 h of exposure to TPL, LM and AM EOs are shown in Figure 3. The EOs of TPL and LM induced a significant (p < 0.05) leakage of potassium and magnesium ions from *S. cerevisiae* cells in a time-dependent manner. In contrast, AM EO caused only a slight non-significant (p > 0.05) leakage of potassium ions and did not affect the magnesium ions during the 6 h treatment period compared to the untreated control. Furthermore, the release of potassium ions from *S. cerevisiae* cells was considerably higher than that of magnesium ions. Overall, the observed significant leakage of potassium and magnesium ions from *S. cerevisiae* cells after the treatment with TPL and LM EOs indicate that both EOs interact with or damage the cytoplasmic membrane affecting its permeability for specific ions.

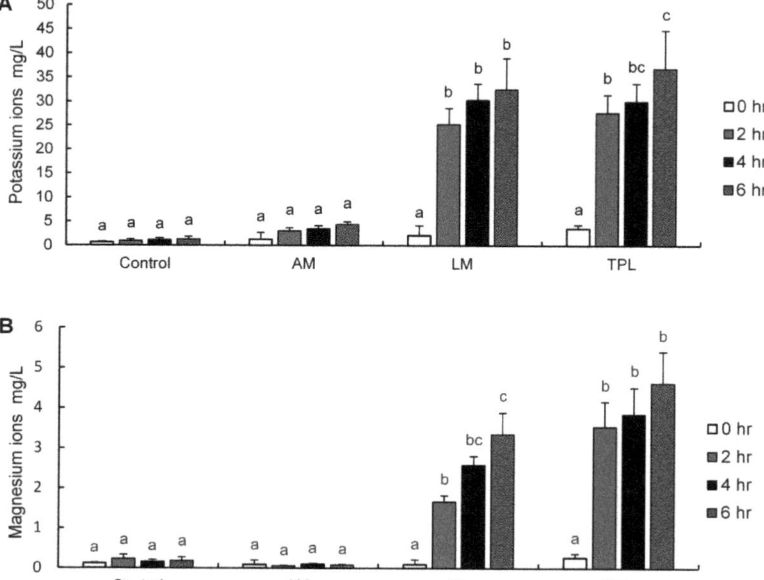

Figure 3. Leakage of potassium (**A**) and magnesium (**B**) ions from *S. cerevisiae* cells treated with anise myrtle, lemon myrtle and Tasmanian pepper leaf essential oils (0.5%, *v/v*) for 0, 2, 4 and 6 h. Data are means ± SD (*n* = 3); means with different letters (a, b, c) in the same column are significantly (*p* < 0.05) different.

3.5. Synergic Effects of EOs

The interactions between the EOs of TPL, LM and AM and their activity against nine yeasts produced different results, ranging from antagonistic to synergistic effects, depending on the EOs when combined at sub-inhibitory concentrations as well as the yeast (Table 3). The strongest synergistic effect was found between LM/TPL EOs effective against seven yeasts, followed by AM/TPL EOs effective against six yeasts and LM/AM EOs only effective against *R. mucilaginosa*. None of the EO combinations had an effect against *D. anomala*.

Table 3. Effects of essential oils against yeasts and food related bacteria using the checkerboard method.

Microorganisms	Combinations											
	TPL EO (A) + LM EO (B)				TPL EO (A) + AM EO (B)				LM EO (A) + AM EO (B)			
	MIC (*v/v*%)	FIC (*v/v*%)	FICI	Action	MIC (*v/v*%)	FIC (*v/v*%)	FICI	Action	MIC (*v/v*%)	FIC (*v/v*%)	FICI	Action
S. pombe	0.016 (A) 0.016 (B)	0.004 0.002	0.38	Synergy	0.016 (A) >2.0 (B)	0.008 0.008	0.5	Synergy	0.016 (A) >2.0 (B)	0.016 0.008	1	Additive
S. cerevisiae	0.031 (A) 0.031 (B)	0.008 0.001	0.28	Synergy	0.031 (A) >2.0 (B)	0.016 0.008	0.5	Synergy	0.031 (A) >2.0 (B)	0.031 0.008	1	Additive
D. anomala	0.002 (A) 0.008 (B)	0.002 0.008	1.95	Indifferent	0.002 (A) >2.0 (B)	0.004 0.0625	1.95	Indifferent	0.008 (A) >2.0 (B)	0.016 0.125	1.95	Indifferent
C. albicans	0.008 (A) 0.063 (B)	0.002 0.016	0.49	Synergy	0.008 (A) >2.0 (B)	0.004 0.063	0.49	Synergy	0.063 (A) >2.0 (B)	– –		Antagonistic
Z. bailii	0.031 (A) 0.031 (B)	0.008 0.002	0.31	Synergy	0.031 (A) >2.0 (B)	0.008 0.063	0.25	Synergy	0.031 (A) >2.0 (B)	0.031 0.0625	0.99	Additive
Z. rouxii	0.004 (A) 0.031 (B)	0.002 0.0005	0.5	Synergy	0.004 (A) >2.0 (B)	0.004 0.008	0.98	Additive	0.016 (A) >2.0 (B)	0.031 0.008	1	Additive
R. mucilaginosa	0.002 (A) 0.016 (B)	0.001 0.004	0.74	Additive	0.002 (A) >2.0 (B)	0.0005 0.031	0.24	Synergy	0.016 (A) >2.0 (B)	0.0005 0.125	0.03	Synergy
R. glutinis	0.008 (A) 0.016 (B)	0.002 0.0005	0.28	Synergy	0.008 (A) >2.0 (B)	0.004 0.008	0.49	Synergy	0.016 (A) >2.0 (B)	0.016 0.063	1	Additive
D. bruxellensis	0.002 (A) 0.008 (B)	0.0005 0.0005	0.31	Synergy	0.002 (A) >2.0 (B)	0.002 0.008	0.98	Additive	0.008 (A) >2.0 (B)	0.016 0.063	1.95	Indifferent

FIC: Fractional Inhibitory Concentration; FICI: Fractional Inhibitory Concentration Index; MIC: minimum inhibitory concentration; TPL: Tasmanian pepper leaf; LM: lemon myrtle; AM: anise myrtle; EO: essential oil.

3.6. Microscopy of Yeast Cell Structure (Untreated and Treated)

The untreated *S. cerevisiae* cells showed a normal oval-shaped cell structure with smooth and intact cell membranes without any signs of damage (Figure 4A,B and Figure 5A) Cells exposed to AM EO (Figure 5B) did not show any visible membrane damage, whereas cells exposed to LM EO (Figure 4C,D and Figure 5C) and TPL EO (Figure 4E,F and Figure 5D) had a deformed and disrupted cell structure that appeared swollen and expanded in size due to damage and lysis of the cell membrane. The organelles of untreated cells (Figure 4A,B) can be clearly seen within the cells' boundary, whereas organelles of damaged cells could not be seen within the cells due to their leakage through the disrupted membrane. The leaked organelles were visible outside the membrane-damaged cells in the bright-field microscopic images (Figure 4D,F). The scanning electron microscope produced detailed 3D images of pore formation on the membranes of cells exposed to LM (Figure 5C) and TPL (Figure 5D) EOs. Furthermore, the damaged cells had a wrinkled and uneven membrane with a larger round-shaped appearance compared to the untreated cells.

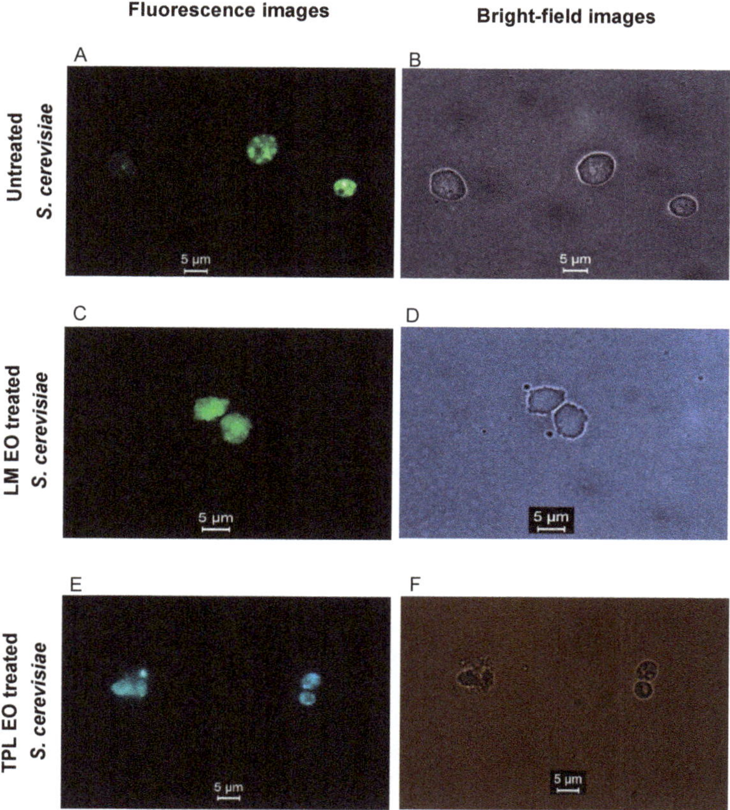

Figure 4. Fluorescence and bright-field electron microscopic images of untreated *S. cerevisiae* cells (**A,B**), and *S. cerevisiae* cells treated with essential oils from lemon myrtle (**C,D**) and Tasmanian pepper leaf (**E,F**).

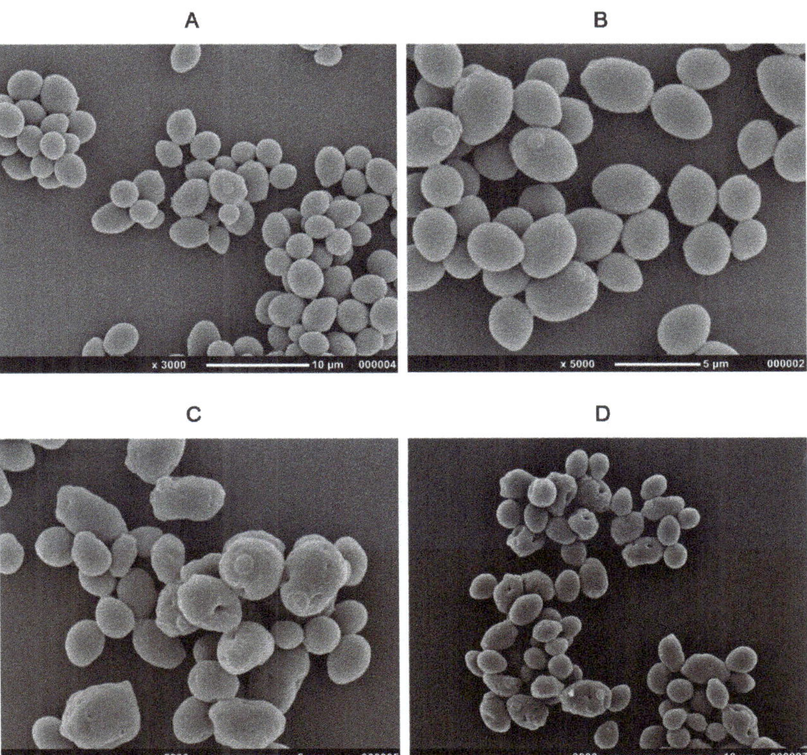

Figure 5. Scanning electron microscopic images of untreated *S. cerevisiae* cells (**A**), and *S. cerevisiae* cells treated with 0.5% (*v/v*) essential oils from anise myrtle (**B**), lemon myrtle (**C**) and Tasmanian pepper leaf (**D**).

3.7. Identification of the Main Bioactive Compounds in the Studied Eos

The following compounds could be identified as the main 'bioactives' in the three Eos by GC–MS (qualitative analysis only): anethole (11.70 min) and methyl chavicol (10.11 min) in AM EO (Figure 6A), polygodial (20.03) and guaiol (18.26 min) in TPL EO (Figure 6B), and neral (Z-isomer of citral; 10.84 min) and geranial (*E*-isomer of citral; 11.30 min) in LM EO (Figure 6C). Full chemical composition tests had been previously reported for commercially produced Eos of AM, TPL and LM and the results confirm our findings for the major identified bioactive compounds [18,22,31,33]. There are two chemotypes of AM that contain the same compounds but at different ratios, especially between anethole and its isomer methyl chavicol. The first chemotype of AM EO consists of anethole (94.97%), methyl chavicol (4.43%), α-pinene (0.09%), 1,8-Cineole (0.02%) and α-farnesene (0.07%), and the second chemotype of AM EO has methyl chavicol (77.54%), anethol (19.95%), 1,8-Cineole (0.80%), α-pinene (0.40%) and α-farnesene (0.11%) [31,33]. C ($C_{10}H_{12}O$) and its isomer methyl chavicol (estragole) are phenylpropane aromatic compounds which were reported to possess antibacterial and antifungal properties [20,41–47]. Our findings showed that the two major compounds in AM EO are similar to the first chemotype where anethole being the major component followed by its isomer methyl chavicol. A full chemical profiling of TPL EO showed the following sesquiterpene ($C_{15}H_{24}$) compounds: polygodial (36.74%), guaiol (4.46%), calamenene (3.42%), spathulenol (1.94%), drimenol (1.91%), cadina-1,4-diene (1.58%), 5-hydroxycalamenene (1.47%) bicyclogerma-crene (1.15%), α-cubebene (0.88%), β-caryophyllene (0.87%), α-copaene (0.48%), cadalene (0.44%), d-cadinol (0.40%), elemol (0.39%), T-muurolol (0.39%), germacrene-D (0.33%), and other sesquiterpenes compounds at levels below 0.16%, camphene, α-gurjunene, and β-cubebene [22]. Our results of GC–MS qualitative analysis were in agreement with Menary et al. [22] in which polygodial was the main compound in TPL EO followed by guaiol. Despite the fact that TPL EO contains some miner compounds in addition to its major constituent polygodial, most reports suggest that polygodial is the main contributor to the antibacterial and antifungal properties [18–20,23,48–54]. Plant Eos often contain several compounds where the antimicrobial activity arises from their major component; however, the minor compounds could form a complex synergistic interaction with the major compound, enhancing its antimicrobial action [55–57]. Previous chemical investigations reveal that LM EO predominantly (82–91%) consists of citral ($C_{10}H_{16}O$) which is a monoterpene aldehyde that exists in two isomers geranial (E-isomer or citral A) and neral (Z-isomer or citral B) [30,58]. According to three separate chemical profiling investigations of LM, EO showed the following constituents: geranial (45–49%), neral (37–42%), 5-Hepten-2-one,6-methyl (1.54–13.82%), 2,3-Dehydro-1,8-cineole (3.52%), nerol (2.66%), Germacrene B (0.2–2.18%), geraniol (0.8–1.26%), linalool (0.5–5.85%), myrcene (0.4–4.39%), and citronellal (0.25–2.19%) [30,59,60]. Our study showed the two major compounds in the tested LM EO are geranial and neral which is supported by the results reported by Forbes-Smith and Paton [30], Southwell et al. [59], and Pengelly [60]. The two major compounds geranial and neral (citral) contribute to the antimicrobial and antifungal activities in LM EO.

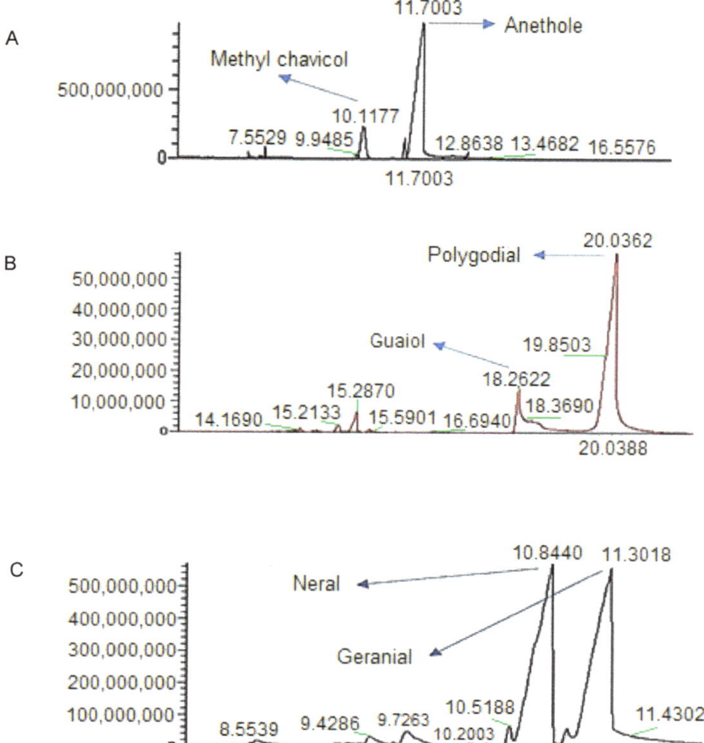

Figure 6. GC–MS chromatograms of anise myrtle (**A**), Tasmanian pepper leaf (**B**) and lemon myrtle (**C**) essential oils showing their main bioactive compounds.

4. Discussion

The EO of AM showed no antimicrobial activity against the tested microorganisms below 20 mg/mL. Brophy and Boland [33] reported that the EO of AM can occur in two different chemotypes having the same chemical composition but at different concentrations: chemotype 1 contained anethole as the main bioactive compound with a relative concentration of 93 to 95% and methyl chavicol between 4.4 and 5.6%, whereas chemotype 2 contained methyl chavicol as the main bioactive compound (66 to 77%) and anethole as the minor one (20 to 33%). The bioactive compound profile of AM EO in the present study is similar to chemotype 1 reported by Brophy and Boland, with anethole as the main bioactive compound followed by methyl chavicol at a much lower level. Wilkinson and Cavanagh [61] evaluated the antimicrobial activity of two AM EO samples: the first sample showed no activity against *E. coli* and *C. albicans*, while the second sample had an inhibitory effect against *E. coli*, *S. aureus* and *C. albicans*. In addition, a study by Hood et al. [62] reported antimicrobial activity of AM EO against *E. coli* and *S. aureus*, whereas Nirmal et al. [63] could not observe any activity against these bacteria. The controversial findings using AM EOs as an antimicrobial agent (effective vs. non-effective) are most likely caused by the reported significant variation in the content and ratio of anethole and methyl chavicol.

In contrast to the AM EOs, four different types of LM EOs studied by Wilkinson et al. [64] showed all a strong antibacterial and antifungal activity, which was positively correlated with citral, the main bioactive compound in these four EOs. Unlike the standard antibiotic drugs which possess either an antifungal or antibacterial effect, the tested EOs of LM and TPL had both antibacterial and anti(weak-acid resistant)-yeast activity. Polygodial, the main bioactive compound in TPL EO, was previously reported [65] to possess a stronger antifungal activity than the standard antifungal drug amphotericin B and also showed synergistic effects with antimicrobial drugs [53]. A study by Sultanbawa et al. [66] evaluated the inhibitory activity of LM EO and reported MICs of 0.313% against *E. coli* and 0.156% against *S. aureus* (complete inhibition after 22 h). In the present study, LM EO completely inhibited the growth of *E. coli* and *S. aureus* after 24 h at MBCs of 0.167% (1.67 mg/mL) for *E. coli* and 0.083% (0.83 mg/mL) for *S. aureus*. Both studies revealed that Gram-positive bacteria were more susceptible to LM EOs than Gram-negative bacteria. Interestingly, the bacterial inhibition concentration of LM EO that prevented the growth of *E. coli* in the present study was almost the same as that reported by Sultanbawa et al. [66] (0.333 vs. 0.313%). Overall, the results of both studies can be used to establish an effective concentration range for LM EO against *E. coli*.

Furthermore, the MICs in the present study were similar to the MBCs/MFCs of TPL EO against *D. bruxellensis*, *D. anomala*, *S. pombe*, *R. glutinis*, *S. cerevisiae*, *C. albicans* and *E. coli* and for LM EO against *C. albicans* and *S. aureus*, which clearly indicates that both EOs exhibited a biocidal activity at their MIC levels. The pronounced biocidal activity of TPL EO against bacteria and yeasts, observed in the present study and many other studies, was attributed to polygodial, the main bioactive compound in TPL EO [20,21,48–54]. However, similar bactericidal and fungicidal properties have also been reported for LM EO and its main bioactive compound citral [60,64,66–72]. It should be mentioned that most available antifungal drugs possess only fungistatic activity and there is an increasing fungal resistance toward these drugs. Therefore, EOs from TPL and LM may offer an alternative strategy to tackle food spoilage offering both fungicidal and bactericidal activity [20,21,49–54,63,68–71,73–76].

The fluorescence and scanning electron microscopic images of yeast cells treated with TPL and LM EOs revealed the fungicidal "mode of action" of these two EOs. Polygodial and citral have been reported to cause structural disruption of cell membranes, increase membrane permeability, and form channels or lesions in cytoplasmic membranes, causing leakage of intracellular components and finally cell death [53,54,77]. The observed yeast cell death in the present study was caused by multiple effects: cell membrane lysis, membrane pore formation and cell content leakage.

However, the addition of exogenous ergosterol had a diminishing effect on the antimicrobial activity of TPL and LM EOs against yeasts, resulting in increased MICs. Interactions between ergosterol and both EOs could be shown in the ergosterol binding assay. This interaction(s) can be described best as the binding of EOs (and their bioactive compounds) to ergosterol and thereby inhibiting or diminishing its crucial biological function in the cell membrane. Ergosterol is the main sterol component of fungal cell membranes and responsible for membrane rigidity, fluidity and permeability, contributing to the functionality of membrane bound enzymes and transporters [78]. Furthermore, the binding of (bioactive) compounds to ergosterol in yeast cell membranes is also affecting the integrity and fluidity of the membrane and can form microspores or channels where ions and cell content can leak through, eventually resulting in cell death. The mechanism of "killing" fungal cells by binding to ergosterol in the cell membrane has also been reported for other EOs and their bioactive compounds, such as thymol from *Thymus vulgaris*, carvacrol, geraniol, nerol and *Coriandrum sativum* leaf oil [35,79–81].

The sorbitol osmotic protection assay was performed to test potential effects of TPL and LM EOs on the integrity of the yeast cell wall. Increased MICs after the addition of sorbitol (0.8 M) indicate an interaction between the EOs and the fungal cell wall. However, the MICs of TPL and LM EOs did not change after the addition of exogenous sorbitol to the test media which is a clear indication that the yeast cell wall was not affected by these two EOs. Our findings agree with literature reports about unchanged MICs of geraniol, thymol, nerol, eugenol, menthol and terpinen-4-ol after the addition of 0.8 M sorbitol [80,82–84].

5. Conclusions

The present study could clearly demonstrate the antifungal activity of Australian native TPL and LM EOs against yeasts that are resistant to weak-acid preservatives and commonly cause serious spoilage problems for the beverage industry. Based on their "anti-yeast" effects, these EOs have the potential to be used as natural preservatives that could substitute, partly or totally, their synthetic counterparts in combating the increasing hazards of resistant yeasts to protect both the food industry and consumers.

Author Contributions: Conceptualization, F.A., R.M., S.W. and Y.S.; methodology, F.A.; software, F.A.; validation, R.M., S.W. and Y.S.; formal analysis, F.A.; investigation, F.A.; resources, R.M., S.W. and Y.S.; data curation, F.A.; writing—original draft preparation, F.A.; writing—review and editing, R.M., S.W., M.E.N. and Y.S.; visualization, F.A.; supervision, R.M., S.W. and Y.S.; project administration, Y.S.; funding acquisition, F.A. and Y.S. All authors have read and agreed to the published version of the manuscript.

Funding: This research was funded by the Public Authority for Applied Education and Training, Kuwait and the Queensland Alliance for Agriculture and Food Innovation, The University of Queensland, Brisbane, QLD, Australia.

Institutional Review Board Statement: Not applicable.

Informed Consent Statement: Not applicable.

Data Availability Statement: All relevant data are reported in the paper; the corresponding author, Yasmina Sultanbawa, should be contacted for furhter details.

Acknowledgments: The authors thank Brian Burren, Kinnari Shelat, Andrew Cusack and Margaret Currie for their expert technical assistance.

Conflicts of Interest: The authors declare no conflict of interest. The funders had no role in the design of the study; in the collection, analyses, or interpretation of data; in the writing of the manuscript, or in the decision to publish the results.

References

1. Burt, S. Essential oils: Their antibacterial properties and potential applications in foods—A review. *Int. J. Food Microbiol.* **2004**, *94*, 223–253. [CrossRef]
2. Rhind, J. From Historical Origins to the Present Day: The Philosophies of the Pioneers and Influential Thinkers. In *Essential Oils: A Handbook for Aromatherapy Practice*; Singing Dragon: London, UK; Philadelphia, PA, USA, 2012; Volume 2, pp. 12–23.
3. Bakkali, F.; Averbeck, S.; Averbeck, D.; Idaomar, M. Biological effects of essential oils—A review. *Food Chem. Toxicol.* **2008**, *46*, 446–475. [CrossRef] [PubMed]

4. Shaaban, H.A.E.; Shibamoto, T.; El-Ghorab, A.H. Bioactivity of essential oils and their volatile aroma components: Review. *J. Essent. Oil Res.* **2012**, *24*, 203. [CrossRef]

5. Van de Braak, S.; Leijten, G. *Essential Oils and Oleoresins: A Survey in The Netherlands and Other Major Markets in the European Union*; CBI, Centre for the Promotion of Imports from Developing Countries: Rotterdam, The Netherlands, 1994.

6. Ndagijimana, M.; Belletti, N.; Lanciotti, R.; Guerzoni, M.; Ardini, F. Effect of Aroma Compounds on the Microbial Stabilization of Orange-based Soft Drinks. *J. Food Sci.* **2004**, *69*, SNQ20–SNQ24. [CrossRef]

7. Tiwari, B.K.; Valdramidis, V.P.; O'Donnell, C.P.; Muthukumarappan, K.; Bourke, P.; Cullen, P.J. Application of natural antimicrobials for food preservation. *J. Agric. Food Chem.* **2009**, *57*, 5987. [CrossRef]

8. Tajkarimi, M.M.; Ibrahim, S.A.; Cliver, D.O. Antimicrobial herb and spice compounds in food. *Food Control* **2010**, *21*, 1199–1218. [CrossRef]

9. Saad, N.Y.; Muller, C.D.; Lobstein, A. Major bioactivities and mechanism of action of essential oils and their components. *Flavour Fragr. J.* **2013**, *28*, 269–279. [CrossRef]

10. Gyawali, R.; Ibrahim, S.A. Natural products as antimicrobial agents. *Food Control* **2014**, *46*, 412–429. [CrossRef]

11. Stratford, M.; James, S.A. Non-alcoholic beverages and yeasts. In *Yeasts in Food: Beneficial and Detrimental Aspects*; Boekhout, T., Robert, V., Eds.; Woodhead Publishing Ltd.: Cambridge, UK, 2003; pp. 309–345.

12. Stratford, M.; Steels, H.; Nebe-von-Caron, G.; Novodvorska, M.; Hayer, K.; Archer, D.B. Extreme resistance to weak-acid preservatives in the spoilage yeast Zygosaccharomyces bailii. *Int. J. Food Microbiol.* **2013**, *166*, 126–134. [CrossRef] [PubMed]

13. Divol, B.; du Toit, M.; Duckitt, E. Surviving in the presence of sulphur dioxide: Strategies developed by wine yeasts. *Appl. Microbiol. Biotechnol.* **2012**, *95*, 601–613. [CrossRef]

14. Pitt, J.I.; Hocking, A.D. Spoilage of stored, processed and preserved foods. In *Fungi and Food Spoilage*, 3rd ed.; Springer: New York, NY, USA, 2009; pp. 401–419.

15. Hayashi, M.; Naknukool, S.; Hayakawa, S.; Ogawa, M.; Ni'matulah, A.-B.A. Enhancement of antimicrobial activity of a lactoperoxidase system by carrot extract and β-carotene. *Food Chem.* **2012**, *130*, 541–546. [CrossRef]

16. Newman, D.J.; Cragg, G.M. Natural products as sources of new drugs over the 30 years from 1981 to 2010. *J. Nat. Prod.* **2012**, *75*, 311–335. [CrossRef]

17. Krisch, J.; Rentsenkhand, T.; Vágvölgyi, C. Essential oils against yeast and moulds causing food spoilage. In *Science against Microbial Pathogens: Communicating Current Research and Technological Advances Microbiology*; Méndez-Vilas, A., Ed.; Formatex Research Center: Badajoz, Spain, 2011; pp. 1135–1142.

18. Pengelly, A. Indigenous and naturalised herbs: *Tasmannia lanceolata*: Mountain pepper. *Aust. J. Med. Herbal.* **2002**, *14*, 71–74.

19. Clarke, M. *Australian Native Food Industry Stocktake*; RIRDC Publication No. 12/066; Union Offset Printing: Canberra, Australia, 2012.

20. Fujita, K.-I.; Kubo, I. Naturally occurring antifungal agents against Zygosaccharomyces bailii and their synergism. *J. Agric. Food Chem.* **2005**, *53*, 5187–5191. [CrossRef]

21. Kubo, I.; Fujita, K.I.; Lee, S.H.; Ha, T.J. Antibacterial activity of polygodial. *Phytother. Res.* **2005**, *19*, 1013–1017. [CrossRef]

22. Menary, R.C.; Dragar, V.A.; Thomas, S.; Read, C.D. *Mountain Pepper Extract, Tasmannia Lanceolata: Quality Stabilisation and Registration: A Report for the Rural Industries Research and Development Corporation*; RIRDC: Barton, Australia, 2003.

23. Sultanbawa, Y. Tasmanian Pepper Leaf (*Tasmannia lanceolata*) Oils. In *Essential Oils in Food Preservation, Flavor and Safety*; Preedy, V.R., Ed.; Elsevier: Amsterdam, The Netherlands, 2016; pp. 819–823.

24. Cock, I.E. The phytochemistry and chemotherapeutic potential of *Tasmannia lanceolata* (Tasmanian pepper): A review. *Pharmacogn. Commun.* **2013**, *3*, 13–25.

25. Buchaillot, A.; Caffin, N.; Bhandari, B. Drying of lemon myrtle (*Backhousia citriodora*) leaves: Retention of volatiles and color. *Dry. Technol.* **2009**, *27*, 445–450. [CrossRef]

26. Jones, G.L. Asia and Australia–Essentially a Well Oiled Connection. In *Aromatic Plants from Asia, their Chemistry and Application in Food and Therapy*; Jirovetz, L., Dung, N.X., Varshney, V.K., Eds.; Har Krishnan Bhalla: Dehradun, India, 2007; p. 247.

27. Scheman, A.; Scheman, N.; Rakowski, E.-M. European directive fragrances in natural products. *Dermatitis* **2014**, *25*, 51–55. [CrossRef]

28. Ress, N.; Hailey, J.; Maronpot, R.; Bucher, J.; Travlos, G.; Haseman, J.; Orzech, D.; Johnson, J.; Hejtmancik, M. Toxicology and carcinogenesis studies of microencapsulated citral in rats and mice. *Toxicol. Sci.* **2003**, *71*, 198–206. [CrossRef]

29. Sultanbawa, Y. Lemon Myrtle (Backhousia citriodora) Oils. In *Essential Oils in Food Preservation, Flavor and Safety*; Preedy, V.R., Ed.; Elsevier: Amsterdam, The Netherlands, 2016; pp. 517–520.

30. Forbes-Smith, M.; Paton, J. Innovative products from Australian native foods. In *Rural Industries Research and Development Corporation*; RIRDC: Barton, Australia, 2002; pp. 1–78.

31. Blewitt, M.; Southwell, I.A. Backhousia anisata Vickery, an alternative source of (E)-anethole. *J. Essent. Oil Res.* **2000**, *12*, 445–454. [CrossRef]

32. Konczak, I.; Zabaras, D.; Dunstan, M.; Aguas, P. Antioxidant capacity and phenolic compounds in commercially grown native Australian herbs and spices. *Food Chem.* **2010**, *122*, 260–266. [CrossRef]

33. Brophy, J.J.; Boland, D.J. The leaf essential oil of two chemotypes of Backhousia anisata Vickery. *Flavour Fragr. J.* **1991**, *6*, 187–188. [CrossRef]

34. Ahmad, A.; Viljoen, A. The in vitro antimicrobial activity of Cymbopogon essential oil (lemon grass) and its interaction with silver ions. *Phytomedicine* **2015**, *22*, 657–665. [CrossRef]

35. Lima, I.O.; Pereira, F.D.O.; Oliveira, W.A.D.; Lima, E.D.O.; Menezes, E.A.; Cunha, F.A.; Diniz, M.D.F.F.M. Antifungal activity and mode of action of carvacrol against Candida albicans strains. *J. Essent. Oil Res.* **2013**, *25*, 138–142. [CrossRef]

36. Frost, D.J.; Brandt, K.D.; Cugier, D.; Goldman, R. A whole-cell Candida albicans assay for the detection of inhibitors towards fungal cell wall synthesis and assembly. *J. Antibiot.* **1995**, *48*, 306–310. [CrossRef]

37. Prashar, A.; Hili, P.; Veness, R.G.; Evans, C.S. Antimicrobial action of palmarosa oil (Cymbopogon martinii) on *Saccharomyces cerevisiae*. *Phytochemistry* **2003**, *63*, 569–575. [CrossRef]

38. Zengin, H.; Baysal, A.H. Antibacterial and antioxidant activity of essential oil terpenes against pathogenic and spoilage-forming bacteria and cell structure-activity relationships evaluated by SEM microscopy. *Molecules* **2014**, *19*, 17773–17798. [CrossRef]

39. Shimada, S.; Andou, M.; Naito, N.; Yamada, N.; Osumi, M.; Hayashi, R. Effects of hydrostatic pressure on the ultrastructure and leakage of internal substances in the yeast Saccharomyces cerevisiae. *Appl. Microbiol. Biotechnol.* **1993**, *40*, 123–131. [CrossRef]

40. Chaliha, M.; Cusack, A.; Currie, M.; Sultanbawa, Y.; Smyth, H. Effect of packaging materials and storage on major volatile compounds in three Australian native herbs. *J. Agric. Food. Chem.* **2013**, *61*, 5738–5745. [CrossRef] [PubMed]

41. Karapinar, M.; Aktuğ, Ş.E. Inhibition of foodborne pathogens by thymol, eugenol, menthol and anethole. *Int. J. Food Microbiol.* **1987**, *4*, 161–166. [CrossRef]

42. Hançer Aydemir, D.; Cifci, G.; Aviyente, V.; Boşgelmez-Tinaz, G. Quorum-sensing inhibitor potential of trans-anethole aganist Pseudomonas aeruginosa. *J. Appl. Microbiol.* **2018**, *125*, 731–739. [CrossRef]

43. Dąbrowska, M.; Zielińska-Bliźniewska, H.; Kwiatkowski, P.; Łopusiewicz, Ł.; Pruss, A.; Kostek, M.; Kochan, E.; Sienkiewicz, M. Inhibitory Effect of Eugenol and trans-Anethole Alone and in Combination with Antifungal Medicines on Candida albicans Clinical Isolates. *Chem. Biodivers.* **2021**, *18*, e2000843. [CrossRef]

44. Kubo, I.; Fujita, K.I.; Nihei, K.I. Antimicrobial activity of anethole and related compounds from aniseed. *J. Sci. Food Agric.* **2008**, *88*, 242–247. [CrossRef]

45. Huang, Y.; Zhao, J.; Zhou, L.; Wang, J.; Gong, Y.; Chen, X.; Guo, Z.; Wang, Q.; Jiang, W. Antifungal activity of the essential oil of Illicium verum fruit and its main component trans-anethole. *Molecules* **2010**, *15*, 7558–7569. [CrossRef] [PubMed]

46. Naksang, P.; Tongchitpakdee, S.; Thumanu, K.; Oruna-Concha, M.J.; Niranjan, K.; Rachtanapun, C. Assessment of antimicrobial activity, mode of action and volatile compounds of Etlingera pavieana essential oil. *Molecules* **2020**, *25*, 3245. [CrossRef]

47. Lis-Balchin, M.; Deans, S.G.; Eaglesham, E. Relationship between bioactivity and chemical composition of commercial essential oils. *Flavour Fragr. J.* **1998**, *13*, 98–104. [CrossRef]

48. Kubo, I.; Himejima, M. Anethole, a synergist of polygodial against filamentous microorganisms. *J. Agric. Food Chem.* **1991**, *39*, 2290–2292. [CrossRef]

49. Kubo, I.; Fujita, K.I.; Lee, S.H. Antifungal mechanism of polygodial. *J. Agric. Food Chem.* **2001**, *49*, 1607–1611. [CrossRef] [PubMed]

50. Machida, K.; Tanaka, T.; Taniguchi, M. Depletion of glutathione as a cause of the promotive effects of polygodial, a sesquiterpene on the production of reactive oxygen species in Saccharomyces cerevisiae. *J. Biosci. Bioeng.* **1999**, *88*, 526–530. [CrossRef]

51. Himejima, M.; Kubo, I. Fungicidal activity of polygodial in combination with anethole and indole against Candida albicans. *J. Agric. Food Chem.* **1993**, *41*, 1776–1779. [CrossRef]

52. Fujita, K.-I.; Kubo, I. Multifunctional action of antifungal polygodial against Saccharomyces cerevisiae: Involvement of pyrrole formation on cell surface in antifungal action. *Bioorg. Med. Chem.* **2005**, *13*, 6742–6747. [CrossRef] [PubMed]

53. Kubo, I.; Taniguchi, M. Polygodial, an antifungal potentiator. *J. Nat. Prod.* **1988**, *51*, 22–29. [CrossRef]

54. Yano, Y.; Taniguchi, M.; Tanaka, T.; Oi, S.; Kubo, I. Protective effects of Ca^{2+} on cell membrane damage by polygodial in Saccharomyces cerevisiae. *Agric. Biol. Chem.* **1991**, *55*, 603–604. [CrossRef]

55. Bassolé, I.H.N.; Lamien-Meda, A.; Bayala, B.; Tirogo, S.; Franz, C.; Novak, J.; Nebié, R.C.; Dicko, M.H. Composition and Antimicrobial Activities of *Lippia multiflora* Moldenke, *Mentha* x *piperita* L. and *Ocimum basilicum* L. Essential Oils and Their Major Monoterpene Alcohols Alone and in Combination. *Molecules* **2010**, *15*, 7825. [CrossRef]

56. Bassolé, I.H.N.; Juliani, H.R. Essential oils in combination and their antimicrobial properties. *Molecules* **2012**, *17*, 3989–4006. [CrossRef]

57. Delaquis, P.J.; Stanich, K.; Girard, B.; Mazza, G. Antimicrobial activity of individual and mixed fractions of dill, cilantro, coriander and eucalyptus essential oils. *Int. J. Food Microbiol.* **2002**, *74*, 101–109. [CrossRef]

58. Kuwahara, Y.; Suzuki, H.; Matsumoto, K.; Wada, Y. Pheromone study on acarid mites. XI. Function of mite body as geometrical isomerization and reduction of citral (the alarm pheromone). *Appl. Entomol. Zool.* **1983**, *18*, 30–39. [CrossRef]

59. Southwell, I.A.; Russell, M.; Smith, R.L.; Archer, D.W. *Backhousia citriodora* F. Muell. (Myrtaceae), a superior source of citral. *J. Essent. Oil Res.* **2000**, *12*, 735–741. [CrossRef]

60. Pengelly, A. Antimicrobial activity of lemon myrtle and tea tree oils. *Aust. J. Med. Herbal.* **2003**, *15*, 9.

61. Wilkinson, J.M.; Cavanagh, H. Antibacterial activity of essential oils from Australian native plants. *Phytother. Res.* **2005**, *19*, 643–646. [CrossRef]

62. Hood, J.R.; Wilkinson, J.M.; Cavanagh, H.M. Evaluation of common antibacterial screening methods utilized in essential oil research. *J. Essent. Oil Res.* **2003**, *15*, 428–433. [CrossRef]

63. Nirmal, N.P.; Mereddy, R.; Li, L.; Sultanbawa, Y. Formulation, characterisation and antibacterial activity of lemon myrtle and anise myrtle essential oil in water nanoemulsion. *Food Chem.* **2018**, *254*, 1–7. [CrossRef] [PubMed]

64. Wilkinson, J.M.; Hipwell, M.; Ryan, T.; Cavanagh, H.M. Bioactivity of Backhousia citriodora: Antibacterial and antifungal activity. *J. Agric. Food Chem.* **2003**, *51*, 76–81. [CrossRef]

65. Andres, M.-I.; Forsby, A.; Walum, E. Polygodial-induced noradrenaline release in human neuroblastoma SH-SY5Y cells. *Toxicol. Vitr.* **1997**, *11*, 509–511. [CrossRef]

66. Sultanbawa, Y.; Cusack, A.; Currie, M.; Davis, C. An innovative microplate assay to facilitate the detection of antimicrobial activity in plant extracts. *J. Rapid Methods Autom. Microbiol.* **2009**, *17*, 519–534. [CrossRef]

67. Aoudou, Y.; Léopold, T.N.; Michel, J.D.P.; Franccedil, E.; Moses, M.C. Antifungal properties of essential oils and some constituents to reduce foodborne pathogen. *J. Yeast Fungal Res.* **2010**, *1*, 001–008.

68. Wuryatmo, E.; Klieber, A.; Scott, E.S. Inhibition of citrus postharvest pathogens by vapor of citral and related compounds in culture. *J. Agric. Food Chem.* **2003**, *51*, 2637–2640. [CrossRef]

69. Wannissorn, B.; Jarikasem, S.; Soontorntanasart, T. Antifungal activity of lemon grass oil and lemon grass oil cream. *Phytother. Res.* **1996**, *10*, 551–554. [CrossRef]

70. Onawunmi, G.O. Evaluation of the antimicrobial activity of citral. *Lett. Appl. Microbiol.* **1989**, *9*, 105–108. [CrossRef]

71. Silva, C.d.B.d.; Guterres, S.S.; Weisheimer, V.; Schapoval, E.E. Antifungal activity of the lemongrass oil and citral against *Candida* spp. *Braz. J. Infect. Dis.* **2008**, *12*, 63–66. [CrossRef]

72. Marques, A.M.; Lima, C.H.; Alviano, D.S.; Esteves, R.L.; Kaplan, M.A.C. Traditional use, chemical composition and antimicrobial activity of Pectis brevipedunculata essential oil: A correlated lemongrass species in Brazil. *Emir. J. Food Agric.* **2013**, *25*, 798–808. [CrossRef]

73. McCallion, R.F.; Cole, A.; Walker, J.; Blunt, J.; Munro, M. Antibiotic substances from New Zealand plants. *Planta Med.* **1982**, *44*, 134–138. [CrossRef]

74. Lim, T.K. Tasmannia lanceolata. In *Edible Medicinal and Non-Medicinal Plants: Fruits*; Springer: Dordrecht, The Netherlands, 2013; Volume 6, pp. 493–499.

75. Kubo, I.; Fujita, K.I. Naturally occurring anti-Salmonella agents. *J. Agric. Food Chem.* **2001**, *49*, 5750–5754. [CrossRef]

76. Banihashemi, Z.; Abivardi, C. Evaluation of fungicidal and fungistatic activity of plant essential oils towards plant pathogenic and saprophytic fungi. *Phytopathol. Mediterr.* **2011**, *50*, 245–256.

77. Taniguchi, M.; Yano, Y.; Tada, E.; Ikenishi, K.; Oi, S.; Haraguchi, H.; Hashimoto, K.; Kubo, I. Mode of action of polygodial, an antifungal sesquiterpene dialdehyde. *Agric. Biol. Chem.* **1988**, *52*, 1409–1414.

78. Abe, F.; Hiraki, T. Mechanistic role of ergosterol in membrane rigidity and cycloheximide resistance in Saccharomyces cerevisiae. *Biochim. Biophys. Acta (BBA) Biomembr.* **2009**, *1788*, 743–752. [CrossRef]

79. de Lira Mota, K.S.; de Oliveira Pereira, F.; de Oliveira, W.A.; Lima, I.O.; de Oliveira Lima, E. Antifungal activity of Thymus vulgaris L. essential oil and its constituent phytochemicals against Rhizopus oryzae: Interaction with ergosterol. *Molecules* **2012**, *17*, 14418–14433. [CrossRef]

80. Miron, D.; Battisti, F.; Silva, F.K.; Lana, A.D.; Pippi, B.; Casanova, B.; Gnoatto, S.; Fuentefria, A.; Mayorga, P.; Schapoval, E.E. Antifungal activity and mechanism of action of monoterpenes against dermatophytes and yeasts. *Rev. Bras. Farmacogn.* **2014**, *24*, 660–667. [CrossRef]

81. de Almeida Freires, I.; Murata, R.M.; Furletti, V.F.; Sartoratto, A.; de Alencar, S.M.; Figueira, G.M.; de Oliveira Rodrigues, J.A.; Duarte, M.C.T.; Rosalen, P.L. *Coriandrum sativum* L.(coriander) essential oil: Antifungal activity and mode of action on Candida spp., and molecular targets affected in human whole-genome expression. *PLoS ONE* **2014**, *9*, e99086.

82. de Castro, R.D.; de Souza, T.M.P.A.; Bezerra, L.M.D.; Ferreira, G.L.S.; de Brito Costa, E.M.M.; Cavalcanti, A.L. Antifungal activity and mode of action of thymol and its synergism with nystatin against Candida species involved with infections in the oral cavity: An in vitro study. *BMC Complement. Altern. Med.* **2015**, *15*, 417. [CrossRef] [PubMed]

83. Rajkowska, K.; Nowak, A.; Kunicka-Styczyńska, A.; Siadura, A. Biological effects of various chemically characterized essential oils: Investigation of the mode of action against Candida albicans and HeLa cells. *RSC Adv.* **2016**, *6*, 97199–97207. [CrossRef]

84. Leite, M.C.; de Brito Bezerra, A.P.; de Sousa, J.P.; de Oliveira Lima, E. Investigating the antifungal activity and mechanism(s) of geraniol against Candida albicans strains. *Med. Mycol.* **2015**, *53*, 275–284. [CrossRef] [PubMed]

Review

Antioxidant and Antimicrobial Properties of Selected Phytogenics for Sustainable Poultry Production

Caven M. Mnisi [1,2,*], Victor Mlambo [3], Akho Gila [1,2], Allen N. Matabane [1,2], Doctor M. N. Mthiyane [1,2], Cebisa Kumanda [4], Freddy Manyeula [5] and Christian S. Gajana [6]

1   Food Security and Safety Focus Area, Faculty of Natural and Agricultural Sciences, North-West University, Mmabatho 2735, South Africa
2   Department of Animal Science, School of Agricultural Sciences, Faculty of Natural and Agricultural Sciences, North-West University, Mmabatho 2735, South Africa
3   School of Agricultural Sciences, Faculty of Agriculture and Natural Sciences, University of Mpumalanga, Mbombela 1200, South Africa
4   Department of Animal Sciences, University of Pretoria, Pretoria 0028, South Africa
5   Department of Agriculture and Animal Health, College of Agriculture and Environmental Science, University of South Africa, Florida 1710, South Africa
6   Department of Livestock and Pasture Sciences, University of Fort Hare, Alice 5700, South Africa
*   Correspondence: 23257539@nwu.ac.za; Tel.: +27-18-389-2738

Abstract: The use of antibiotic growth promoters (AGP) in poultry production not only promotes the emergence of pathogenic multi-drug resistant bacteria, but it also compromises product quality, threatens animal and human health, and pollutes the environment. However, the complete withdrawal of AGP without alternatives could result in uncontrollable disease outbreaks that would jeopardize large-scale poultry intensification. Thus, the use of phytogenic products as potential alternatives to in-feed AGP has attracted worldwide research interest. These phytogenic products contain numerous biologically active substances with antioxidant and antimicrobial activities that can enhance poultry health, growth performance, and meat quality characteristics. In addition, the incorporation of phytogenic products as feed additives in poultry diets could result in the production of high-quality, drug-free, and organic poultry products that are safe for human consumption. Thus, this review examines the current evidence on the antioxidant and antimicrobial properties of a selection of phytogenic products, their effects on nutrient utilization, and physiological and meat quality parameters in poultry. The paper also reviews the factors that could limit the utilization of phytogenic products in poultry nutrition and proposes solutions that can deliver efficient and sustainable poultry production systems for global food and nutrition security.

Keywords: antibiotic growth promoters; bioactive compounds; phytogenics; poultry; sustainability

Citation: Mnisi, C.M.; Mlambo, V.; Gila, A.; Matabane, A.N.; Mthiyane, D.M.N.; Kumanda, C.; Manyeula, F.; Gajana, C.S. Antioxidant and Antimicrobial Properties of Selected Phytogenics for Sustainable Poultry Production. *Appl. Sci.* **2023**, *13*, 99. https://doi.org/10.3390/app13010099

Academic Editors: Georgeta Pop and Călin Jianu

Received: 8 November 2022
Revised: 30 November 2022
Accepted: 13 December 2022
Published: 21 December 2022

1. Introduction

Poultry is a major contributor to human nutrition through the provision of high-quality protein, lipids, and other nutrients. However, the future of the South African poultry industry is faced with a lot of uncertainties and sustainability challenges owing to the unsustainability and high market cost of some conventional feed ingredients, including synthetic antibiotics. Thus, to deliver sustainable production systems, phytogenic products can be used to enrich poultry diets with nutrients and bioactive compounds (phytochemicals). These phytogenics are naturally endowed with a milieu of bioactive chemicals that account for their antioxidant and antimicrobial activities in poultry. Most of these phytochemicals are components of essential oils, which include carvacrol, capsaicin, cineole, cinnamaldehyde, eugenol, flavonoids, isoflavones, polyphenols, resveratrol, thymol, and many others that are well known for their potent antioxidant and antimicrobial properties [1,2]. Consequently, dietary supplementation with plant essential oils or their botanical components has the potential to boost the birds' antioxidant and antimicrobial systems, thus improving feed utilization, growth performance, and meat quality and stability. Some studies have demonstrated positive effects of phytogenic products on antioxidant systems that are critical for the protection of birds from the negative effects of lipid peroxidation [3,4]. However, in a few instances, phytogenic feed additives have been unable to prevent meat lipid peroxidation in chickens [5]. Another demonstrable benefit of phytogenic bioactive compounds in poultry diets is their ability to improve the quality and shelf-life of poultry meat through their ability to modify its fatty acid composition [6,7]. The fatty acid composition of meat is an important quality parameter in this era of prevalent non-communicable chronic metabolic diseases afflicting humanity in southern Africa [8] and globally [4]. On the other hand, Puvača et al. [9] reported that phytogenics used as natural antimicrobial growth promoters act as biochemical elements of numerous pathways involved in growth, development, and health, where they participate in the modulation of physiological and immunological processes in poultry. Through these pathways, phytochemicals exhibit various antimicrobial activities [10]. This paper, therefore, presents a review of the antioxidant and antimicrobial properties of phytogenic products used in poultry production. It further presents an overview of selected plant sources of beneficial compounds and their contribution to food and nutrition security and sustainable poultry production. We also explore the challenges and possible solutions surrounding their use in practical poultry diets. Unlike existing reviews on this topic, this offering provides the reader with broader perspectives on the utility of phytogenics as mitigators of contemporary economic, environmental, and social challenges that face poultry producers. This review not only characterizes the beneficial compounds and their modes of action but also addresses how they impact the entire poultry production chain and associated sustainability hurdles.

2. Antioxidant Properties of Phytogenic Products Used in Poultry Production

Phytogenics are very rich in bioactive compounds with potent antioxidant properties that can modulate the immune system, reduce oxidative stress, and thus improve growth performance and meat quality [11]. Due to their ability to shield lipids and proteins, antioxidants are frequently and effectively used as natural additives in poultry feed. They are increasingly becoming popular as alternatives to synthetic antioxidant additives, such as butylated hydroxytoluene and tocopheryl acetate. Their beneficial effects are mediated through the improvement of the antioxidant status of poultry [12,13], including increasing the activities and gene expression of antioxidant enzymes [3] whilst decreasing lipid peroxidation [14]. In addition, the phytochemicals can modulate the antioxidant status of poultry through their alteration of meat fatty acid composition from the saturated (lauric, myristic, palmitic, and stearic acids) to the monounsaturated (e.g., oleic acid) and polyunsaturated (PUFA) (long-chain) fatty acid lineages [4,7]. Saturated fatty acids are undesirable in meat due to their association with hypercholesterolemia and other chronic metabolic diseases in humans [4,7] whilst PUFAs, particularly omega-3 fatty acids, are associated with good health [15]. Furthermore, some of the antioxidant bioactive compounds in phytogenics, such as anthocyanins, polyphenols, retinol, and tocopherol, have good pigmentation-imparting effects [16,17], with the potential to enhance meat color. These compounds also have the potential to enhance meat stability by delaying the oxidation of stored meat.

According to Dyshlyuk et al. [18], the antioxidant defense and immunity-boosting qualities of dietary plant polyphenolics help maintain a balance between the generation of free radicals and their neutralization by supporting the antioxidant system along the gut lining. The significant antioxidant activity of polyphenolics is attributed to the presence of hydroxyl groups, which function as hydrogen donors to surrogate radicals created during the initial stage of lipid oxidation, thus delaying the generation of hydroxyl peroxide. Indeed, the intake of phytogenic products in poultry results in an increase in serum antioxidant enzyme activities and a decrease in the malondialdehyde level [15–18]. The antioxidant properties of phytogenic compounds, such as α-tocopheryl acetate or butylated hydroxytoluene, are useful in the protection of dietary lipids from oxidative damages [19]. Plant oils containing natural antioxidants contribute to the improved oxidative stability of meat and meat products containing higher levels of polyunsaturated fatty acid. Accordingly, antioxidants in phytogenic products have the potential to improve meat quality [16]. This is achieved when plant-derived antioxidants scavenge excess oxygen free radicals and chelating metals [20] or by increasing the activity of antioxidant enzymes, thereby reducing oxidative damage [21] in meat products. Furthermore, the protective effects of most dietary phytochemicals are a result of multiple distinct mechanisms [22], and hence, meat quality is affected by protecting muscle function in a live animal body. In addition, phytochemicals have been shown to benefit animals in terms of improved performance, meat quality [23], and an enhanced endogenous antioxidant system. This is possibly done by directly affecting specific molecular targets or indirectly as stabilized conjugates affecting metabolic pathways [24]. As a result, both external and endogenous stimuli activate and deactivate critical events in intracellular relays, allowing proper signaling to diverse downstream target molecules in a highly sophisticated manner to fine-tune cellular homeostasis [25]. Thus, the use of these plant-derived phytochemicals (e.g., carvacrol, thymol, catechins, quercetin, oregano, curcumin, and cinnamaldehyde) is regarded as an ingenious strategy to adopt in livestock production as antioxidant agents for enhancing the antioxidant capacity of the body and relieving oxidative stress, thus improving their growth performance and meat quality [26]. The ability of phytoproducts to act as antioxidants is mostly attributed to the existence of antioxidant compounds. However, their level mostly depends on the type of plant [27] as well as the growth environment. Since the primary roles of polyphenols in the plant relate to how it responds to challenging environmental conditions, Borguini et al. [28] demonstrated a stronger antioxidant potential of plants produced in an organic system compared to conventional ones.

3. Antimicrobial Properties of Phytogenic Products in Poultry

Although there is still much to learn about the mechanisms by which plant-derived phytochemicals exert their antimicrobial effects on various microbial populations, on target sites, in a feed matrix, and in the presence of other phytochemicals with opposing effects, their antimicrobial effects in animal health and nutrition have been extensively studied. Phytogenics are widely known for their ability to combat infections on a microbiological level. Additionally, purging of the digestive tract helps reduce ATP or energy losses during inflammation and immune responses, prevents diseases, promotes feed and nutrient digestion, and enhances growth performance and production. Transit time, digestive secretions, and the activity of digestive enzymes are some of the mechanisms known to affect gut function, and their combined effects have an impact on nutritional digestibility. Indeed, poultry performance is directly correlated with gut health and function, which is influenced by constant interactions amongst nutrition, intestinal integrity, gut flora, and the immune system. A favorable enhancement of the eubiosis could be how phytogenic compounds might selectively affect microorganisms. This leads to better utilization and absorption of nutrients, resulting in higher performance. Numerous phytogenics have positive effects on the digestive system, including spasmolytic, laxative, and antiflatulent properties. Phytogenics have attracted a lot of attention because they contain a variety of beneficial compounds, such as flavonoids and isoprene glucosinolate derivatives, which have antimicrobial properties. Lee et al. [29] showed that phytogenic compounds enhanced the intestinal activities of trypsin, lipase, and amylase in broilers. At a dose of 0.6 mL/L, oregano and thyme essential oils were found to have a bactericidal action on *E. coli* [30]. Mathis et al. [31] showed that walnut leaves (*Juglandaceae*) slowed the spread of *Clostridium perfrengens* in hens while improving weight gain. Likewise, pomegranate (*Punica granatum*) and green tea (*Camellia sinensis*) products were shown to alter the intestinal microbiota by promoting the proliferation of non-pathogenic bacteria in the digestive system [32]. Broilers fed diets supplemented with the extract of green tea leaves and pomegranate rinds promoted a greater relative abundance of lactic acid-producing bacteria [32]. The in vitro effectiveness of certain berry, date, and thyme extracts against chicken-derived *E. coli* and Salmonella isolates was demonstrated in a 2009 study by Dhifi et al. [33]. Cinnamon extract, thyme, and clove stimulated the digestive secretions of bile, mucus, and saliva and improved enzyme activities, which are of great nutritional interest [34]. In addition, some oils extracted from plants positively influenced the activity of trypsin and amylase in chickens and had a stimulatory effect on the intestinal mucus in chickens, maintaining an equilibrium in the microflora present in their gut [35]. Oladeji et al. [36] claimed that phytogenics significantly increased poultry production performance in terms of weight gain and the feed conversion ratio. This improvement was attributed to the high nutrient availability due to changes in the intestinal ecosystem, such as the increase in the population of lactic acid-producing bacteria.

Lactic acid-producing bacteria have a positive impact on the lower gastrointestinal tract by regulating the composition of intestinal microflora, promoting intestinal immunity, and developing intestinal health. This could be the reason why several researchers have concluded that phytogenic feed additives improve nutrient digestibility, growth performance, and gut health in poultry [37]. Aksit et al. [38] reported earlier that phytogenics' antibacterial effect may help to improve the carcass's microbiological freshness. According to Ganguly [39], the phytogenic growth promoter remains active throughout the gastrointestinal tract, where it exerts broad-spectrum antimicrobial action, improves broiler overall growth performance, and further improves nutrient utilization by enhancing gastrointestinal histomorphology and host immunity. Mohebodini et al. [40] reported that the inclusion of resveratrol from grape by-products in broiler diets increased the levels of immunoglobulin G (IgG) and immunoglobulin M (IgM). These antibodies provide specific immunity to prevent the adhesion of pathogenic microbes and some viruses on the intestinal wall [41]. Therefore, a healthy gut composed of beneficial microbiota is very important for poultry performance and welfare and can be achieved by incorporating phytogenic products into their diet.

4. Selected Plant Sources of Antimicrobial and Antioxidant Compounds

4.1. Moringa oleifera

Moringa oleifera is a medium-sized perennial, evergreen, and deciduous tree from the family *Moringaceae* that grows quickly in a variety of soil types and hot, humid, and dry tropical and subtropical climates [42]. Out of the 13 species in this family, *M. oleifera* is the most popular and widely utilized species [43]. Owing to its nutritional profile, *M. oleifera* is prized globally for its contribution to the livestock industry and food and pharmaceutical sectors [44]. *M. oleifera* is rich in essential nutrients, such as lipids, proteins, minerals, and vitamins [45]. Reports have indicated that the leaves have the highest protein value (250 to 270 g/kg CP) [42], with some protein being available in the seeds, roots, and flowers [46]. Furthermore, *M. oleifera* is also a great source of well-balanced amino acids and vitamins, of which the amount of vitamin A is about 10 times more than in carrots and pumpkins and the amount of vitamin C is 7 times more than in oranges [47]. According to Mbikay [48], moringa leaves contain vitamins A, B and B-Complexes, C (ascorbic acid), E (α-tocopherol), K, and pro-vitamin A in the form of beta-carotene, along with minerals such as sulfur, calcium, phosphorus, potassium, manganese, and iron. Moringa products are also rich in monounsaturated fatty acids [45] and polyunsaturated fatty acids, such as linoleic acid, α-linolenic acid, arachidic acid, stearic acid, oleic acid, and palmitic acids [49].

In addition, *M. oleifera* contains a myriad of phytochemicals, such as flavonols, quercetin, apigenin, kaempferol, luteolin, myricetin glycosides, and polyphenols, which exhibit antioxidant, antimicrobial, anticarcinogenic, anti-inflammatory, and medicinal activities [50,51]. However, the leaves also contain antinutritional substances, such as alkaloids, carotenoids, glucosinolates, isothiocyanates, saponins, tocopherols, tannins, oxalates, phytates, trypsin and amylase inhibitors, lectins, and cyanogenic glucosides [52], which may affect their utility in poultry diets and induce different responses. For example, Kakengi et al. [53] found that adding 10% and 20% *M. oleifera* leaf meal in place of sunflower seed meal enhanced feed intake and dry matter intake while lowering the egg mass output in laying hens. However, Khan et al. [54] also reported that the inclusion of 1.2% *M. oleifera* leaf powder in broiler feeds improved the intestinal microarchitecture and cellular count, which in turn promoted gut health.

4.2. Lippia javanica

Lippia is a genus that belongs to the family *Verbenaceae* [55], which is distributed in the tropics of Africa and South America [56]. There are more than 200 species in the genus *Lippia* (*Lamiales: Verbenaceae*), many of which are aromatic [55]. According to Viljoen et al. [57], *Lippia javanica* (*Burm. f.*) *Spreng.*, often known as fever tea, is a perennial, erect, woody shrub that is rich in nutrients and bioactive compounds. The plant has been used to treat various ailments such as colds, cough, fever, influenza, malaria, measles, rashes, stomach problems, and headaches [58]. Nutritionally, *L. javanica* possesses varying amounts of essential nutrients, such as protein, fats, carbohydrates, minerals, and vitamins [59]. It is a good source of cobalt, cadmium copper, chromium, iron, magnesium, calcium, manganese, zinc, selenium, potassium, phosphorus, and lead [60,61]. Bio-compounds, such as caryophyllene, carvone, ipsenone, ipsdienone, limonene, linalool, myrcene, myrcenone, ocimenone, p-cymene, piperitenone, sabinene, and tagetenone, are abundant in the volatile oil of *L. javanica* [59]. *L. javanica* also contains alkaloids, amino acids, flavonoids, iridoids, triterpenes, and other volatile and non-volatile secondary metabolites. Furthermore, *L. javanica* contains phytochemicals, such as anthocyanins, anthraquinones, coumarins, saponins, flavones, alkaloids, flavonoids, tannins, phenols, cardiac glycosides [55], and phenylethanoid glycosides (isoverbascoside and verbascoside) [62]. From the essential oils extracted from *L. javanica*, Pascual et al. [63] reported the presence of p-cymene, camphor, limonene, β-caryophyllene, linalool, α-pinene, thymol, carvone, ipsenone, myrcenone, myrcene, piperitenone, caryophyllene, and tagetenone, which give the plant antiseptic, antibacterial, and antiviral properties [64]. Varied results have been reported from feeding trials involving *L. javanica* leaf meals depending on the amount used and the bird species. For example, the inclusion of *L. javanica* in broiler diets at a rate of 5 g/kg had positive effects on the growth performance, carcass characteristics, and fatty acid profiles of broiler meat [65]. However, Mnisi et al. [66] reported that the inclusion of *L. javanica* at a rate of 25 g/kg feed promoted similar growth performance, health status, and carcass and meat quality traits as the commercial grower diet containing antibiotics in Japanese quail.

4.3. Camellia sinensis

Camellia sinensis, commonly known as green tea, is an evergreen shrub that belongs to the family *Theaceae* [67]. The *Camellia* genus comprises over 200 species, with a majority being native and adapted to China [68]. The green tea plant is the most grown for its buds and leaves, which are used to make tea [69]. Green tea leaves have a crude protein content that ranges from 18.15 to 22.9% and metabolizable energy between 11.3 and 14.6 MJ/kg [70]. The leaves are rich in vitamins (A, C, E, K, and B complex), lipids (linolenic and linoleic acids), pigments (carotenoids and chlorophyll), and minerals [71]. Green tea contains beneficial compounds that have stress-reducing functions and antioxidant activities with neurological effects and anti-inflammatory properties [72]. Its polyphenols account for up to 35% of the dry weight of the leaves [73]. However, the majority of the antioxidants are catechins, which make up about 10 to 20% of dry leaves. The major catechins identified in the plant are (-)-epigallocatechin gallate, (-)-epicatechin (EC), (-)-epigallocatechin (EGC), and (-)-epicatechin gallate (ECG) [63], which have potential health benefits [74].

Furthermore, it contains 26 different amino acids, alkaloids (caffeine, theobromine, and theophylline), carotenoids, lipids, L-ascorbic acid, carbohydrates, methylxanthines, minerals, chlorophyll, saponins, organic compounds, and volatile organic compounds [74,75]. Other important bioactive compounds that have been isolated in the plant are flavonoids such as kaempferol, myricetin, and quercetin, which account for 2–3% of the dry leaves [76].

Mahlake et al. [77] indicated that replacing the antibiotic zinc-bacitracin with green tea leaf powder in the diets of Jumbo quail boosted the overall feed intake but had no effect on weight gain or feed conversion efficiency. Additionally, 10% aqueous green tea extract and other antibiotics had comparable antibacterial effects on the antibiotic-resistant *S. pyogenes*, *P. mirabilis*, and *S. aureus* species [78].

4.4. Allium sativum L.

Garlic (*Allium sativum* L.) is a bulbous plant that belongs to the *Anaryllidaceae* family [79]. It is believed to have originated in Asia, although it is now utilized throughout the world [80]. Garlic is globally used as a nutraceutical, containing carbohydrates, protein, amino acids, lipids, fiber, vitamins, minerals, organic acids, saponins, phenolic compounds, and a large group of organo-sulfur compounds [81], along with phytochemicals (saponins, tannins, alkaloids, and flavonoids) [82]. Additionally, garlic has enzymes, minerals, vitamins [83], and moderate sulfur content but little B vitamin [84]. Garlic bioactive compounds are organosulfur compounds, which are responsible for its smell, flavor, and antioxidant and medicinal properties [85]. Sulfur compounds, such as trisulfides, thiosulfinates, tetrathiol, sulfates, scordinine, pseudoscordinine, methyl sulfides, methionine, glutathione, disulfides, dimethyl sulfides, diallyl sulfides, cystine, cysteine sulfoxides, cysteine, cycloalliin, allyl trisulfides, allyl sulfides, allyl disulfides, alliin, allicin (thiosulfonate), and ajoene, have all been reported in garlic [80,86].

Other reported phytochemicals are flavonoids (flavones and quercetins) [87], phenolic compounds, saponins, and sapogenines [88]. These bioactive compounds have antibacterial, hypo-cholesterolemic, antioxidant, and growth-promoting qualities, which are beneficial to livestock and poultry [83]. For example, the addition of 0.5 kg/ton garlic meal in diets as an alternative to antibiotics promoted the high live weight in broilers [89]. According to Eltazi et al. [90], supplementing a standard diet with 3% garlic powder promoted higher body weight gain, higher feed intake, and an improved feed conversion ratio, along with higher dressing and breast percentages in broilers.

4.5. Allium cepa

Allium cepa, commonly known as an onion, is a member of the *Liliaceae* family, which is currently grown all over the world and has its roots in central Asia [91]. The onion is an important source of nutritional components, such as proteins, carbohydrates, sugars (arabinose, fructose, galactose, glucose), vitamins, lipids, minerals, and some flavonoids and polyphenols components [92]. Onions are a rich source of vitamin A (β-carotenoid), vitamin B (B1, B2, B3), vitamin E (α-tocopherols), and vitamin C (ascorbic acid) [93,94], which exhibit antioxidant activities. They are also a good source of diverse dietary flavonoids, phospholipids, and glycolipids, along with organosulfur compounds, such as allicin, alliin, diallyl disulfide, S-methyl-L-cysteine S-oxide, propanethial S-oxide, and 3-mercapto-2-methypentan-1-ol [95]. Furthermore, Metrani et al. [96] reported that onions contain various sulfoxides, which include (+)-S-(1-propenyl)-L-cysteine sulfoxide, (+)-S-methyl-L-cysteine sulfoxide, S-propyl-L-cysteine sulfoxide, S-methyl-L-cysteine sulfoxide, and S-propenyl-L-cysteine sulfoxide, which exhibit antibacterial and antioxidant activities. In addition, onion bulbs are said to be a good source of phytonutrients and antioxidants due to the presence of organosulfur compounds, such as cysteine sulfoxides, quercetin, quercetin glucosides, and allicin [97,98].

Onion bulbs' saponins have been shown to have biological effects, including antifungal and anti-inflammatory properties [97]. These substances stimulate the digestive and immune systems, which in turn enhance growth performance and general health in birds [99]. It has been claimed that onion extracts or powder used in drinking water or livestock feed have both growth-promoting and anti-pathogenic properties [100]. The phytochemicals in onions alter the gut flora and immune system and encourage the growth of colonic and mucosal microflora, which function as a barrier to prevent microbes from entering the gastrointestinal tract [101]. Aditya et al. [102] found that adding 5 or 7.5 g of onion to broiler diets increased the overall feed intake and body weight of the hens without affecting the feed conversion ratio. Supplementing drinking water with up to 1% onion extracts during the starter and grower periods increased the average daily feed without compromising broiler chickens' feed conversion ratio [100]. Muscovy ducks fed diets containing 1% onion meal had an enhanced feed conversion ratio, live body weight, and weight gain [103]. Dosoky et al. [104] found that Japanese quail hens fed diets with 800 g of dried onion per kg DM had increased body weight.

4.6. Mentha piperita

Peppermint (*Mentha piperita*) is a perennial herb that belongs to the *Lamiaceae* family [105]. *Mentha piperita* is one of the most commonly used medicinal plants in the world [106]. This is mainly because of the presence of phytochemicals that have anti-inflammatory, anti-aging, antimicrobial, antioxidant, emmenagogue, antinociceptive, and rubefacient properties [107]. Bio-compounds, such as proteins, carbohydrates, lipids, minerals, vitamins, alkaloids, saponins, glycosides, steroids, and tannins, have been extracted from the plant [105,108]. Peppermint products have high concentrations of minerals, such as P, K, Na, Ca, Mg, and Zn, and vitamins A, C, and E [105]. The leaves contain flavonoids and phenolic acids, along with acetaldehyde, amyl alcohol, cadinene, caffeic acid, cardiac glycosides, dimethyl sulfide, limonene, menthol, menthone, menthyl esters, phellandrene, pinene, and pugelone [2,109], which have antioxidant and antimicrobial activities.

Furthermore, the plant contains mint oil, which is composed primarily of linoleic, palmitic, and linolenic palmitic acid [110]. About 0.5 to 4% of the essential oils found in peppermint leaves are composed of menthol (25–78%), menthone (14–36%), isomenthone (1.5–10%), menthyl acetate (2.8–10%), and cineol (3.5–14%) [111]. Due to its bioactive substances, peppermint is frequently used in the poultry industry to boost the immune system, in addition to its potent antibacterial and antioxidant capabilities [112]. According to Khempaka et al. [113], peppermint leaves had positive impacts on broiler ammonia generation, abdominal fat deposition, and antioxidant activity. Early in a broiler's life, peppermint leaves are effective at promoting growth [114]. When added to broiler diets in various concentrations (0, 5, 10, or 15 g/kg), peppermint leaves significantly boosted their body weight and daily body weight gain compared to a control diet [115].

4.7. Aloe vera

Aloe vera is one of over 420 species in the genus *Aloe*, which has been variously categorized as belonging to the *Asphodelaceae*, *Liliaceae*, or *Aloaceae* families [116]. It is a succulent, perennial xerophyte with thick, fleshy, pointed leaves that cluster at the stem. *Aloe vera* is ubiquitous in components with biological activity, such as polysaccharides, phenolic compounds, minerals, water- and lipid-soluble vitamins, organic acids, and lipids [117]. It is made up of 96% water and 4% dry matter, containing protein, fat, dietary fiber, and 75 other biologically active compounds [118,119]. Anthraquinones, saccharides, vitamins, enzymes, and low-molecular-weight compounds are the main components of *A. vera* and are responsible for its anti-inflammatory, immunomodulatory, wound-healing, antibacterial, antiviral, antifungal, anti-tumor, anti-diabetic, and antioxidant properties [120].

In addition, other phytochemicals, such as alkaloids, anthraquinones, anthrones, chromones, coumarins, flavonoids, lignin, naphthalene, saponins, sterols, pyrans, and pyrones, have been isolated from the plants of the genus [121]. In poultry nutrition, the incorporation (1.5, 2, 2.5%) of *A. vera* gel in broiler feed reduced the *Escherichia coli* count [122]. Similarly, Dai et al. [123] found that *A. vera* products increased *Lactobacillus spp.* and *Bifidobacteria* while decreasing the amount of *E. coli*. In another study, Darabighane and Zarei [124] reported improved feed utilization efficiency upon adding 1.5%, 2%, and 2.5% of *A. vera* gel to the diet of broilers with coccidiosis. The authors stated that this could be attributed to the potential of *A. vera* to enhance the intestinal health and immune system response in the birds.

4.8. Seaweeds

Seaweeds are macroalgae that grow in the littoral zone of aquatic systems. They consist of green (*Chlorophyceae*), red (*Rhodophyta*), and brown (*Phaeophyceae*) algae. There are 10,100 different kinds of seaweeds that have been identified worldwide. Seaweeds can be found in all types of marine settings; hence, their nutritional value varies greatly depending on the species, habitat, geographical origin, production region, season, harvest time, environmental and physiological fluctuations, and water temperature [125]. Seaweeds contain polysaccharides, proteins, amino acids, minerals, vitamins, and antioxidant chemicals [125]. Stengel et al. [126], reported that seaweeds are a rich source of bioactive compounds, such as polyphenols, which have antioxidant and antimicrobial properties and thus, can be used as a nutraceutical additive in poultry production [127].

Choi et al. [128] demonstrated that adding seaweed supplements at a rate of 5 g/kg improved the feed conversion efficiency. According to Abudabos et al. [129], adding seaweed to broiler diets improved the quality of the meat by reducing the number of microorganisms in the digestive tract and boosting immunological function. Nhlane et al. [130] investigated the impact of green seaweed (*Ulva* sp.) meal on native chickens and found that its dietary addition boosted native chickens' feed intake and overall body weight gain but not their ability to convert feed into energy. Broiler hens supplemented with green seaweed at doses of 10 or 30 g/kg did not alter the growth or feed efficiency; however, birds given the higher dose (30 g/kg) showed increased dressing percentage and breast muscle yield compared to those given the control or 10 g/kg dose [131]. In another study, the inclusion of seaweed up to 150 g/kg in the diets of ducks had no negative impacts on the growth rate or carcass quality [132].

5. Contribution of Phytogenics to Environmental Health and Food Security

Food security has four dimensions, including (1) availability—national, (2) accessibility —household, (3) utilization—individual, and (4) stability. To achieve full food security, all four dimensions must be intact [133], which is still a major challenge in most low- and middle-income countries (LMICs). Rapid human population growth, poverty, the COVID-19 pandemic, and climate change, with its concomitant recurrent droughts [134], are among the major factors that exacerbate food insecurity. Ironically, over 80% of households in Sub-Saharan Africa (SSA) practice poultry farming with predominantly indigenous poultry breeds [135]. However, they have not yet been sufficiently exploited to alleviate food insecurity. This current scenario, therefore, provides an opportunity to explore sustainable strategies to ensure poultry intensification in the sub-continent. To meet the increasing global demand for nutritious food, the poultry industry, which contributed 40.6% of total meat production (337.3 million tons) in 2020 [136], must be sustainably intensified.

However, poultry intensification raises significant concerns about environmental sustainability. This is because intensive poultry production depends on external inputs, primarily comprising monoculturally produced feed ingredients such as maize and soybean, whose production is associated with serious environmental impact issues, including climate change, deforestation, and loss of biodiversity [137]. In addition, broiler chickens excrete enormous amounts of unutilized nitrogen and phosphorus into the environment annually, which, if mismanaged, leads to environmental manure [138], atmospheric ammonia pollution [139], and eutrophication and its induction of aquatic hypoxia and harmful algal blooms [140]. Thus, improved nutrient utilization efficiency in birds consuming diets with phytogenic products could have positive environmental consequences by reducing the nutrient pollution of poultry wastes. Furthermore, the use of phytogenic plants as feed additives results in improvements in protein digestibility, which further leads to the greater utilization of dietary amino acids and a reduction in the excretion of nitrogenous compounds [141]. As a result, the use of phytogenic feed additives lowers the emissions from poultry production. In addition, the anti-microbial effects of phytogenic feed additives result in lower ammonia release. This is achieved by a reduction of Gram-negative bacteria and an increase in the beneficial bacterial activities in the hindgut of poultry, leading to increased formation of volatile fatty acids in excreta, a lower slurry pH, and consequently, the reduction of ammonia emissions [142].

In addition, the usage of antibiotics in intensified poultry farms can accelerate the spread of antibiotic resistance genes in the environment [9,10,142,143]. Therefore, sustainable poultry intensification could be a strategy to enhance food security, particularly in SSA, without causing catastrophic damage to the ecosystem, such as the feeding of phytogenic feed additives. The use of phytogenic compounds could provide a more environmentally friendly, lower-cost solution for improving food and nutrition security while also benefiting poultry consumers' health. This will, in turn, meet the demand for animal protein and, ultimately, reduce hunger and food insecurity, even in low-income countries. This would also ensure that consumers have access to healthy poultry products that are free of antibiotic residues, and as such, comply with the goal of achieving good health and well-being. The antimicrobial properties of these phytogenics could also reduce the amount of carbon and methane that is released into the environment, which could contribute to the goal of combatting climate change. Moreover, phytogenics have probiotic, prebiotic, and antibiotic activities that can reduce pathogenic bacterial infections and help reduce disease outbreaks, thus promoting food safety. This is very important, given that most of the recent pandemics have been of animal origin [144].

6. Anti-Nutritional Factors as Constraints to the Use of Phytogenics in Poultry Nutrition

The incorporation of phytogenic products in poultry diets is known to improve the performance and health of birds, but their widespread adoption could be limited by the presence of anti-nutritional factors (ANFs) (Table 1). Anti-nutritional factors have been reported to reduce nutrient utilization or feed intake when included in animal feeds [145]. Several studies have reported the toxicological effects of ANFs in poultry [146,147]. For example, phytates, tannins, and phytosterols in *M. oleifera* reduced weight gain in Arbor Acres broiler chickens by reducing the digestibility of amino acids [148]. In addition, reductions in crude protein digestibility, metabolizable energy utilization, feed intake, weight gain, and feed conversion efficiency were observed in broilers fed a diet with cassava pellets containing cyanogenic glycosides in place of maize [149,150]. Feed intake, weight gain, and carcass traits were reduced in chickens fed a diet containing 40 g/kg of polyphenol-rich *C. sinensis* [151]. Hassan [152] observed poor growth in broiler chicks fed 3 g saponins/kg diet due to a depressed feed intake. Farhadi et al. [153] reported that tannins from eucalyptus leaf powder reduced body weight and the overall feed conversion ratio in broilers because tannins bind proteins and form indigestible complexes. With phytogenics used as whole plant parts, it was shown that the high fiber content reduced the performance of the birds by increasing the passage rate [154]. Similarly, Nduku et al. [155] found that feeding one-week-old broiler chicks a diet containing *M. oleifera* high in fiber reduced the daily weight gain and impaired the feed conversion ratio. This is because simple non-ruminants such as chickens have a limited capacity to utilize high-fiber diets due to a lack of microbial digestive enzymes.

Table 1. Anti-nutritional factors present in selected plants used in poultry diets.

Phytogenics	Anti-Nutritional Factors	References
Moringa	Tannins, oxalates, phytate, saponins, cyanogenic glycosides, alkaloids, flavonoids	[156]
Lippia	Tannins, oxalates, saponins, phytate, alkaloids	[157]
Camellia sinensis	Flavonoids, phenolics	[158]
Garlic	Alkaloids, tannins, saponins	[159]
Onion	Alkaloids, tannins, flavonoids, total phenolics	[159]
Peppermint	Alkaloids, saponins, glycosides, tannins, flavonoids	[108]
Aloe	Flavonoids, tannins, alkaloids	[160]
Artemisia afra	Flavonoids, alkaloids, phenolic acids, lignans, proanthocynidins	[161]
Carpobrotus edulis	Saponins, flavonoids, alkaloids, cyanogenic glycosides, tannins	[162]

Table 1. *Cont.*

Phytogenics	Anti-Nutritional Factors	References
Eucalyptus	Tannins, phytate, oxalates, saponins	[163]
Seaweeds	Flavonoids, phlorotannins	[164]

7. Amelioration of Antinutritional Effects in Poultry Consuming Phytogenics

As aforementioned, the utility of phytogenics is hindered by the presence of ANFs, which could be ameliorated using a variety of techniques. Solid-state fermentation, alkali treatments, tannin-binding agents, thermal processes, exogenous feed enzymes, and plant breeding are some of the strategies that can be used to ameliorate ANFs. Solid-state fermentation (SSF) is a process that takes place in a solid matrix in the absence or near absence of free water with the aid of the metabolic activity of microorganisms [165]. When fermenting phytogenics to reduce ANFs, it is also important to use microorganisms, such as lactic acid bacteria, that will promote beneficial intestinal microbiota while acting against pathogenic enteric microbes [166]. Microbial activity during SSF can reduce the levels of ANFs in potential phytogenic substrates. Indeed, Olukomaiya et al. [167] reported a significant reduction in the phytic acid content in fermented lupins flour using *Aspergillus sojae* and *A. ficuum*. Similarly, Shi et al. [168] reported a reduction in the concentrations of crude fiber, tannins, hemicellulose, and phytate in fermented *M. oleifera* inoculated with *A. niger*, *Candida utilis*, and *Bacillus subtilis* at a ratio of 1:1:2. Similarly, *M. oliefera* seed flour was reported to have reduced tannin, phytic, phenol, flavonoid, saponin, and alkaloid contents after fermentation and germination processes compared to unprocessed flour [169].

Thermal processing is also an effective means of inactivating thermo-labile ANFs in phytogenic products [170]. Heat-based methods include autoclaving, pressure cooking, microwaving, extrusion cooking, and toasting. Batista et al. [171] reported that autoclaving reduced the contents of trypsin and α-amylase inhibitors and resistant starch in *Phaseolus vulgaris*. Toasting the seeds from *Senna obtusifolia* at 80 °C for one hour significantly lowered the hemagglutinin, phytate, oxalate, saponin, and tannin contents [172]. Thermal application has been shown to partially hydrolyze tannic acid and release gallic acid molecules, and these newly produced gallic acid and galloyl groups had enhanced antimicrobial and antioxidant effects compared to fresh tannic acid [172]. Moreover, boiling, steaming, stir-frying, and roasting were reported to decrease the total polyphenols contents of *Capsicum annuum* L. [173].

Exogenous enzymes have the potential to improve the utility of phytogenics for sustainable poultry production [174]. Proteases and carbohydrases (xylanases and cellulases) are exogenous enzymes commonly used in poultry feeds. Costa et al. [175] successfully used exogenous *ulvan lyase* to boost the meat nutritional value of birds by feeding them *Ulva lactuca* supplemented with carbohydrases. Abou-Arab and Abu-Salem [176] treated *Jatropha curcas* seeds with sodium bicarbonate ($NaHCO_3$) and sodium hydroxide ($NaOH$) and successfully inactivated the phytic acid, trypsin inhibitors, total phenols, and saponins. This review shows that ANFs in phytogenic products could hinder their utility as sources of beneficial bioactive compounds. However, there are several methods that could be used to reduce the ANF content of phytogenic products and improve their utilization in poultry nutrition. Nonetheless, factors such as costs and the ease of application often guide the preferable processing method of farmers.

8. Prospects and Conclusions

The use of phytogenic products as functional feed ingredients in poultry diets could allow for AGP-free large-scale sustainable production of poultry products in the agro-food value chain. This paper presents evidence that the incorporation of phytogenic products in poultry diets can boost feed utilization efficiency, reduce feed-food competition, and enhance their contribution to eradicating hunger and food insecurity. However, there is also evidence that their utilization can be limited by the presence of plant secondary metabolites, particularly when higher dietary inclusion levels are used. Thus, it is important to establish their maximum tolerance level in poultry diets and, thereafter, develop strategies to ameliorate any antinutritional activities should higher dietary inclusion levels be desired. Such practical strategies would ensure the large-scale adoption of phytogenic products by farmers and feed manufacturing companies.

Author Contributions: Conceptualization, C.M.M.; writing—original draft preparation, C.M.M., V.M., A.G. D.M.N.M., C.K., and F.M.; writing—review and editing, C.M.M., V.M., A.G., A.N.M., D.M.N.M., C.K., F.M., and C.S.G.; administration, C.M.M. All authors have read and agreed to the published version of the manuscript.

Funding: The funding received from the University of Fort Hare for open-access publishing is hereby acknowledged.

Institutional Review Board Statement: Not applicable.

Informed Consent Statement: Not applicable.

Data Availability Statement: Not applicable.

Conflicts of Interest: The authors declare no conflict of interest.

References

1. Farag, M.R.; Alagawany, M.; Tufarelli, V. In vitro antioxidant activities of resveratrol, cinnamaldehyde and their synergistic effect against cyadox-induced cytotoxicity in rabbit erythrocytes. *Drug Chem. Toxicol.* **2016**, *17*, 1–10.
2. Patil, S.R.; Patil, R.S.; Godghate, A.G. *Mentha piperita* Linn: Phytochemical, antibacterial and dipterian adulticidal approach. *Int. J. Pharm. Pharm. Sci.* **2016**, *8*, 352–355.
3. Reis, J.H.; Gebert, R.R.; Barreta, M.; Boiago, M.M.; Souza, C.F.; Baldissera, M.D.; Santos, I.D.; Wagner, R.; Laporta, L.V.; Stefani, L.M.; et al. Addition of grape pomace flour in the diet on laying hens in heat stress: Impacts on health and performance as well as the fatty acid profile and total antioxidant capacity in the egg. *J. Therm. Biol.* **2019**, *80*, 141–149. [CrossRef]
4. Galli, G.M.; Gerbet, R.R.; Griss, L.G.; Fortuoso, B.F.; Petrolli, T.G.; Boiago, M.M.; Souza, C.F.; Baldissera, M.D.; Mesadri, J.; Wagner, R.; et al. Combination of herbal components (curcumin, carvacrol, thymol, cinnamaldehyde) in broiler chicken feed: Impacts on response parameters, performance, fatty acid profiles, meat quality and control of coccidia and bacteria. *Microb. Pathog.* **2020**, *139*, 1–11. [CrossRef] [PubMed]
5. Reis, J.H.; Gebert, R.R.; Barreta, M.; Baldissera, M.D.; dos Santos, I.D.; Wagner, R.; Campigotto, G.; Jaguezeski, A.M.; Gris, A.; de Lima, J.L.F.; et al. Effects of phytogenic feed additive based on thymol, carvacrol and cinnamic aldehyde on body weight, blood parameters and environmental bacteria in broilers chickens. *Microb. Pathog.* **2018**, *125*, 168–176. [CrossRef]
6. Starčević, K.; Krstulović, L.; Brozić, D.; Maurić, M.; Stojević, Z.; Mikulec, Ž.; Bajić, M.; Mašek, T. Production performance, meat composition and oxidative susceptibility in broiler chicken fed with different phenolic compounds. *J. Sci. Food Agric.* **2014**, *95*, 1172–1178. [CrossRef]
7. Ahmed, S.T.; Islam, M.M.; Bostami, A.B.M.R.; Mun, H.S.; Kim, Y.J.; Yang, C.J. Meat composition, fatty acid profile and oxidative stability of meat from broilers supplemented with pomegranate (*Punica granatum* L.) by-products. *Food Chem.* **2015**, *188*, 481–488. [CrossRef]
8. Gouda, H.N.; Charlson, F.; Sorsdahl, K.; Ahmadzada, S.; Ferrari, A.J.; Erskine, H.; Leung, J.; Santamauro, D.; Lund, C.; Aminde, L.N.; et al. Burden of non-communicable diseases in sub-Saharan Africa, 1990–2017: Results from the Global Burden of Disease Study 2017. *Lancet. Glob. Health* **2019**, *7*, 1375–1387. [CrossRef]
9. Puvača, N.; Stanaćev, V.; Glamočić, D.; Lević, J.; Perić, L.; Milić, D. Beneficial effects of phytoadditives in broiler nutrition. *World Poult. Sci. J.* **2013**, *69*, 27–34. [CrossRef]
10. Scalbert, A. Antimicrobial properties of tannins. *Phytochemistry* **1991**, *30*, 3875–3883. [CrossRef]
11. Mehdi, Y.; Létourneau-Montminy, M.P.; Gaucher, M.L.; Chorfi, Y.; Suresh, G.; Rouissi, T.; Brar, S.K.; Côté, C.; Ramirez, A.A.; Godbout, S. Use of antibiotics in broiler production: Global impacts and alternatives. *Anim. Nutr.* **2018**, *4*, 170–178. [CrossRef]
12. Polat, U.; Yesilbag, D.; Eren, M. Serum biochemical profile of broiler chickens fed diets containing rosemary and rosemary volatile oil. *J. Biol. Environ. Sci.* **2011**, *5*, 23–30.
13. Paraskeuas, V.; Fegeros, K.; Palamidi, I.; Hunger, C.; Mountzouris, K.C. Growth performance, nutrient digestibility, antioxidant capacity, blood biochemical biomarkers and cytokines expression in broiler chickens fed different phytogenic levels. *Anim. Nutr.* **2017**, *3*, 114–120. [CrossRef]
14. Rahman, M.M.; Kim, S.J. Effects of dietary *Nigella sativa* seed supplementation on broiler productive performance, oxidative status and qualitative characteristics of thighs meat. *Ital. J. Anim. Sci.* **2016**, *15*, 241–247. [CrossRef]
15. Siroma, T.K.; Machate, D.J.; Zorgetto-Pinheiro, V.A.; Figueiredo, P.S.; Marcelino, G.; Hiane, P.A.; Bogo, D.; Pott, A.; Cury, E.R.J.; Guimarães, R.D.C.A.; et al. Polyphenols and ω-3 PUFAs: Beneficial outcomes to obesity and its related metabolic diseases. *Front. Nutr.* **2022**, *8*, 781622. [CrossRef]

16. Zdanowska-Ssiadek, A.; Lipínska-Palka, P.; Damaziak, K.; Michalczuk, M.; Marchewka, J. Antioxidant effects of phytogenic herbal-vegetable mixtures additives used in chicken feed on breast meat quality. *Anim. Sci. Pap. Rep.* **2018**, *36*, 393–408.

17. Tian, X.Z.; Lu, Q.; Paengkoum, P.; Paengkoum, S. Short communication: Effect of purple corn pigment on change of anthocyanin composition and unsaturated fatty acids during milk storage. *J. Dairy Sci.* **2020**, *103*, 7808–7812. [CrossRef]

18. Dyshlyuk, L.; Babich, O.; Ivanova, S.; Vasilchenco, N.; Atuchin, V.; Korolkov, I.; Prosekov, A. Antimicrobial potential of ZnO, TiO_2 and SiO_2 nanoparticles in protecting building materials from biodegradation. *Int. Biodeterior.* **2020**, *146*, 104821. [CrossRef]

19. Nakatani, N. Phenolic antioxidants from herbs and spices. *Biofactors* **2000**, *13*, 141–146. [CrossRef]

20. De Castilho, T.S.; Matias, T.B.; Nicolini, K.P.; Nicolini, J. Study of interaction between metal ions and quercetin. *Food Sci. Hum. Well.* **2018**, *7*, 215–219. [CrossRef]

21. Na, H.K.; Surh, Y.J. Modulation of Nrf2-mediated antioxidant and detoxifying enzyme induction by the green tea polyphenol EGCG. *Food Chem. Toxicol.* **2008**, *46*, 1271–1278. [CrossRef] [PubMed]

22. Calabrese, V.; Cornelius, C.; Mancuso, C.; Pennisi, G.; Calafato, S.; Bellia, F.; Bates, T.E.; Giuffrida Stella, A.M.; Schapira, T.; Dinkova Kostova, A.T.; et al. Cellular stress response: A novel target for chemoprevention and nutritional neuroprotection in aging, neurodegenerative disorders and longevity. *Neurochem. Res.* **2008**, *33*, 2444–2471. [CrossRef] [PubMed]

23. Brewer, M.S. Natural antioxidants: Sources, compounds, mechanisms of action, and potential applications. *Compr. Rev. Food Sci. Food Saf.* **2011**, *10*, 221–247. [CrossRef]

24. Mahfuz, S.; Shang, Q.; Piao, X. Phenolic compounds as natural feed additives in poultry and swine diets: A review. *J. Anim. Sci. Biotechnol.* **2021**, *12*, 48. [CrossRef] [PubMed]

25. Kuralkar, P.; Kuralkar, S.V. Role of herbal products in animal production—An updated review. *J. Ethnopharmacol.* **2021**, *278*, 114246. [CrossRef] [PubMed]

26. Rossi, B.; Toschi, A.; Piva, A.; Grilli, E. Single components of botanicals and nature-identical compounds as a non-antibiotic strategy to ameliorate health status and improve performance in poultry and pigs. *Nutr. Res. Rev.* **2020**, *33*, 218–234. [CrossRef]

27. Wojdyło, A.; Oszmiański, J.; Czemerys, R. Antioxidant activity and phenolic compounds in 32 selected herbs. *Food Chem.* **2007**, *105*, 940–949. [CrossRef]

28. Borguini, R.G.; Markowicz, C.H.; Bastos, J.M.; Moita-neto, F.S.; Capasso, E.A.; Ferraz da Silva, T. Antioxidant potential of tomatoes cultivated in organic and conventional systems. *Braz. Arch. Biol. Technol.* **2013**, *56*, 521–529. [CrossRef]

29. Lee, K.W.; Everts, H.; Kappert, H.J.; van der Kuilen, J.; Lemmers, A.G.; Frehner, M.; Beynen, A.C. Growth performance, intestinal viscosity, fat digestibility and plasma cholesterol in broiler chickens fed a rye-containing diet without or with essential oil components. *Int. J. Poult. Sci.* **2004**, *3*, 613–618.

30. Burt, S.; Reindeers, R.D. Antibacterial activity of selected plant essential oils against *E. coli* O157:H7. *Lett. Appl. Microbiol.* **2003**, *36*, 162–167. [CrossRef]

31. Mathis, G.F.; Hofacre, C.; Scicutella, N. Performance improvement with a feed added coated blend of essential oils, a coated blend of organic and inorganic acids with essential oils, or virginiamycin in broilers challenged with *Clostridium perfringens*. In Proceedings of the International Poultry Scientific Forum, Strasbourg, France, 26–30 August 2007.

32. Chen, D.; Chen, G.; Sun, Y.; Zeng, X.; Ye, H. Physiological genetics, chemical composition, health benefits and toxicology of tea (*Camellia sinensis* L.) flower: A review. *Food Res. Int.* **2020**, *137*, 109584. [CrossRef]

33. Dhifi, W.; Hamrouni, I.; Ayachi, S.; Chahed, T.; Saidani, M.; Marzouk, B. Biochemical characterization of some Tunisian olive oils. *J. Food Lipids* **2004**, *11*, 287–296. [CrossRef]

34. Alloui, M.N.; Agabou, A.; Alloui, N. Application of herbs and phytogenic feed additives in poultry production—A Review. *Glob. J. Anim. Sci.* **2014**, *2*, 234–243.

35. Nantapo, C.W.T.; Marume, U. Exploring the potential of *Myrothamnus flabellifolius* Welw. (resurrection tree) as a phytogenic feed additive in animal nutrition. *Animals* **2022**, *12*, 1973. [CrossRef]

36. Oladeji, I.S.; Adegbenro, M.; Osho, I.B.; Olarotimi, O.J. The efficacy of phytogenic feed additives in poultry production: A review. *Turk. J. Agric. Food Sci. Technol.* **2019**, *7*, 2038–2041. [CrossRef]

37. Perricone, V.; Comi, M.; Giromini, C.; Rebucci, R.; Agazzi, A.; Savoini, G.; Bontempo, V. Green tea and pomegranate extract administered during critical moments of the production cycle improves blood antiradical activity and alters cecal microbial ecology of broiler chickens. *Animals* **2020**, *10*, 785. [CrossRef]

38. Aksit, M.; Goksoy, E.; Kok, F.; Ozdemir, D.; Ozdogan, M. The impacts of organic acid and essential oil supplementations to diets on the microbiological quality of chicken carcasses. *Arch. Geflugelkd.* **2006**, *70*, 168–173.

39. Ganguly, S. Phytogenic growth promoter as replacers for antibiotic growth promoter in poultry birds. *Adv. Pharmacoepidem. Drug Saf.* **2013**, *2*, 119. [CrossRef]

40. Mohebodini, H.; Jazi, V.; Bakhshalinejad, R.; Shabani, A.; Ashayerizadeh, A. Effect of dietary resveratrol supplementation on growth performance, immune response, serum biochemical indices, cecal microflora, and intestinal morphology of broiler chickens challenged with *Escherichia coli*. *Livest. Sci.* **2019**, *229*, 13–21. [CrossRef]

41. Vanmarsenille, C.; Elseviers, J.; Yvanoff, C.; Hassanzadeh-Ghassabeh, G.; Garcia Rodriguez, G.; Martens, E.; Depicker, A.; Martel, A.; Haesebrouck, F. In planta expression of nanobody-based designer chicken antibodies targeting Campylobacter. *PLoS ONE* **2018**, *13*, e0204222. [CrossRef]

42. Ferreira, P.M.P.; Farias, D.F.; Oliveira, J.T.D.A.; Carvalho, A.D.F.U. *Moringa oleifera*: Bioactive compounds and nutritional potential. *Rev. Nutr.* **2008**, *21*, 431–437. [CrossRef]

43. Manzoor, M.; Anwar, F.; Iqbal, T.; Bhanger, M.I. Physico-chemical characterization of *Moringa concanensis* seeds and seed oil. *J. Am. Oil Chem. Soc.* **2007**, *84*, 413–419. [CrossRef]

44. Abd El-Hack, M.E.; Alagawany, M.; Elrys, A.S.; Desoky, E.S.M.; Tolba, H.M.; Elnahal, A.S.; Elnesr, S.S.; Swelum, A.A. Effect of forage *Moringa oleifera* L.(moringa) on animal health and nutrition and its beneficial applications in soil, plants and water purification. *Agriculture* **2018**, *8*, 145. [CrossRef]

45. Ravani, A.; Prasad, R.V.; Gajera, R.R.; Joshi, D.C. Potentiality of *Moringa oleifera* for food and nutritional security—A review. *Agric. Rev.* **2017**, *38*, 228–232. [CrossRef]

46. Dalei, J.; Rao, V.M.; Sahoo, D.; Rukmini, M.; Ray, R. Review on nutritional and pharmacological potencies of *Moringa oleifera*. *Eur. J. Pharm. Med. Res.* **2016**, *3*, 150–155.

47. Gopalakrishnan, L.; Doriya, K.; Kumar, D.S. *Moringa oleifera*: A review on nutritive importance and its medicinal application. *Food Sci. Hum. Wellness* **2016**, *5*, 49–56. [CrossRef]

48. Mbikay, M. Therapeutic potential of *Moringa oleifera* leaves in chronic hyperglycaemia and dyslipidaemia: A review. *Front. Pharm.* **2012**, *3*, 1–12. [CrossRef]

49. Saini, R.K.; Shetty, N.P.; Giridhar, P. GC-FID/MS analysis of fatty acids in Indian cultivars of *Moringa oleifera*: Potential sources of PUFA. *J. Am. Oil Chem. Soc.* **2014**, *91*, 1029–1034. [CrossRef]

50. Maizuwo, A.I.; Hassan, A.S.; Momoh, H.; Muhammad, J.A. Phytochemical constituents, biological activities, therapeutic potentials and nutritional values of *Moringa oleifera* (Zogale): A review. *J. Drug Des. Med. Chem.* **2017**, *3*, 60–66.

51. Rodríguez-Pérez, C.; Quirantes-Piné, R.; Fernández-Gutiérrez, A.; Segura-Carretero, A. Optimization of extraction method to obtain a phenolic compounds-rich extract from *Moringa oleifera Lam* leaves. *Ind. Crops Prod.* **2015**, *66*, 246–254. [CrossRef]

52. Makkar, H.P.S.; Becker, K. Nutritional value and antinutritional components of whole and ethanol extracted *Moringa oleifera* leaves. *Anim. Feed Sci. Technol.* **1996**, *63*, 211–228. [CrossRef]

53. Kakengi, A.; Kaijage, J.; Sarwatt, S.; Mutayoba, S.; Shem, M.; Fujihara, T. Effect of *Moringa oleifera* leaf meal as a substitute for sunflower seed meal on performance of laying hens in Tanzania. *Livest. Res. Rural Dev.* **2007**, *19*, 446.

54. Khan, I.; Zaneb, H.; Masood, S.; Yousaf, M.S.; Rehman, H.F.; Rehman, H. Effect of *Moringa oleifera* leaf powder supplementation on growth performance and intestinal morphologyin broiler chickens. *J. Anim. Physiol. Anim. Nutr.* **2017**, *101*, 114–121. [CrossRef]

55. Wafula, M.C.; Mutisya, M.D.; Malombe, I. Non-volatile chemical composition and botanical extracts from *Lippia javanica* (Burm. F) *Spreng. in control of cowpea aphids. J. Appl. Sci.* **2019**, *19*, 325–330.

56. Ombito, J.O.; Salano, E.N.; Yegon, P.K.; Ngetich, W.K.; Mwangi, E.M. A review on the chemistry of some species of genus *Lippia* (Verbenaceae family). *J. Sci. Innov. Res.* **2014**, *3*, 460–466. [CrossRef]

57. Viljoen, A.M.; Subramoney, S.V.; van Vuuren, S.F.; Başer, K.H.C.; Demirci, B. The composition, geographical variation and antimicrobial activity of *Lippia javanica* (Verbenaceae) leaf essential oils. *J. Ethnopharmacol.* **2005**, *96*, 271–277. [CrossRef]

58. Coopoosamy, R.M.; Naidoo, K.K. An ethnobotanical study of medicinal plants used by traditional healers in Durban, South Africa. *Afr. J. Pharm. Pharmacol.* **2012**, *6*, 818–823. [CrossRef]

59. Maroyi, A. *Lippia javanica* (Burm. F.) Spreng.: Traditional and commercial uses and phytochemical and pharmacological significance in the african and indian subcontinent. *Evid. Based Complem. Altern. Med.* **2017**, *2017*, 1–42.

60. Mahlangeni, N.T.; Moodley, R.; Jonnalagadda, S.B. Elemental composition of *Cyrtanthus obliquus* and *Lippia javanica* used in South African herbal tonic, Imbiza. *Arab. J. Chem.* **2018**, *11*, 128–136. [CrossRef]

61. Chawafambira, A. The effect of incorporating herbal (*Lippia javanica*) infusion on the phenolic, physicochemical, and sensorial properties of fruit wine. *Food Sci. Nutr.* **2021**, *9*, 4539–4549. [CrossRef]

62. Olivier, D.K.; Shikanga, E.A.; Combrinck, S.; Krause, R.W.M.; Regnier, T.; Dlamini, T.P. Phenylethanoid glycosides from *Lippia javanica*. *S. Afr. J. Bot.* **2010**, *76*, 58–63. [CrossRef]

63. Pascual, M.E.; Slowing, K.; Carretero, M.E.; Villar, Á. Antiulcerogenic activity of *Lippia alba* (Mill.) NE Brown (Verbenaceae). *Farmaco* **2001**, *56*, 501–504. [CrossRef] [PubMed]

64. Chagonda, L.S.; Chalchat, J.C. Essential oil composition of *Lippia javanica* (Burm. f.) spreng chemotype from Western Zimbabwe. *J. Essent. Oil-Bear. Plants* **2015**, *18*, 482–485. [CrossRef]

65. Mpofu, D.A.; Marume, U.; Mlambo, V.; Hugo, A. The effects of *Lippia javanica* dietary inclusion on growth performance, carcass characteristics and fatty acid profiles of broiler chickens. *Anim. Nutr.* **2016**, *2*, 160–167. [CrossRef] [PubMed]

66. Mnisi, C.M.; Matshogo, T.B.; van Niekerk, R.; Mlambo, V. Growth performance, haemo-biochemical parameters and meat quality characteristics of male Japanese quails fed a *Lippia javanica*-based diet. *S. Afr. J. Anim. Sci.* **2017**, *47*, 661–671. [CrossRef]

67. Namita, P.; Mukesh, R.; Vijay, K.J. *Camellia sinensis* (green tea): A review. *Glob. J. Pharmacol.* **2012**, *6*, 52–59.

68. Chen, L.; Zhou, Z.X.; Yang, Y.J. Genetic improvement and breeding of tea plant (*Camellia sinensis*) in China: From individual selection to hybridization and molecular breeding. *Euphytica* **2007**, *154*, 239–248. [CrossRef]

69. Tai, Y.; Liu, C.; Yu, S.; Yang, H.; Sun, J.; Guo, C.; Huang, B.; Liu, Z.; Yuan, Y.; Xia, E.; et al. Gene co-expression network analysis reveals coordinated regulation of three characteristic secondary biosynthetic pathways in tea plant (*Camellia sinensis*). *BMC Genom.* **2018**, *19*, 1–13. [CrossRef]

70. Abdo, Z.M.A.; Hassan, R.A.; Amal, A.E.; Shahinaz, A.H. Effect of adding *Camellia sinensis* and its aqueous extract as natural antioxidants to laying hen diet on productive, reproductive performance and egg quality during storage and its content of cholesterol. *Egypt. Poult. Sci. J.* **2010**, *30*, 1121–1149.

71. Wang, Y.; Ho, C.T. Polyphenolic chemistry of tea and coffee: A century of progress. *J. Agric. Food Chem.* **2009**, *57*, 8109–8114. [CrossRef]

72. Reygaert, W.C. The antimicrobial possibilities of green tea. *Front. Microbiol.* **2014**, *5*, 434. [CrossRef]

73. Cheng, Y.; Sun, J.; Zhao, H.; Guo, H.; Li, J. Functional mechanism on stem cells by tea (*Camellia sinensis*) bioactive compounds. *Food Sci. Hum. Wellness* **2022**, *11*, 579–586. [CrossRef]

74. Chen, D.; Ding, Y.; Chen, G.; Sun, Y.; Zeng, X.; Ye, H. Components identification and nutritional value exploration of tea (*Camellia sinensis* L.) flower extract: Evidence for functional food. *Int. Food Res. J.* **2020**, *132*, 109100. [CrossRef]

75. Chen, G.J.; Yuan, Q.X.; Saeeduddin, M.; Ou, S.Y.; Zeng, X.X.; Ye, H. Recent advances in tea polysaccharides: Extraction, purification, physicochemical characterization and bioactivities. *Carbohydr. Polym.* **2016**, *153*, 663–678. [CrossRef]

76. Gramza-Michałowska, A.; Kobus-Cisowska, J.; Kmiecik, D.; Korczak, J.; Helak, B.; Dziedzic, K.; Górecka, D. Antioxidative potential, nutritional value and sensory profiles of confectionery fortified with green and yellow tea leaves (*Camellia sinensis*). *Food Chem.* **2016**, *211*, 448–454. [CrossRef]

77. Mahlake, S.K.; Mnisi, C.M.; Lebopa, C.K.; Kumanda, C. The effect of green tea (*Camellia sinensis*) leaf powder on growth performance, selected haematological indices, carcass characteristics and meat quality parameters of Jumbo quail. *Sustainability* **2021**, *13*, 7080. [CrossRef]

78. Al-Kayali, K.K.; Razooqi, B.M.; Mtaab, A.S. Antibacterial activity of aqueous extract of green tea on bacteria isolated from children with impetigo. *Diyala J. Med.* **2011**, *1*, 37–43.

79. Friesen, N.; Fritsch, R.M.; Blattner, F.R. Phylogeny and new intrageneric classification of *Allium* L. (*Alliaceae*) based on nuclear ribosomal DNA ITS sequences. *Aliso* **2002**, *22*, 372–395. [CrossRef]

80. Morales-González, J.A.; Madrigal-Bujaidar, E.; Sánchez-Gutiérrez, M.; Izquierdo-Vega, J.A.; Valadez-Vega, M.D.C.; Álvarez-González, I.; Morales-González, Á.; Madrigal-Santillán, E. Garlic (*Allium sativum* L.): A brief review of its antigenotoxic effects. *Foods* **2019**, *8*, 343. [CrossRef]

81. Tesfaye, A.; Mengesha, W. Traditional uses, phytochemistry and pharmacological properties of garlic (*Allium Sativum*) and its biological active compounds. *Int. J. Sci. Res. Eng. Technol.* **2015**, *1*, 142–148.

82. Indrasanti, D.; Indradji, M.; Hastuti, S.; Aprilliyani, E.; Rosyadi, K. The administration of garlic extract on *Eimeria stiedai* oocysts and the hematological profile of the *Coccidia* infected rabbits. *Media Peternak.* **2017**, *3*, 158–164. [CrossRef]

83. Milosevic, N.; Stojcic, M.D.; Stanacev, V.; Peric, L.; Veljic, M. The performance and carcass traits of broilers feed with garlic (*Allium sativum*) additive. *Eur. Poult. Sci.* **2013**, *77*, 254–259.

84. Ogbuewu, I.P.; Mbajiorgu, C.A.; Okoli, I.C. Antioxidant activity of ginger and its effect on blood chemistry and production physiology of poultry. *Comp. Clin. Pathol.* **2019**, *28*, 655–660. [CrossRef]

85. Clement, F.; Pramod, S.N.; Venkatesh, Y.P. Identity of the immunomodulatory proteins from garlic (*Allium sativum*) with the major garlic lectins or agglutinins. *Int. Immunopharmacol.* **2010**, *10*, 316–324. [CrossRef] [PubMed]

86. Țigu, A.B.; Moldovan, C.S.; Toma, V.A.; Farcaș, A.D.; Moț, A.C.; Jurj, A.; Fischer-Fodor, E.; Mircea, C.; Pârvu, M. Phytochemical analysis and in vitro effects of *Allium fistulosum* L. and *Allium sativum* L. extracts on human normal and tumor cell lines: A comparative study. *Molecules* **2021**, *26*, 574. [CrossRef]

87. Kimura, S.; Tung, Y.C.; Pan, M.H.; Su, N.W.; Lai, Y.J.; Cheng, K.C. Black garlic: A critical review of its production, bioactivity, and application. *J. Food Drug Anal.* **2017**, *25*, 62–70. [CrossRef]

88. Sobolewska, D.; Michalska, K.; Podolak, I.; Grabowska, K. Steroidal saponins from the genus *Allium*. *Phytochem. Rev.* **2016**, *15*, 1–35. [CrossRef]

89. Pagrut, N.; Ganguly, S.; Tekam, S.; Bhainsare, P. Effect of supplementation of garlic extract on the productive performance of broiler chicks. *Ratio* **2018**, *11*, 12.

90. Eltazi, S.; Ka, M.; Ma, M. Effect of using garlic powder as natural feed additive on performance and carcass quality of broiler chicks. *Assiut. Vet. Med. J.* **2014**, *60*, 45–53. [CrossRef]

91. Brewster, J.L. *Onions and Other Vegetable Alliums*, 2nd ed.; CABI: Wallingford, UK, 2008.

92. Sami, R.; Elhakem, A.; Alharbi, M.; Benajiba, N.; Almatrafi, M.; Helal, M. Nutritional values of onion bulbs with some essential structural parameters for packaging process. *Appl. Sci.* **2021**, *11*, 2317. [CrossRef]

93. Mahmood, N.; Muazzam, M.A.; Ahmad, M.; Hussain, S.; Javed, W. Phytochemistry of *Allium cepa* L. (Onion): An overview of its nutritional and pharmacological importance. *Sci. Inq. Rev.* **2021**, *5*, 41–59. [CrossRef]

94. Goodarzi, M.; Nanekarani, S. Effect of onion extract in drink water on performance and carcass traits in broiler chickens. *IERI Procedia* **2014**, *8*, 107–112. [CrossRef]

95. Pareek, S.; Sagar, N.A.; Sharma, S.; Kumar, V. Onion (*Allium cepa* L.). In *Fruit and Vegetable Phytochemicals: Chemistry and Human Health*, 2nd ed.; Yahia, E.M., Ed.; John Wiley & Sons Ltd.: Chichester, UK, 2018; Volume 2, pp. 1145–1162.

96. Metrani, R.; Singh, J.; Acharya, P.K.; Jayaprakasha, G.S.; Patil, B. Comparative metabolomics profiling of polyphenols, nutrients and antioxidant activities of two red onion (*Allium cepa* L.) cultivars. *Plants* **2020**, *9*, 1077. [CrossRef]

97. Kothari, D.; Lee, W.D.; Niu, K.M.; Kim, S.K. The genus *Allium* as poultry feed additive: A review. *Animals* **2019**, *9*, 1032. [CrossRef]

98. Zhao, X.X.; Lin, F.J.; Li, H.; Li, H.B.; Wu, D.T.; Geng, F.; Gan, R.Y. Recent advances in bioactive compounds, health functions, and safety concerns of onion (*Allium cepa* L.). *Front. Nutr.* **2021**, *8*, 669805. [CrossRef]

99. Guo, A.; Bao, K.; Sang, S.; Zhang, X.; Shao, B.; Zhang, C.; Yang, X. Soft-chemistry synthesis, solubility and interlayer spacing of carbon nano-onions. *RSC Adv.* **2021**, *11*, 6850–6858. [CrossRef]

100. Goodarzi, M.; Nanekarani, S.; Landy, N. Effect of dietary supplementation with onion (*Allium cepa* L.) on performance, carcass traits and intestinal microflora composition in broiler chickens. *Asian Pac. J. Trop. Dis.* **2014**, *4*, S297–S301. [CrossRef]

101. Slavin, J.L. Carbohydrates, dietary fiber, and resistant starch in white vegetables: Links to health outcomes. *Adv. Nutr.* **2013**, *4*, 351S–355S. [CrossRef]

102. Aditya, S.; Ahammed, M.; Jang, S.H.; Ohh, S.J. Effects of dietary onion (*Allium cepa*) extract supplementation on performance, apparent total tract retention of nutrients, blood profile and meat quality of broiler chicks. *Asian-Australas. J. Anim. Sci.* **2017**, *30*, 229–235. [CrossRef]

103. Ibrahiem, I.A.; Elam, T.A.; Mohamed, F.F.; Awadalla, S.A.; Yousif, Y.I.; Ibrahim, I.A.; Elam, T.A.; Mohamed, F.F.; Awadalla, S.A.; Yousif, Y.I. Effect of onion and/or garlic as feed additives on growth performance and immunity in broiler muscovy ducks. In Proceedings of the First Scientific Conference of Faculty of Veterinary Medicine, Moshtohor, Egypt, 1–4 September 2004; pp. 1–4.

104. Dosoky, W.M.; Zeweil, H.S.; Ahmed, M.H.; Zahran, S.M.; Shaalan, M.M.; Abdelsalam, N.R.; Abdel-Moneim, A.M.E.; Taha, A.E.; El-Tarabily, K.A.; Abd El-Hack, M.E. Impacts of onion and cinnamon supplementation as natural additives on the performance, egg quality, and immunity in laying Japanese quail. *Poult. Sci.* **2021**, *100*, 101482. [CrossRef]

105. Mainasara, M.M.; Bakar, M.F.A.; Waziri, A.H.; Musa, A.R. Comparison of phytochemical, proximate and mineral composition of fresh and dried peppermint (*Mentha piperita*) leaves. *J. Sci. Technol.* **2018**, *10*, 85–91. [CrossRef]

106. Trevisan, S.C.C.; Menezes, A.P.P.; Barbalho, S.M.; Guiguer, É.L. Properties of *Mentha piperita*: A brief review. *World J. Pharm. Med. Res.* **2017**, *3*, 309–313.

107. Farnad, N.; Heidari, R.; Aslanipour, B. Phenolic composition and comparison of antioxidant activity of alcoholic extracts of peppermint (*Mentha piperita*). *J. Food Meas. Charact.* **2014**, *8*, 113–121. [CrossRef]

108. Sujana, P.; Sridhar, T.M.; Josthna, P.; Naidu, C.V. Antibacterial activity and phytochemical analysis of *Mentha piperita* L.(Peppermint)—An important multipurpose medicinal plant. *Sci. Res. J.* **2013**, *4*, 77–83.

109. Badal, R.M.; Badal, D.; Badal, P.; Khare, A.; Shrivastava, J.; Kumar, V. Pharmacological action of *Mentha piperita* on lipid profile in fructose-fed rats. *Iran. J. Pharm. Res.* **2011**, *10*, 843–847.

110. Uribe, E.; Marín, D.; Vega-Gálvez, A.; Quispe-Fuentes, I.; Rodríguez, A. Assessment of vacuum-dried peppermint (*Mentha piperita* L.) as a source of natural antioxidants. *Food Chem.* **2016**, *190*, 559–565. [CrossRef]

111. Beigi, M.; Torki-Harchegani, M.; Ghasemi Pirbalouti, A. Quantity and chemical composition of essential oil of peppermint (*Mentha × piperita* L.) leaves under different drying methods. *Int. J. Food Prop.* **2018**, *21*, 267–276. [CrossRef]

112. Dorman, H.D.; Koşar, M.; Başer, K.H.C.; Hiltunen, R. Phenolic profile and antioxidant evaluation of *Mentha x piperita* L.(peppermint) extracts. *Nat. Prod. Commun.* **2009**, *4*, 535–545. [CrossRef]

113. Khempaka, S.; Pudpila, U.; Molee, W. Effect of dried peppermint (*Mentha cordifolia*) on growth performance, nutrient digestibility, carcass traits, antioxidant properties, and ammonia production in broilers. *J. Appl. Poult. Res.* **2013**, *22*, 904–912. [CrossRef]

114. Toghyani, M.; Toghyani, M.; Gheisari, A.; Ghalamkari, G.; Mohammadrezaei, M. Growth performance, serum biochemistry and blood hematology of broiler chicks fed different levels of black seed (*Nigella sativa*) and peppermint (*Mentha piperita*). *Livest. Sci.* **2010**, *129*, 173–178. [CrossRef]

115. Abdel-Wareth, A.A.; Kehraus, S.; Südekum, K.H. Peppermint and its respective active component in diets of broiler chickens: Growth performance, viability, economics, meat physicochemical properties, and carcass characteristics. *Poult. Sci.* **2019**, *98*, 3850–3859. [CrossRef]

116. Dagne, E.; Bisrat, D.; Viljoen, A.; Van Wyk, B.E. Chemistry of Aloe species. *Curr. Org. Chem.* **2000**, *4*, 1055–1078. [CrossRef]

117. Minjares-Fuentes, R.; Rodríguez-González, V.M.; González-Laredo, R.F.; Eim, V.; González-Centeno, M.R.; Femenia, A. Effect of different drying procedures on the bioactive polysaccharide acemannan from *Aloe vera* (*Aloe barbadensis Miller*). *Carbohydr. Polym.* **2017**, *168*, 327–336. [CrossRef]

118. Boudreau, M.D.; Mellick, P.W.; Olson, G.R.; Felton, R.P.; Thorn, B.T.; Beland, F.A. Clear evidence of carcinogenic activity by a whole-leaf extract of aloe barbadensis miller (*Aloe vera*) in F344/n rats. *Toxicol. Sci.* **2013**, *131*, 26–39. [CrossRef]

119. Zhang, Y.; Bao, Z.; Ye, X.; Xie, Z.; He, K.; Mergens, B.M.; Zheng, Q.B. Chemical investigation of major constituents in Aloe vera leaves and several commercial Aloe juice powders. *J. AOAC Int.* **2018**, *101*, 1741–1751. [CrossRef]

120. Christaki, E.V.; Florou-Paneri, P.C. *Aloe vera*: A plant for many uses. *J. Food Agric. Environ.* **2010**, *8*, 245–249.

121. Aida, P.U.I.A.; Cosmin, P.U.I.A.; Emil, M.O.I.Ş.; Graur, F.; Fetti, A.; Florea, M. The phytochemical constituents and therapeutic uses of genus Aloe: A review. *Not. Bot. Horti Agrobot. Cluj Napoca* **2021**, *49*, 12332.

122. Darabighane, B.; Zarei, A.; Shahneh, A.Z.; Mahdavi, A. Effects of different levels of *Aloe vera* gel as an alternative to antibiotic on performance and ileum morphology in broilers. *Ital. J. Anim. Sci.* **2011**, *10*, 189–194. [CrossRef]

123. Dai, B.S.; Jiang, L.; Chen, S.X. Effects of medicinal herb and polysaccharide from Aloe on gut microflora, immune function and growth performance in broiler. *China Poult.* **2007**, *29*, 21–24. (In Chinese)

124. Darabighane, B.; Zarei, A. The effects of the different levels of *Aloe vera* gel on oocysts shedding in broilers with coccidiosis. *Planta Med.* **2012**, *78*, 496. [CrossRef]

125. Michalak, I.; Mahrose, K. Seaweeds, intact and processed, as a valuable component of poultry feeds. *J. Mar. Sci. Eng.* **2020**, *8*, 620. [CrossRef]

126. Stengel, D.B.; Connan, S.; Popper, Z.A. Algal chemodiversity and bioactivity: Sources of natural variability and implications for commercial application. *Biotech. Adv.* **2011**, *29*, 483–501. [CrossRef] [PubMed]

127. van Hal, J.W.; Huijgen, W.J.J.; López-Contreras, A.M. Opportunities and challenges for seaweed in the biobased economy. *Trends Biotechnol.* **2014**, *32*, 231–233. [CrossRef] [PubMed]

128. Choi, Y.J.; Lee, S.R.; Oh, J.W. Effects of dietary fermented seaweed and seaweed fusiforme on growth performance, carcass parameters and immunoglobulin concentration in broiler chicks. *Asian-Australas. J. Anim. Sci.* **2014**, *27*, 862–870. [CrossRef]

129. Abudabos, A.M.; Okab, A.B.; Aljumaah, R.S.; Samara, E.M.; Abdoun, K.A.; Al-Haidary, A.A. Nutritional value of green seaweed (*Ulva lactuca*) for broiler chickens. *Ital. J. Anim. Sci.* **2013**, *12*, 177–181. [CrossRef]

130. Nhlane, L.T.; Mnisi, C.M.; Mlambo, V.; Madibana, M.J. Nutrient digestibility, growth performance, and blood indices of Boschveld chickens fed seaweed-containing diets. *Animals* **2020**, *10*, 1296. [CrossRef]

131. Mohammadigheisar, M.; Shouldice, V.L.; Sands, J.S.; Lepp, D.; Diarra, M.S.; Kiarie, E.G. Growth performance, breast yield, gastrointestinal ecology and plasma biochemical profile in broiler chickens fed multiple doses of a blend of red, brown and green seaweeds. *Brit. Poult. Sci.* **2020**, *61*, 590–598. [CrossRef]

132. El-Deek, A.A.; Brikaa, M.A. Nutritional and biological evaluation of marine seaweed as a feedstuff and as a pellet binder in poultry diet. *Int. J. Poult. Sci.* **2009**, *8*, 875–881. [CrossRef]

133. Peng, W.; Berry, E.M. The concept of food security. *Encycl. Food Sec. Sustain.* **2019**, *2*, 1–7.

134. Ogundeji, A.A.; Danso-Abbeam, G.; Jooste, A. Climate information pathways and farmers' adaptive capacity: Insights from South Africa. *Environ. Dev.* **2022**, *44*, 100743. [CrossRef]

135. Mthiyane, D.M.N.; Mhlanga, B.S. The nutritive value of marula (*Sclerocarya birrea*) seed cake for broiler chickens: Nutritional composition, performance, carcass characteristics and oxidative and mycotoxin status. *Trop. Anim. Health Prod.* **2017**, *49*, 835–842. [CrossRef]

136. Food and Agriculture Organization of the United Nations (FAO). Meat Market Review: Emerging Trends and Outlook. Available online: https://www.fao.org/publications/card/en/c/CB2423EN/#:~{}:text=World%20meat%20production%20in%202020,import%20demand%20from%20East%20Asia (accessed on 1 December 2020).

137. Andretta, I.; Hickmann, F.M.W.; Remus, A.; Franceschi, C.H.; Mariani, A.B.; Orso, C.; Kipper, M.; Létourneau-Montminy, M.P.; Pomar, C. Environmental impacts of pig and poultry production: Insights from a systematic review. *Front. Vet. Sci.* **2021**, *8*, 750733. [CrossRef]

138. Lovatto, P.A.; Hauschild, L.; Hauptli, L.; Lehnen, C.R.; Carvalho, A.A. Modelagem da ingestão, retenção e excreção de nitrogênio e fósforo pela suinocultura Brasileira. *J. Ver. Bras. Zootec.* **2005**, *34*, 2348–2354. [CrossRef]

139. Russ, A.; Schaeffer, E. Ammonia Emissions from Broiler Operations Higher Than Previously Thought. 2. Environmental Integrity Project. Available online: https://www.environmentalintegrity.org/wp-content/uploads/2017/02/Ammonia-Report.pdf (accessed on 22 January 2018).

140. Glibert, P.M. From hogs to HABs: Impacts of industrial farming in the US on nitrogen and phosphorus and greenhouse gas pollution. *Biogeochemistry* **2020**, *150*, 139–180. [CrossRef]

141. de Mesquita Souza Saraiva, M.; Lim, K.; do Monte, D.F.M.; Givisiez, P.E.N.; Alves, L.B.R.; de Freitas Neto, O.C.; Kariuki, S.; Júnior, A.B.; de Oliveira, C.J.B.; Gebreyes, W.A. Antimicrobial resistance in the globalized food chain: A One Health perspective applied to the poultry industry. *Braz. J. Microbiol.* **2022**, *53*, 465–486. [CrossRef]

142. Torres, A.G. Airborne particulate matter from livestock production systems: A review of an air pollution problem. *Environ. Pollut.* **2010**, *158*, 1–17.

143. Forgetta, V.; Rempel, H.; Malouin, F.; Vaillancourt, R., Jr.; Topp, E.; Dewar, K.; Diarra, M.S. Pathogenic and multidrug-resistant *Escherichia fergusonii* from broiler chicken. *Poult. Sci.* **2012**, *91*, 512–525. [CrossRef]

144. IFAH. *The Costs of Animal Disease*; White Paper; Oxford Analytica: Oxford, UK, 2012.

145. Soetan, K.O.; Oyewole, O.E. The need for adequate processing to reduce the anti-nutritional factors in plants used as human foods and animal feeds: A review. *Afr. J. Food Sci.* **2009**, *3*, 223–232.

146. Emiola, I.A.; Ologhobo, A.; Akinlade, J.; Adedeji, S.O.; Bamgbabe, O. Effect of inclusion of differently processed *Mucuna utilis* seed meal on performance characteristics of broilers. *Trop. Anim. Health Prod.* **2003**, *6*, 13–21.

147. Erdaw, M.M.; Wu, S.; Iji, P.A. Growth and physiological responses of broiler chickens to diets containing raw, full-fat soybean and supplemented with a high-impact microbial protease. *Asian-Australas. J. Anim. Sci.* **2017**, *30*, 1303–1313. [CrossRef]

148. Kavoi, B.M.; Gakuya, D.W.; Mbugua, P.N.; Kiama, S.G. Effects of dietary *Moringa oleifera* leaf meal supplementation on chicken intestinal structure and growth performance. *J. Morphol. Sci.* **2016**, *33*, 186–192. [CrossRef]

149. Tang, D.F.; Ru, F.; Song, S.Y.; Choct, M. The effect of cassava chips, pellets, pulp and maize based diets on performance, digestion and metabolism of nutrients for broilers. *J. Anim. Vet. Adv.* **2014**, *11*, 1332–1337.

150. Akapo, A.O.; Oso, A.O.; Bamgbose, A.M.; Sanwo, K.A.; Jegede, A.V.; Sobayo, R.A.; Idowu, O.M.; Fan, J.; Li, L.; Olorunsola, R.A. Effect of feeding cassava (*Manihot esculenta Crantz*) root meal on growth performance, hydrocyanide intake and haematological parameters of broiler chicks. *Trop. Anim. Health Prod.* **2014**, *46*, 1167–1172. [CrossRef] [PubMed]

151. Jelveha, K.; Rasoulia, B.; Seidavia, A.; Diarra, S.S. Comparative effects of Chinese green tea (*Camellia sinensis*) extract and powder as feed supplements for broiler chickens. *J. Appl. Anim. Res.* **2018**, *46*, 1114–1117. [CrossRef]

152. Hassan, S.M. Effects of Guar meal, Guar gum and saponin rich Guar meal extract on productive performance of starter broiler chicks. *Afr. J. Agric. Res.* **2013**, *8*, 2464–2469.

153. Farhadi, D.; Karimi, A.; Sadeghi, G.; Sheikhahmadi, A.; Habibian, M.; Raei, A.; Sobhani, K. Effects of using eucalyptus (*Eucalyptus globulus* L.) leaf powder and its essential oil on growth performance and immune response of broiler chickens. *Iran. J. Vet. Res.* **2017**, *18*, 60–62.

154. Kulshreshtha, G.; Rathgeber, B.; Stratton, G.; Thomas, N.; Evans, F.; Critchley, A.; Prithiviraj, B. Feed supplementation with red seaweeds, *Chondrus crispus* and *Sarcodiotheca gaudichaudii*, affects performance, egg quality, and gut microbiota of layer hens. *Poult. Sci.* **2014**, *93*, 2991–3001. [CrossRef]

155. Nduku, X.P.; Mabusela, S.P.; Nkukwana, T.T. Growth and meat quality of broiler chickens fed *Moringa oleifera* leaf meal, a probiotic and an organic acid. *S. Afr. J. Anim. Sci.* **2020**, *50*, 710–718. [CrossRef]

156. Igwilo, I.O.; Okonkwo, J.C.; Ugochukwu, G.C.; Ezekwesili, C.N.; Nwenyi, V. Comparative studies on the nutrient composition and anti-nutritional factors in different parts of *Moringa oleifera* plant found in Awka, Nigeria. *Bioscientist* **2017**, *5*, 1–12.

157. Narzarya, H.; Basumatary, S. Amino acid profiles and anti-nutritional contents of traditionally consumed six wild vegetables. *Curr. Chem. Lett.* **2019**, *8*, 137–144. [CrossRef]

158. Mahlake, S.K.; Mnisi, C.M.; Kumanda, C.; Mthiyane, D.M.N.; Montso, P.K. Green tea (*Camellia sinensis*) products as alternatives to antibiotics in poultry nutrition: A Review. *Antibiotics* **2022**, *11*, 565. [CrossRef]

159. Salawu, K.; Owolarafe, T.A.; Ononamadu, C.J.; Ihegboro, G.O.; Lawal, T.A.; Aminu, M.A.; Oyekale, A.J. Phytochemical, nutritional composition and heavy metals content of *Allium cepa* (Onion) and *Allium sativum* (Garlic) from Wudil Central Market, Kano State, Nigeria. *Biokemistri* **2019**, *33*, 311–317.

160. Nalimu, F.; Oloro, J.; Kahwa, I.; Ogwang, P.E. Review on the phytochemistry and toxicological profiles of *Aloe vera* and *Aloe ferox*. *Future J. Pharm. Sci.* **2021**, *7*, 1–21. [CrossRef]

161. Trendafilova, A.; Moujir, L.M.; Sousa, P.M.C.; Seca, A.M.L. Research advances on health effects of edible Artemisia species and some sesquiterpene lactones constituents. *Foods* **2021**, *10*, 65. [CrossRef]

162. Castañeda-Loaiza, V.; Placines, C.; Rodrigues, M.J.; Pereira, C.; Zengin, G.; Uysal, A.; Jeko, J.; Cziáky, Z.; Reis, C.P.; Gaspar, M.M. If you cannot beat them, join them: Exploring the fruits of the invasive species *Carpobrotus edulis* (L.) N.E. Br as a source of bioactive products. *Ind. Crop. Prod.* **2020**, *144*, 112005. [CrossRef]

163. Nnenna, A.O.; Okwudil, A.; Kelechi, A.K.; Kenechukwu, O.C.; Izuchukwu, U.I.; Mayer, E.M. Nutritional profile, bioactive compound content and antioxidant activity of ethanol leaf extract of *Eucalyptus tereticornis*. *Eur. J. Biomed. Pharm. Sci.* **2020**, *7*, 61–73.

164. Feng, Y.; Carroll, A.R.; Addepalli, R.; Fechner, G.A.; Avery, V.M.; Quinn, R.J. Vanillic acid derivatives from the green algae (*Cladophora socialis*) as potent protein tyrosine phosphatase 1B inhibitors. *J. Nat. Prod.* **2007**, *70*, 1790–1792. [CrossRef]

165. Lizardi-Jimenez, M.A.; Hernandez-Martınez, R. Solid state fermentation (SSF): Diversity of applications to valorize waste and biomass. *3 Biotech* **2017**, *7*, 1–9. [CrossRef]

166. de Pereira, G.V.M.; de Coelho, B.O.; Júnior, A.I.M.; Thomaz-Soccol, V.; Soccol, C.R. How to select a probiotic? A review and update of methods and criteria. *Biotechnol. Adv.* **2018**, *36*, 2060–2076. [CrossRef]

167. Olukomaiya, O.O.; Adiamo, O.Q.; Fernando, W.C.; Mereddy, R.; Li, X.; Sultanbawa, Y. Effect of solid-state fermentation on proximate composition, anti-nutritional factor, microbiological and functional properties of lupin flour. *Food Chem.* **2020**, *315*, 126238. [CrossRef]

168. Shi, H.; Su, B.; Chen, X.; Pian, R. Solid state fermentation of *Moringa oleifera* leaf meal by mixed strains for the protein enrichment and the improvement of nutritional value. *PeerJ* **2020**, *8*, e10358. [CrossRef] [PubMed]

169. Ijarotimi, O.S.; Adeoti, O.A.; Ariyo, O. Comparative study on nutrient composition, phytochemical, and functional characteristics of raw, germinated, and fermented *Moringa oleifera* seed flour. *Food Sci. Nutr.* **2013**, *1*, 452–463. [CrossRef] [PubMed]

170. Choi, J.; Kim, W.K. Dietary application of tannins as a potential mitigation strategy for current challenges in poultry production: A Review. *Animals* **2020**, *10*, 2389. [CrossRef] [PubMed]

171. Batista, K.A.; Pereira, K.W.J.; Moreira, B.R.; Silva, C.N.S.; Fernandes, K.F. Effect of autoclaving on the nutritional quality of hard-to-cook common beans (*Phaseolus vulgaris*). *Int. J. Environ. Agric. Biotechnol.* **2020**, *5*, 22–30. [CrossRef]

172. Tarimbuka, L.I.; Yusuf, H.B.; Wafar, R.J. Response of weaner rabbits fed toasted sickle pod (*Senna occidentilis*) seed meal. *Asian J. Agric. Res.* **2017**, *1*, 1–7. [CrossRef]

173. Hwang, I.G.; Shin, Y.J.; Lee, S.; Lee, J.; Yoo, S.M. Effects of different cooking methods on the antioxidant properties of red pepper (*Capsicum annuum L*). *Prev. Nutr. Food Sci.* **2012**, *17*, 286–292. [CrossRef]

174. Dosković, V.; Bogosavljević-Bosković, S.; Pavlovski, Z. Enzymes in broiler diets with special reference to protease. *World Poult. Sci. J.* **2013**, *69*, 343–360. [CrossRef]

175. Costa, M.M.; Pestana, J.M.; Carvalho, P.; Alfaia, C.M.; Martins, C.F.; Carvalho, D.; Mourato, M.; Gueifão, S.; Delgado, I.; Coelho, I.; et al. Effect on broiler production performance and meat quality of feeding *Ulva lactuca* supplemented with carbohy-drases. *Animals* **2022**, *12*, 1720. [CrossRef]

176. Abou-Arab, A.A.; Abu-Salem, F.M. Nutritional quality of *Jatropha curcas* seeds and effect of some physical and chemical treatments on their anti-nutritional factors. *Afr. J. Food Sci.* **2010**, *4*, 93–103.

Systematic Review

Anti-Breast Cancer Activity of Essential Oil: A Systematic Review

Mohammad Adam Mustapa [1], Ikhsan Guswenrivo [2], Ade Zuhrotun [3], Nur Kusaira Khairul Ikram [4] and Muchtaridi Muchtaridi [1,5,*]

[1] Department of Pharmaceutical Analysis and Medicinal Chemistry, Faculty of Pharmacy, Universitas Padjadjaran, Sumedang 45363, Indonesia
[2] Research Center for Applied Zoology, National Research and Innovation Agency (BRIN), Jakarta Pusat 10340, Indonesia
[3] Department of Biological Pharmacy, Faculty of Pharmacy, Universitas Padjadjaran, Bangdung 40132, Indonesia
[4] Institute of Biological Sciences, Faculty of Science, Universiti Malaya, Kuala Lumpur 50603, Malaysia
[5] Research Collaboration Center for Radiopharmaceuticals Theranostic, National Research and Innovation Agency (BRIN), Sumedang 45363, Indonesia
* Correspondence: muchtaridi@unpad.ac.id; Tel.: +62-22-8428-8888 (ext. 3210)

Abstract: Breast cancer is the second highest cancer-related death worldwide. The treatment for breast cancer is via chemotherapy; however, occurrences of multidrug resistance, unselective targets, and physicochemical problems suggest that chemotherapy treatment is ineffective. Therefore, there is a need to find better alternatives. Essential oil is a plant secondary metabolite having promising bioactivities and pharmacological effects, including anti-breast cancer capabilities. This review intends to discuss and summarize the effect of essential oils on anti-breast cancer from published journals using keywords in PubMed, Scopus, and Google Scholar databases. Our findings reveal that the compositions of essential oils, mainly terpenoids, have excellent anti-breast cancer pharmacological effects with an IC_{50} value of 0.195 µg/mL. Hence, essential oils have potential as anti-breast cancer drugs candidates with the highest efficacy and the fewest side effects.

Keywords: breast cancer; essential oils; terpenoids

Citation: Mustapa, M.A.; Guswenrivo, I.; Zuhrotun, A.; Ikram, N.K.K.; Muchtaridi, M. Anti-Breast Cancer Activity of Essential Oil: A Systematic Review. *Appl. Sci.* **2022**, *12*, 12738. https://doi.org/10.3390/app122412738

Academic Editors: Călin Jianu and Georgeta Pop

Received: 4 November 2022
Accepted: 29 November 2022
Published: 12 December 2022

Publisher's Note: MDPI stays neutral with regard to jurisdictional claims in published maps and institutional affiliations.

1. Introduction

Breast cancer is a complex condition triggered by abnormalities in the the proliferation of breast cells. Although the disease is present worldwide, there are regional variations in death and survival rates due to differences in the population structure, way-of-life, genetics, and the environment [1,2]. Breast cancer is the most common cancer and the leading cause of mortality among women [3,4]. In the United States, there were around 252,710 new instances of invasive breast cancer and 6341 new cases of ductal carcinoma. The Asia–Pacific area accounts for over 24% of all breast cancer cases, with China, Japan, Vietnam, and Indonesia having the greatest incidence rates [5,6]. The prevalence of this condition is still very high globally, and in Indonesia, and it is predicted that the incidence would rise to 85 per 100,000 women by 2022 [7–9]. Thus, the treatment of breast cancer is still receiving special attention. The treatment of breast cancer to date consists of three methods, namely, surgical therapy, radiation therapy, and chemotherapy [10]. However, these therapies cause a lot of tissue damage and other unwanted side effects [6,11]. Therefore, materials with low toxicity are needed as an alternative that can be taken for therapy. Besides the toxic effect produced being much smaller [12], the effectiveness is also not inferior to drugs used in chemotherapy. One of the natural ingredients widely used in the treatment of breast cancer is essential oil [13]. Essential oils are secondary metabolites in several plants with volatile properties and have many pharmacological effects, including the treatment of cancers [14], such as lung cancer [15], colon cancer [16–18], prostate cancer [19,20], breast cancer [21,22], cervical cancer [23,24], and many more. Currently, many studies are regarding the activity of essential oil for the treatment of breast cancer which has also been summarized in review articles. However, a systematic review of the use of essential oil in the treatment of breast cancer has not been specifically carried out. Therefore, this review aims to summarize and discuss studies related to the use of essential oils for breast cancer therapy arranged based on the rules of a systematic review.

2. Methodology

The arrangement of this systematic review was based on the results of the collected journals and reviews using Scopus, Pubmed, and Google Scholar databases with the keywords "essential oil for breast cancer", "monoterpene for breast cancer", "sesquiterpene for breast cancer", and so on. The inclusion criteria were in the form of research articles and journals for the last 10 years (2011–2021), while the exclusion criteria were journals without identification of compounds (Figure 1).

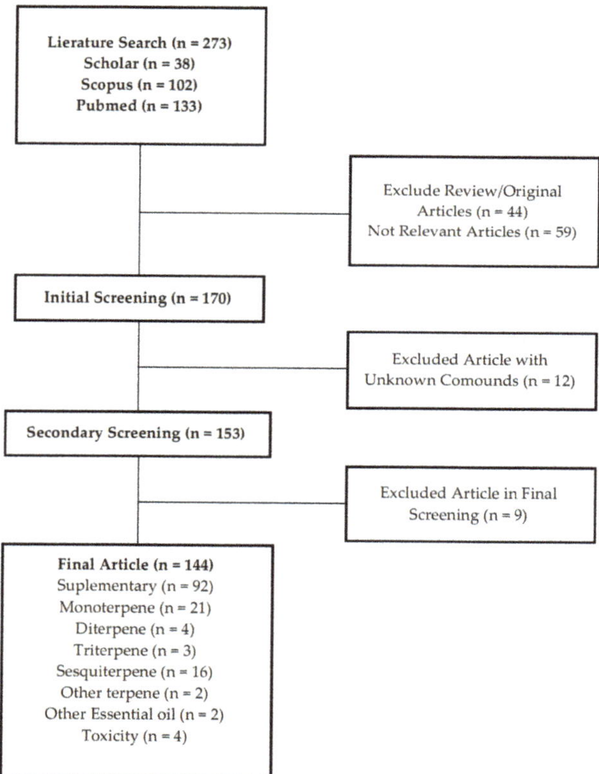

Figure 1. Flowchart of the methodology.

3. Result

In the literature search, 135 articles were listed. After screening the title and abstract, 48 articles were selected for a detailed study. Finally, 46 articles were chosen for this review.

3.1. Breast Cancer

Breast cancer is the most common cancer with a high mortality in women [3]. The malignant illness which causes this cancer starts in the breast cells. Factors including the population structure, lifestyle, genetics, and the environment, can increase the risk of acquiring this cancer, similar to other malignant tumors [1,2,25].

According to breast cancer oncology, neoplastic cells are different from other normal body cells. Normal tissues in the body have limited growth regulation keeping their structure and function working as usual. Cancer cells, however, have prolonged and chronic proliferation without external stimulation [26]. A breast tumor usually begins with ductal hyperproliferation which will develop into a benign tumor or metastatic carcinoma if stimulated continuously by various carcinogenic factors. The tumor microenvironment, such as the influence of the stroma or macrophages, also plays an important role in the initiation and progression of breast cancer [27–29].

The two theories put forth as the basis for the initiation and progression of breast cancer are the cancer stem cell theory and the stochastic theory (Figure 2). According to the cancer stem cell theory, all tumor subtypes develop from the same stem cell or progenitor cell. Different phenotypes of a tumor are caused by genetic and epigenetic alterations that are acquired in stem cells or progenitor cells. The stochastic theory describes that each tumor subtype originates from a single type of cell (stem cells, progenitor cells, or differentiated cells). Any breast cell may progressively develop random mutations, and if these mutations result in malicious behavior, the breast cells are categorized as tumor cells.

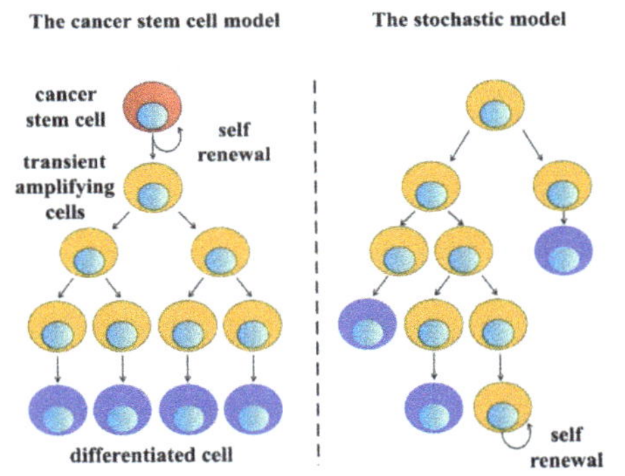

Figure 2. Stem cell and stochastic model.

According to the cancer sites, there are many types of breast cancer, including invasive and non-invasive cancer. When abnormal cells originating from milk ducts or lobules spread out closer to the breast tissue, invasive carcinoma develops [30]. These type of cancer cells can pass through the breast tissue and migrate to various parts of the body by the immune system or through systemic circulation [31]. The most frequent malignancy in women is invasive breast cancer. Non-invasive breast cancer, on the other hand, refers to cancer that has not spread from the lobules [32]. Atypical cells can form and progress into invasive breast cancer, even when they have not yet spread to other tissues outside of the lobules or ducts.

Treatment of breast cancer currently consists of three methods, including surgical therapy, radiation therapy, and chemotherapy. Every technique has positives and negatives aspects, starting with surgical therapy, which aims to prevent, diagnose, stage, and remove malignant tissue. In the case of radiation therapy, the irradiated rays can often kill more than 40% of cancer cells. If chemotherapy is used, it can lessen the quantity of cancer cells and stop them from spreading. However, each approach has a number of disadvantages, such as pain, infection, bleeding, blood clots, and gastrointestinal issues for surgical therapy. The mechanism of action for radiation is typically harmful to normal cells, which can occasionally be destroyed by the delivery of high radiation doses throughout the radiation process. Chemotherapy, on the other hand, is not selective and also has a negative effect on healthy cells [6]. In addition, the limitations of chemotherapy are side effects and multiple drug resistance issues, prompting researchers to explore treatments utilizing natural compounds such as essential oils. Essential oils are plant-defense compounds (secondary metabolites), which contain active components with therapeutic action, one of which is anti-radical; hence, they can be used as an alternative cancer treatment.

3.2. Essential Oil

The original name of essential oil was Quinta Essential, given by Paracelsus von Hohenheim in the sixteenth century [33]. It is a combination of volatile secondary metabolites produced by plants, mostly used for pollinator attraction and self-defense against predators. Hydrocarbons and volatile terpenes make up most of the essential oils. Plant tissues' glandular cells produce essential oil, which is then accumulated in the resin vessel [34]. Ethereal oil and cooking oil are additional names for essential oil. Depending on the type of plant, the oil has a harsh flavor, is volatile at room temperature without decomposing, and often dissolves in organic solvents, though it is insoluble in water [35,36]. Essential oil can be extracted via distillation of plant extract. At high concentrations, essential oil can be used as local anesthetics. For example, clove oil is used to treat toothache, although it has the side effect of damaging mucous membranes [37]. Most essential oils have strong antibacterial and antifungal properties. Some of the most renewable activities of essential oils are controlling Alzheimer's disease, neurodegeneration, and anti-cancer activity [12,38–40].

Essential oils can be grouped based on several classifications and types. In this review, the oil is classified into two major parts: (1) based on the method of extraction and (2) based on the oil content.

3.2.1. Classification Based on the Method of Extraction

Several extraction methods have been reported and steam distillation is the most preferred method in producing essential oils at large quantities with well-maintained purity. However, the method is not suitable for all plant parts because it is not stable at high temperatures [41]. To overcome this problem, methods such as cold pressing and solvent extraction methods have been developed.

a.   Steam Distillation

Steam distillation is a method that is often used for essential oil extraction [42]. The principle of steam distillation is using steam as a separation agent to separate the various component of the mixture [43], such as the separation of essential oil in the stems, leaves, and flowers. The flow of steam around these parts will cause the oil to be evaporated and carried away with the steam, which is then condensed and separated by decantation. This method is often used to make traditional or aromatherapy oils that are pure and free from impurities [44]. The quality and purity of the oil depend on various factors, such as the pressure of the steam passing through the plant material, the refrigerant used, the temperature of the closed system during oil production, as well as the skills of the distiller. The oil produced by steam distillation is of high value because of the high quality and purity of the extracts [37,45].

b.   Cold Pressing Method

The cold pressing method is used to obtain high purity oils [46]. The principle of this method is applying pressure to extract the essential oil substances present in the plant. This method is mostly used by the citrus family to extract oil from fruit peels, such as tangerines, grapefruit, lemons, oranges, and others [47]. The oil is forced out of the plant parts by mechanical pressure, and the extracted results are in liquid form or mixed with water. Thus, a filtering or distillation step is required to separate the oil and the water [48–50].

c.   Extraction Using Solvent

Some plant materials are unsuitable for steam distillation due to high temperatures (in the form of steam) or cold pressing. The resulting oils following this method can be contaminated or have low purity. To avoid this, plants such as jasmine, rose, orange blossom (neroli), tuberose, and oak were extracted using solvents. This process works by passing plant materials through a hydrocarbon solvent, such as ethanol, ether, methanol, hexane, alcohol, and petroleum [38,51]. The solvent mixture is then filtered and distilled under low pressure to obtain the essential oils [52].

d. Microwave-Assisted Hydrodistillation (MAHD)

Microwave-assisted hydrodistillation (MAHD), a complex distillation technique that combines conventional hydrodistillation with microwave heating, has lately gained popularity for the extraction of essential oils from medicinal plants and herbs due to its cost-effectiveness and environmentally friendly nature. A method's efficiency can be increased by increasing the yield, among other factors, through optimization of its parameter conditions. Microwave-assisted hydrodistillation (MAHD) was developed and utilized to extract some plants' essential oils in an effort to make use of microwave heating with the traditional HD. Using the MAHD approach for this extraction has several benefits. Even though the distillation takes less time than the traditional extraction method, the oil yield is slightly higher, and this would help to meet the steadily rising demand for essential oil from medicinal plants. Less time is needed to complete the extraction, resulting in less electricity being used, which lowers operating costs as well. Furthermore, MAHD does not use any chemicals. Because of this, the essential oil obtained using this technique is virtually pure and secure. Given their widespread use in both food preparation and medicine, these requirements are crucial for medicinal essential oils [53].

e. Ohmic-Assisted Hydrodistillation (OAHD)

Ohmic-assisted hydrodistillation (OAHD) is a newly proposed extraction technique that has been utilized to separate essential oils and makes use of the benefits of ohmic heating. Over the past ten years, interest in ohmic hydrodistillation, which combines ohmic heating with hyrodistillation, has increased. In comparison to the traditional hydrodistillation procedure, ohmic-aided hydrodistillation (OAHD) has a shorter extraction time, uses less energy, and produces a greater yield. The extraction time for the OAHD method was 24.75 min, but the HD approach required 1 h to extract the thyme essential oil. In comparison to HD, there were no differences in the essential oil molecules derived by OAHD [54].

3.2.2. Classification Based on Contents

Most essential oils are composed of secondary metabolite components of the terpenoid group (monoterpenoids, triterpenoids, and sesquiterpenoids) and phenylpropanoids. Other names for essential oils are culinary oil and ethereal oil. The oil can have a strong flavor, be volatile at ambient temperature without disintegrating, and frequently dissolve in organic solvents, while being insoluble in water, depending on the plant species [55].

a. Terpenes

Terpenes are the main constituents of essential oils derived from various types of plants or flowers. They are naturally occurring, volatile, unsaturated or open-chain, or cyclic compounds. Terpenes can be classified according to the number of isoprene units, a 5-carbon compound that gives off a scent or taste as a defense mechanism [56]. Based on the isoprene, terpenes are divided into monoterpenes, sesquiterpenes, diterpenes, triterpenes, tetraterpenes, and polyterpenes. Monoterpenes and sesquiterpenes are the largest components in essential oils [57].

Most types of terpenes, excluding clove oil, have a lower specific density than water and are soluble in organic solvents like ether and alcohol. It has a distinctive odor that characterizes the essential oils and high refractive index [57]. Terpenes play an important role in the taste, fragrance, and pigment of a plant. It has been reported that terpene possesses a variety of bioactivities, including anticancer, antibacterial, antihyperglycemic, antifungal, antiviral, analgesic, anti-inflammatory, and antiparasitic [56].

b. Monoterpenes

Almost all essential oils are monoterpenes, a 10 carbons compound with at least one double bond. Monoterpenes have a basic structure consisting of two linked isoprene units. These compounds can undergo cyclization and oxidation in various ways [58]. The high hydrocarbon content and fast reaction to air and heat sources making it not durable [34,59]. Due to their low molecular weight, many of these compounds exist in the form of essential oils. Examples of monoterpenes are geraniol, terpineol, limonene, myrcene, linalool and pinene [58].

It is reported that monoterpenes have antimicrobial, anti-inflammatory, antioxidant, antipruritic, hypotensive, and analgesic pharmacological properties [58].

c. Sesquiterpene

Sesquiterpene consists of 15 carbon atoms and has complex pharmacological action, such as that of chamazulene, which is found in German chamomile. The most prevalent kind of functional group identified in essential oils is the oxygenated groups. Similar to terpenes, it is vital to know the different classes of oxygenated compounds since each class has a distinct potential for health benefits [59–61].

Three linked isoprene units combine to form sesquiterpene lactones with one of the methanol groups oxidized to the lactone form. These sesquiterpenes are vital for plant defense since they are insecticides, antibacterial, antiviral, and antifungal agents. In addition, this compound can also provide biological activities, such as antibacterial, antiviral, antifungal, and anticancer [62].

3.3. Bioactivity of Essential Oils as Cancer Agents

Secondary metabolites found in essential oils can be used as active components in cancer treatments, such as the terpene group. Terpenes have been studied for anti-cancer activity. The following is the anti-cancer activity of essential oils:

3.3.1. Prostate Cancer

Prostate cancer is a cancer of the prostate gland that occurs in men [63]. A study of essential oils for anti-prostate cancer reported that essential oil containing jacaric acid selectively induces apoptosis in hormone-dependent (LN-CaP) and independent human prostate cancer cells (PC-3). The essential oil from *Solanum erianthum* and *Pinus wallichiana* has also been reported to exhibit significant anti-proliferative activity in prostate cancer [64]. The mechanism of essential oils in prostate cancer is to specifically inhibit ROS and have apoptotic activity in cancer cells [65]. In other research, the essential oil of Panax ginseng was shown to have saponins with antimutagenic and anti-tumor effects. *Guatteria pogonopus* leaves significantly inhibit PC-3M metastatic prostate malignancy both in vitro and in vivo [66]. Terpene essential oil has activity against three human cancer cells, one of which is prostate cancer cells. In addition, essential oil activity from the Mentha species was also reported in another prostate cancer cell line, namely LNCaP [67].

3.3.2. Glioblastoma

Glioblastoma multiforme (GBM) is a type of glioma (tumor of brain tissue) that grows and develops rapidly. These tumors are formed from star-shaped glial cells (astrocytes) that support nerve tissue in the brain [68]. Uncontrolled tumor development will lead to cancer [69]. Essential oil from *Hypericum hircinum* has antiproliferative activity on human glioblastoma tumor cells (T98G) [70]. Additionally, the cytosolic $Ca2^+$ content of T98G cells is increased by the essential oil from *Zanthoxylum tinguassuiba*, which contains bisabolol and sesquiterpenes, and the viability of human glioblastoma cells is altered by causing apoptosis. According to a recent study, glioblastoma cell line SF-767 was sensitive to *Ocimum basilicum* L. and *Lippia multiflora* EOs, whereas SF-763 cells were most responsive to *Ageratum conyzoides* L.'s essential oils with potent anti-tumor effects [71].

3.3.3. Colon Cancer

Colon cancer is cancer that develops in the large intestine or at the bottom of the large intestine connected to the anus (rectum) [72]. Several studies reported the activity of essential oils, mainly the terpene group for alternative colon cancer treatment. Geraniol, a monoterpene found in essential oils of various fruits and herbs has been proposed to represent a new class of cancer agents for chemoprevention, because it has significant antiproliferative activity on colon cancer cells (Caco-2) [73]. Essential oils from *Afrostyrax lepidophyllus, Scotonycteris zenkeri,* and *Athanasia brownii Hochr* showed a strong inhibitory effect on human colon carcinoma cell line HCT116 [74]. The essential oil of *Satureja khuzistanica* significantly reduced the viability of SW480 colon cancer cells in a dose-dependent manner [75]. The essential oil of *Artemisia campestris* exhibits significant antitumor activity against HT-29 colon cancer cells [76] while Thymoquinone inhibited the proliferation of a series of human colon cancer cells (Caco-2, HCT-116, LoVo, DLD-1, and HT-29) [77].

3.3.4. Liver Cancer

The aberrant growth of liver tissue that mutates and develops a tumor is known as liver cancer [78]. Essential oils from Thymus citriodorus, Artemisia indica, and Pituranthos tortuosus (Desf.) have substantial cytotoxic effects on HepG2 liver cancer cells when tested for anti-cancer activity [79]. The HepG2 liver cancer cell line undergoes apoptosis in response to zanthoxylum schinifolium essential oil, but not in response to caspase activation. [80]. Moreover, essential oil isolated from the leaves of *N. variabillima* also showed cytotoxic activity on human liver cancer cells [81].

3.3.5. Uterus and Cervix Cancer

Uterus cancer is a malignant tumor that develops commonly in the uterus in women with menopause or over 50 years of age [82]. Several studies have been reported on the activity of essential oils against uterus and cervix cancer. The uterine carcinoma cell line Siha and the cervical cancer cells HeLa, were both sensitive to the essential oils from the leaves of *Casearia sylvestris* and *Liquidambar styracifua* L. [83] The essential oil of *Aristolochia mollissima's* rhizome and aerial increased the cytotoxicity of the human cervical cancer cell line HeLa [84]. Essential oil furanodiene from the rhizome of *Curcuma wenyujin* also demonstrated growth inhibition in an in vivo study in uterine cervical tumors (U14) of rats [85].

3.3.6. Lung Cancer

Lung cancer is a malignancy in the lung tissue originating from cells inside and outside of the lungs (metastasis) [85]. Essential oils from *Xylopia frutescens* leaves were reported to have a cytotoxic effect both in vitro and in vivo in NCI-H358M and PC-3M lung carcinoma cell lines [86]. The essential oil i n *Tridax procumbens* also showed a significant effect on preventing lung cancer cell metastasis on B16F-10 cell lines [87]. The essential oil obtained from *Litsea cubeba* seeds has activity on human NSCLC cells, A549, through the induction of apoptosis and cell cycle arrest [88], while the essential oil of *Solanium spirale* Roxb. showed significant cytotoxicity against NCI-H187 cells [89].

3.3.7. Leukimia

Leukemia is a health condition whereby the body produces excess white blood cells, also called abnormal leukocytes [90]. The THP-1 cell line showed concentration-dependent growth inhibition in the studies of *A. indica* essential oils in leukemia [91]. The cytotoxic efficacy of the essential oils of pine wood, *Cedrus libani, Juniperus excelsa,* and *Juniperus oxycedrus* against drug-sensitive CCRF-CEM and leukemia CEM/ADR5000 expressing multidrug-resistant P-glycoprotein was also demonstrated [92].

3.4. Bioactivity of Essential Oils as Anti-Breast Cancer Agents

Essential oils have been widely studied to treat breast cancer. From the articles collected, it is known that compounds that act as cancer agents are mostly terpene and its derivatives (Table 1).

Table 1. Bioactivity of Essential Oils as Anti-Breast Cancer Agent.

No	Plant Name	Compounds	Methods	Activities	Ref.
1	*Zataria multiflora*	Monoterpenes and triterpenes	MCF-7 and MDA-MB-231 cells	Triggers apoptosis by inducing ROS, mitochondrial membrane potential (MMP) loss, DNA damage, G2 and S-phase arrest in MDA-MB-231 cells and spheroids.	[93]
2	*Zataria multiflora*	Monoterpenes	In vitro using 4T1 cells	Inhibition of proliferation and apoptosis of 4T1 and TC1 cells.	[94]
3	*Cymbopogon citratus*	Monoterpenes	In vivo using 54 Holtzman female rats	Shows tumor reduction as well as necrosis and mitosis.	[95]
4	*Oliveria decumbens*	Monoterpenes	In vitro using 4T1 cells	Induces apoptosis through ROS generation, mitochondrial membrane potential disruption, caspase-3 activation, and DNA damage.	[38]
5	*Pinus sylvetris*	Monoterpenes	In vitro using MCF-7 cells	Inhibits the growth of MCF-7 cells by 45.3% and 99.7%.	[96]
6	*Erythrina corallodendron* L.	MonotMonoterpenes erpen	In vitro using MDA-MB-231, MCF-7 and HMLE cells	Inhibits the proliferation, migration, and invasion of breast cancer cells in a dose-dependent manner.	[97]
7	*Cyphostemma juttae*	Diterpenes	In vitro using MDA-MB 231 and SUM 149 cells	*C. juttae* oil substantially reduces the activation of NF-κB transcription factors, resulting in a significant decrease in several NF-κB target genes.	[98]
8	*Juniperus oxycedrus* L.	Monoterpenes	In vitro using MCF-7 cells	ALEO shows an IC_{50} value of 31% (v/v) against MCF-7 cells after 36 h of treatment.	[99]
9	*Pallines spinosa*	Monoterpenes and triterpenes	In vitro using MCF-7 and MDA-MB-231 cells	Induces apoptosis in MCF-7 and MDA-MB-231 cell lines, and alters Bcl-2 and Bax protein levels.	[60]
10	oleo-gum-resin and its essential oil of *Ferula assa-foetida* and ferulic acid	Monoterpenes	In vitro 4T1 cells	The results show that the three constituents can inhibit the proliferation of 4T1 cells. Our MTT assay results demonstrated a significant cytotoxicity effect in a time and concentration-dependent manner.	[100]
11	*Decatropis bicolor* (Zucc.)	Monoterpenes	In vitro using MDA MB 231 cells	Cytotoxic effect on MDA-MB-231 cells in a dose- and time-dependent manner with an IC_{50} of 53.81 ± 1.691 µg/mL, but independent of the breast epithelial cell line MCF10A (207.51 ± 3.26 µg/mL).	[101]
12	*Cinnamomum longepaniculatum*	Monoterpenes	A549 and MCf-7 cells	The essential oil derived from C. longepaniculatum with the main compounds of terpinen-4-ol, α-terpineol, and safrole induces apoptosis or substantial necrosis in human A549 lung cancer and MCF-7 breast cancer cells.	[61]

Table 1. *Cont.*

No	Plant Name	Compounds	Methods	Activities	Ref.
13	*Nigela sativa*	Thymoquinone (p-benzoquinones)	In vitro using MCF-7 cells	*Nigela sativa* essential oil significantly reduced breast cancer cell survival (MCF-7). The nucleo-cytoplasmic morphological features of NSEO-NE-treated cells were cell membrane blistering, cytoplasmic vacuolation, chromatin marginalization, and nuclear fragmentation. The results demonstrated that NSEO-NE induces apoptosis in MCF-7 cells.	[102]
14	*Opoponax (Commiphora guidottii)*	Sesquiterpenes	MCF-7 cells	The loss of viability was due to the induction of apoptosis as demonstrated by the Annexin V-propidium iodide and caspase-3/7 activity assays/test.	[103]
15	*Hypericum perforatum*	Sesquiterpenes	MCF-7 cells	The essential oil of Hypericum perforatum also showed anticancer activity against MCF-7 cells. The IC_{50} values of essential oil, MTX, and MTX essential oil were 0.78, 6.25, and 0.195 µg/mL, respectively. However, Hypericum perforatum essential oil was found to be non-cytotoxic for MDBK cells.	[104]
16	*Pinus corainensis*	Monoterpens	MDA-MB-231 cells (TNBC)	Inhibits TNF α-induced MDA-MB-231 cell invasion as determined by a three-dimensional spheroid invasion assay.	[105]
17	*Chenopodium ambrosioides* L.	Monoterpenes	MCF-7 cells	Inhibits the growth of MCF-7 cells within 24 h ($p < 0.05$), which is consistent with the results of fluorescent staining of live/dead cells.	[106]
18	*Boswelia sacra*	Triterpenes	MDA-MB-231 cells (TNBC)	Induces cancer cell death, prevent the formation of cellular tissue (MDA-MB-231) cells in Matrigel, and cause multicellular tumor spheroid damage (T47D cells), and regulates molecules involved in apoptosis, signal transduction, and cell cycle development.	[107]
20	*Syzigium aromaticum*	Diterpenes	BSLT (brine shrimp lethality test) method	Tests on BSLT and MTT showed essential oils had the highest cytotoxic effect, followed by ethanol and water extracts. The LD_{50} concentration of essential oil in 24 h of BSLT was 37 µg/mL.	[108, 109]
21	*Laurus nobilis* L.	Monoterpenes	MCF-7 and T47D cells	Shows a strong antiproliferative activity for both leaves and fruit; however, the fruit remained more potent against both breast cancer cell models (MCF7 and T47D).	[110]
22	*Lycopus lucidus Turcz. var. hirtus Regel*	Sesquiterpenes	MCF-7 cells	The essential oil can induce apoptosis of carcinoma cell lines and decrease the level of intracellular GSH.	[111]
23	*Cordia africana Lam.*	Sesquiterpenes	MCF-7 cells	Has an apoptotic mechanism with IC_{50} inhibition of 4.55 µg/mL.	[112]
24	*Lemon volatile oil*	Monoterpenes	MCF-7 cells	Oil derived from lime leaves shows cytotoxic activity against breast cancer cells (MCF-7) at IC_{50} 10% (v/v).	[113]

Table 1. *Cont.*

No	Plant Name	Compounds	Methods	Activities	Ref.
25	*Murraya koenigii*	Sesquiterpenes	MCF-7 cells	Essential oil in particular exhibits strong antibacterial and cytotoxic effects with a dose-dependent trend ($\leq$5.0 µg/mL).	[114]
26	*Cedrelopsis grevei*	Sesquiterpenes	MCF-7 cells	The main constituents are (E)-β-farnesene (27.61%), δ-cadinene (14.48%), α-copaene (7.65%), and β-elemene (6.96%). Grevei essential oil is active against MCF-7 cells (IC_{50} = 21.5 mg/L).	[115]
27	*Asteraceae family*	Sesquiterpenes	MCF-7 and MDA-MB-468 cells	*Britannin* can induce apoptosis in MCF-7 and MDA-MB-468 cells. Western blot analysis shows that Bcl-2 expression markedly decreased in response to *Britannin* treatment, whereas Bax protein expression increased, which is positively correlated with increased p53 expression.	[116]
28	*Tanacetum polycephalum* L. Schultz-Bip	Sesquiterpenes	In vivo (white rats)	Histopathological examination showed that TPHE significantly suppressed the carcinogenic effect of LA7 tumor cells. Tumor sections from TPHE-treated mice showed a significantly decreased expression of Ki67 and PCNA compared to the control group.	[117]
29	*Gaillardin, was isolated from the chloroform extract of Inula oculus-christi aerial*	Sesquiterpenes	MCF-7 cells	Gaillardin is able to induce apoptosis in the breast cancer cell lines of MCF-7 and MDA-MB-468 and determine the mechanism underlying their anticancer effects. Induction of apoptosis with Gaillardin treatment is confirmed by annexin V-FITC/PI staining, and activation of caspase-3, -6, and -9.	[118]
30	*Opoponax (Commiphora guidottii)*	Sesquiterpenes	In vivo (white rats)	Opoponax essential oil shows specific cytotoxicity for mammary tumor cells of humans and mice in vitro and in vivo, and this warrants further investigation into the use of β-bisabolene in medicine.	[103]
31	*Eupatorium lindleyanum DC.*	Sesquiterpenes	MDA-MB-231 cells (TNBC)	Inhibits migration, invasion, and motility of MDA-MB-231 cells in a concentration-dependent manner by invasion and apoptosis.	[119]
32	*Eupatorium lindleyanum DC.*	Sesquiterpenes	MDA-MB-468 cells (TNBC)	EO-induced cytotoxicity is mediated by the induction of apoptosis.	[120]
33	*Ulva fasciata*	Sesquiterpenes	MDA-MB 231cells (TNBC)	Computational study shows the interaction of guai-2-en-10α-ol with the Asp855 residue of the EGFR kinase domain in the active conformation. All these results suggest the anticancer potential of guai-2-en-10α-ol via the EGFR/PI3K/Akt pathway.	[121]
34	Plants such as *Inula helenium*	Sesquiterpenes	MDA-MB 231cells (TNBC)	ALA can inhibit the proliferation, motility, migration, and tube formation of human umbilical vein endothelial cells. ALA also inhibits angiogenesis at the chorioallantoic membrane of chicken embryos and slows the growth of the xenograft of human breast cancer of MDA-MB-231.	[61]

3.4.1. Monoterpenes

Nearly all essential oils contain monoterpenes, which have a structure of 10 carbon atoms with at least one double bond [59]. Monoterpene has been linked to breast cancer in numerous studies. The oxygenated monoterpene 1,8-cineole is a major component of the oil from *L. nobilis* fruits and leaves. The leaves and fruits' crude ethanol fractions, including solvent extracts, show strong antiproliferative activity. However, the fruits are more potent against both breast cancer cell models (MCF7 and T47D). At the IC_{50} value, the mechanisms of apoptosis are different when the proapoptotic efficacy of *L. nobilis* fruits is not regulated by p53 or p21, and the component of the leaves extract substantially increased p53 level. In both extracts, apoptosis is independent of caspase-8 or Fas ligand [110].

Essential oils derived from *Zataria multiflora* were investigated using MCF-7 and MDA-MB-231 cells. It is known that the main component of these essential oils are monoterpenes. The mechanism of monoterpenes is apoptosis by inducing reactive oxygen species (ROS), mitochondrial membrane potential (MMP) disruption, DNA damage, G2 and S-phase [122]. The essential oils were also investigated in vitro using cell lines of 4T1 and TC1 and tested on white winstar mice. As compared to controls, the results showed that the tested essential oils were effective at reducing tumor weight and inhibiting 4T1 and TC1 cell growth and death. They also enhanced the secretion of TNF-, IFN-, and IL-2 while decreasing IL-4. During the treatment with *Zataria multiflora* oil, the biochemical factors of mice did not change significantly [94].

In addition, *Cymbopogon citratus* essential oils include monoterpenes which reduce tumors, necrosis, and mitosis. Carvacrol-treated test animals did not exhibit necrosis, mitosis, or infiltration. The cumulative tumor volume was significantly reduced by carvacrol at a dose of 100 mg/kg/day BW, dropping to 0.11–0.05 cm^3 from 0.38–0.04 cm^3 in the DMBA group (p 0.01). Thus, it may be concluded that carvacrol and *Cymbopogon citratus* extracts exhibited antitumor effects on female rats having DMBA-induced breast cancer [95].

Anti-breast cancer activity was also analyzed by in vitro and in vivo for *Oliveria decumbens* essential oils by enhancing apoptotic and immunomodulatory effects. Based on the MTT test, *Oliveria decumbens* essential oils inhibit viability on 4T1 cancer cells without significant effects on normal cells L929 in 2D analysis, as well as the anti-proliferative effect on 4T1 spheroid (3D analysis). The results showed that *Oliveria decumbens* essential oils induce apoptosis through ROS generation, mitochondrial membrane potential disruption, caspase-3 activations, and DNA damage. On in vivo testing, the effectiveness of *Oliveria decumbens* essential oils were evaluated in tumor-induced 4T1 mice and cytokine to confirm the antitumor effect and development of immune response associated with Th1 expansion [38].

The monoterpene of *Pinus sylvestris'* essential oils demonstrated minimal inhibitory doses of 0.125, 0.1, 3.0, and 10.0 µg/mL, against *B. subtilis, S. cerevisiae, S. aureus,* and *E. coli,* respectively. The proliferation of MCF-7 cells was decreased by 45.3% and 99.7% by the oils at 50 µg/mL and 100 µg/mL, respectively, and 14,36 ± 0.28 µg/mL was the inhibitory concentration of DPPH (2,2-diphenyl-1-picrylhydrazyl) for radical scavenging [96].

Additionally, the essential oils from the leaves of *Erythrina corallodendron* L. reduced the growth, migration, and invasion of breast cancer cells in a dose-dependent manner. Further research on the essential oils from leaves of *E. corallodendron* should be performed to elucidate their potential as a clinical drug or adjuvant to treat migration and invasion of breast cancer [97].

Moreover, essential oils from the *Oxycedrus* L. fruit showed higher efficacy against MCF-7 cells as compared to the extract derived from the leaves. According to reports, *Oxycedrus* L. triggered caspase-9 activation and mitochondrial potential loss in ER + breast cancer cells, indicating that the fruit oils activated the intrinsic mechanism of death in these cells [99].

According to a different study, the essential oils of *Pallines spinosa's* flowers were substantially more effective than those of the oil leaves against MCF-7 (IC_{50} 0.25 ± 0.03 µg/mL) and MDA-MB-231 (IC_{50} 0.21 ± 0.03 µg/mL), respectively. When compared to breast cancer and cell hematology, the toxicity of flower oils was five to eight times lower in normal MCF-10-2A (IC$_{50}$ 1.3 $\pm$ 0.2 μg/mL) and blood mononuclear cells (2.80 $\pm$ 0.45 μg/mL), respectively. Both oils change the levels of the proteins Bcl-2 and Bax and cause caspase-dependent and MCF-7 and MDA-MB-231 cell-dependent apoptosis. Additionally, the oils control the production of cyclin D1, CDK4, and p21 proteins to suppress the cell cycle in both cancer cell lines at the G0/G1 phase [60].

Ferula assa-foetida and *Decatropis bicolor* (Zucc.) also contain monoterpenes having anti-breast cancer effectiveness. The cytotoxic impact on MDA-MB-231 cells is dosage- and time-dependent, with an IC$_{50}$ of 53.81 $\pm$ 1.691 μg/mL, but is independent of the breast epithelial cell line MCF10A (207.51 $\pm$ 3.26 μg/mL). However, ferulic acid did not show any significant effect until 500 μg/mL [100,101].

The essential oils from *Cinnamomum longepaniculatum* were mainly terpinen-4-ol, α-terpineol, and safrole, which are known to induces apoptosis or substantial necrosis in human A549 lung cancer and MCF-7 breast cancer cells [106]. Among the four main components of Korean pine essential oils, d-limonene, 3-carene, α-pinene, and β-myrcene demonstrated the highest suppression of nuclear factor κB (NF-κB) triggered by the tumor necrosis factor (TNFα) [105]. The active substance in the essential oils could inhibit the growth of MCF-7 cells within 24 h ($p < 0.05$), which is consistent with the results of live/dead cell fluorescent staining.

Chenopodium ambrosioides L. inhibited essential oils by 58.98%, 1-isopropyl-4-methylbenzene by 37.8%, and α-terpinene by 32.09%. In comparison to the control, the relative MDA content increased considerably (p 0.05) up to concentrations of 1.25, 0.21, and 0.17 μg/mL for essential oils 1-isopropyl-4-methylbenzene, and α-terpinene, respectively, and subsequently declined ($p < 0.05$) [61].

The molecular profile of lemon essential oils against breast cancer was examined using GC/MS analysis. Limonene (47.24% and 55.23%), geranial (14.48% and 7.94%), and neral (12.1% and 6.1%) made up most of the mixture. At an IC$_{50}$ of 10% (v/v), the oils from the leaves exhibited cytotoxic action against breast cancer cells (MCF-7). According to the results (compared to untreated cells; $p < 0.05$), there was a significant increase in the expression level of the apoptotic protein caspase-8, a significant decrease in the expression level of the anti-apoptotic protein BcL-2, and a significant increase in the expression level of the proliferative marker Ki-67 [113,123].

Another test conducted on red algae *Plocamium* extract with monoterpenes as the dominant compound was against MDA-MB-231 triple-negative breast cancer cells. The results showed that it is able to disrupt mitochondria, activate caspase-3/7, externalize phosphatidylserine, reduce the number of polyploid cells, and is DNA fragmentation consistent with the induction of apoptosis with an IC$_{50}$ value of 16 $\pm$ 2.2 μM, 7.3 $\pm$ 0.4 μM, and 3.3 $\pm$ 0.5 μM after 24 h, 48 h, dan 72 h, respectively [124].

3.4.2. Sesquiterpene

Sesquiterpenes have been widely studied for their activity against breast cancer. Yoe et al. reported that β-Bisabolene and α-Bisabolene from opoponax (*Commiphora guidottii*) showed specific cytotoxicity for mammary tumor cells of humans and mice both In vitro and In vivo. In addition, this compound also showed good selectivity in human cancer cells of MCF-7, MDA-MB-231, and SKBR3 [103]. Therefore, further investigation into the use of β-bisabolene in medicine is needed [103].

Additionally, *Hypericum perforatum essential* oil demonstrated that sesquiterpene such as germacrene D, δ-cadinene, γ-muurolene, germacrene B, α-copaene, bicyclogermacrene, and (E)-caryophyllene was the dominant compound. Cytotoxic activity of the essential oils on MCF7 breast cancer cells resulted in the suppression of cancer growth with an IC$_{50}$ of 0.78 μg/mL. However in normal MDBK cells no cytotoxic activity was observed [125].

Essential oils of *Lycopus lucidus* Turcz. var. hirtus Regel was also reported to contain sesquiterpene; -Humulene (15.97%) was the dominant compound. These essential oils are reported to exhibit cytotoxic activity by reducing the cell viability of breast cancer [111].

Essential oils from *Cordia africana* Lam. contain 32.0% of β-caryophyllene having inhibition on MCF7 breast cancer cells with an IC_{50} of 4.55 μg/mL. In addition, this essential oils also exhibit apoptotic activity by the increase in protein caspase-8 which induces apoptosis in cells [112].

Essential oils from *Murraya koenigii* are also reported to be sesquiterpenes as the dominant compounds [114]. The essential oils inhibit the growth of MCF7 breast cancer cells. Essential oils from *Cedrelopsis grevei* leaves contained (E)-β-farnesene (27.61%), α-copaene (7.65%), δ-cadinene (14.48%), and β-elemene (6.96%) also inhibits the proliferation of MCF7 breast cancer cells with an IC_{50} value of 21.5 mg/L [115].

Herbal medicine derived from *Chloranthus serratus* essential oils was reported to suppress LIM kinase-1 (LIMK1) which plays an important role in invasion and metastasis of tumor cells by regulating actin cytoskeleton architecture, cofilin1 phosphorylation, F-actin polymerization and the cell migration of human breast cancer MDA-MB-468 and MDA-MB-231 cells with IC_{50} values of 4.64 μM and 3.14 μM respectively [126].

Another study reported that sesquiterpene from the *Pimpinella haussknechtii* fruit increased protein aggregation and mRNA expression of ATF-4, CHOP, GADD34, and TRIB3 in MCF7 breast cancer cells, with IC_{50} values of 45 and 25 μM [127]. Alantolactone, a sesquiterpene compound, were reported to inhibit angiogenesis, suppress phosphorylation of vascular endothelial growth factor receptor 2 and its downstream protein kinases including PLC-γ1, FAK, Src, and Akt in endothelial cells, and delay the growth of human breast cancer MDA MB 231 xenografts in mice. When tested on MDA MB 231 cells, the antiangiogenic activity of this chemical indicated that alantolactone is a prospective therapeutic candidate for antiangiogenic cancer therapy, with an IC_{50} of 40.4 μM [128].

Based on these In vitro and In vivo studies, sesquiterpenes were active against breast anticancer cells. In vitro studies on anti-breast cancer activity using MCF-7, MDA MB 231 and MDA MB 468 cells [60,120] shows significant results in inhibiting breast cancer [61,116,120,121]. In addition, the histopathological In vivo studies performed on white mice showed that sesquiterpenes significantly suppressed the carcinogenic effect of LA7 tumor cells in mice [103].

The mechanism of the sesquiterpenes is to inhibit the migration, invasion, and motility of MDA-MB-231 cells in a concentration-dependent manner by invasion and apoptosis [119]. Moreover, studies on the inhibition of the Akt pathway in MDA-MB-468 cells showed that it might stop the growth of those cells, by causing a cell cycle arrest in the G2/M phase and inducing caspase-dependent death [120].

According to a recent study, allantoin can prevent human umbilical vein endothelial cells from proliferating, migrating, moving, and forming tubes. Additionally, allantoin inhibited angiogenesis at the chorioallantoic membrane of chicken embryos and, via blocking angiogenesis, reduced the growth of the breast cancer xenograft MDA-MB-231 in mice [128].

3.4.3. Triterpenes

Triterpene has been reported to be active against breast cancer cells [129,130]. In comparison to the essential oils from the leaf (IC_{50} 2.4 ± 0.5 μg/mL and 1.5 ± 0.1 μg/mL), triterpenes from *Pallines spinosa* demonstrated a considerable cytotoxic effect on MCF-7 (IC_{50} 0.25 ± 0.03 μg/mL) and MDA-MB-231 (IC_{50} 0.21 ± 0.03 μg/mL) cells. When evaluated on normal MCF-10-2A (IC_{50} 1.3 ± 0.2 μg/mL) and blood mononuclear cells (2.80 ± 0.45 μg/mL) as opposed to breast cancer cell hematology, flower essential oils' toxicity was five to eight times lower. Both essential oils change the levels of the proteins Bcl-2 and Bax and cause caspase-dependent apoptosis in MCF-7 and MDA-MB-231 cells. Additionally, via modifying the expression of cyclin D1, CDK4, and p21 proteins in both cancer cell lines, essential oils suppressed the cell cycle at the G0/G1 phase [60].

Boswellia sacra essential oils which were distilled at 100 °C were more effective than essential oils prepared at 78 °C. This is because the extract has higher triterpene content. These compounds have the ability to cause the death of cancer cells, stop the growth of

MDA-MB-231 cells in Matrigel, harm multicellular tumor spheroids (T47D cells), and control molecules that control apoptosis, signal transduction, and cell cycle progression [107].

Xue et al. reported that the ethanolic extract of *Pleurotus eryngii* inhibits MCF7 cells. Three triterpene components, 2,3,6,23-tetrahydroxy-urs-12-en-28 oic acid, 2,3,23-trihydroxyurs-12-en-28 oic acid, and lupeol, make up most of this extract's composition. The ability of these substances to stop the growth of breast cancer cells was demonstrated by their respective IC_{50} values of 15.71, 48, and 66.89 μM, repsectively [131].

3.4.4. Diterpenes

Diterpenes are a group of hydrocarbons that are widely present in essential oils. Several studies reported that diterpenes are good anti-breast-cancer agents. Zito et al. [98] reported that the essential oils of *Cyphostemma juttae* have a dominant diterpene content, which is phytol compound (30%) tested against triple negative breast cancer cells (MDA-MB-231, SUM 149). Phytol substantially reduces the activation of NF-κB transcription factors, resulting in a significant decrease in several NF-κB target genes.

The extract of *Tinospora cordifolia* also contains diterpenes with breast anticancer activity against MCF7 cells. The results showed that the IC_{50} values at 24 and 48 h were 3.2 mM and 2.4 mM respectively. The cytotoxicity of this compound was specific for MCF-7 cells and had no toxic effect on normal Vero cells and V79 cells. This compound also significantly stimulated the formation of intracellular ROS, even at lower doses of 0.6 and 1.2 mM. Thus, it can be concluded that the diterpene compounds from this extract are effective and selective against cancer cells, especially breast cancer cells [132]. The diterpene myrsinol compound J196-10-1, derived from the roots of *Euphorbia prolifera*, can reverse multidrug resistance, namely daunorubicin, vincristine, and topotecan, with the IC_{50} value of daunorubicin from 29.65 μM to 0.55 μM, vincristine from 13.85 μM to 0.063 μM, and topotecan from 4.61 μM to 0.65 μM [106].

Myrsinane-type diterpenes, such as 3,7,10,14,15-tetraacetyl-5-propanoyl-13,17-epoxy-8,10,18-myrsinadiene and 3,7,10,14,15-pentaacetyl-5-butanoyl-13,17-epoxy-8-myrsinane, are present in *Euphorbia connata* Boiss. acetone:chloroform extracts. In MDA-MB cells, these compounds had an IC_{50} value of 24.53 ± 3.39 and 26.67 ± 1.41 μM, while in MCF-7 cells, the value was 37.73 ± 3.41 and 34.57 ± 2.12 μM, respectively, indicating a moderately inhibitory effect on breast cancer. Other diterpenes, such as 5,6-epoxy-8,9,15-triacetyl-3-benzoyl-14-oxo-jatropha-11 (E)-ene (3), exhibited weak cytotoxicity in MDA-MB cells (IC_{50} = 55.67 ± 7.09 μM) and moderate cytotoxicity in MCF7 cells (IC_{50} = 24.33 ± 3.21 μM) [133].

3.4.5. Other Terpenes

Essential oils from *Nepeta cataria* L. have also been tested on PC3 and MCF7 breast cancer cells. The largest content of these essential oils is the stereoisomer of nepetalactone, a class of terpene compounds. Essential oils of *N. cataria* were more effective against PC3 triple negative breast cancer than other breast cancer cells. In addition, western blot also showed the expression of apoptosis-inducing proteins. Therefore, it can be postulated that these plant compounds have the mechanism to treat cancer by inducing cell apoptosis [134]. The dominant compounds in essential oils from *Achillea fragrantissima* comprise 1-terpinen-4-Ol (30.90%) and p-cymen-3-Ol (21.22%), and these essential oils were effective against MCF7 breast cancer cells, with an IC_{50} value of 0.51 μg/mL extracted via hydrodistillation and 0.80 μg/mL extracted via volatile solvent extraction [135].

3.5. Types of Cells Used in Cytotoxic Studies

Cytotoxic studies of essential oils were performed on several types of breast cancer cell lines (Figure 3). Throughout this review, MCF-7 and MDA-MB-231 are the most common cancer cells used in cytotoxic studies, with 19 studies on MCF-7 and 11 on MDA-MB-231. (Figures 3 and 4) The breast cancer cell line MCF-7 contains glucocorticoid, progesterone, and estrogen receptors. The pleural effusion of a 69-year-old Caucasian woman with metastatic breast cancer (adenocarcinoma) was used as the source of this substance in 1970 by Dr. Soule of the Michigan Cancer Foundation in Detroit, Michigan. Because MCF-7 cells still possessed several desirable traits unique to the mammary epithelium, such as the ability to process estrogen in the form of estradiol via estrogen receptors (ER) in the cell cytoplasm, it is valuable for in vitro breast cancer investigations. It is the first hormone to react to a breast cancer cell line. Its special qualities are advantageous for experimental therapies, and the cells are also cytokeratin-sensitive. The epithelial-like cells develop in monolayers when cultured in vitro, and the cells have the ability to form domes.

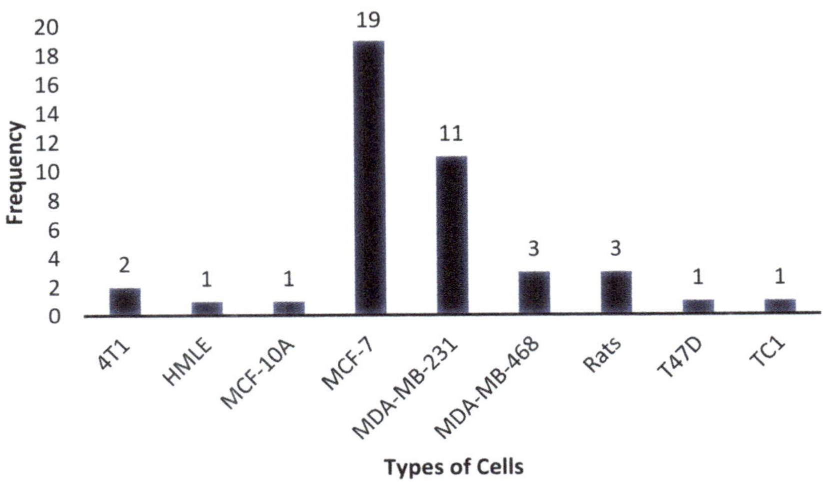

Figure 3. Types of cells used in the cytotoxicity study.

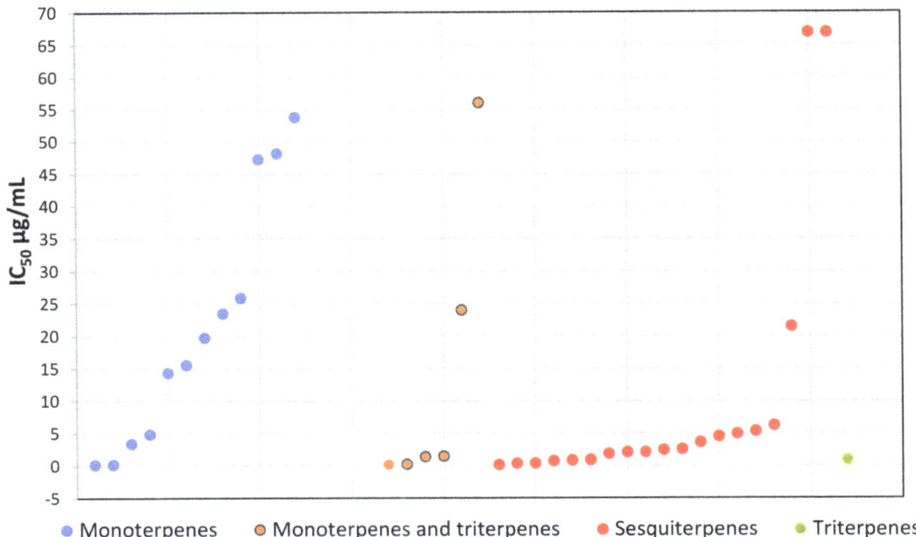

Figure 4. IC$_{50}$ Value of essential oils in breast cancer cells.

The triple-negative breast cancer cell line MDA-MB-231 exhibits an indicative epithelial to mesenchymal transition (EMT), which is linked to BC metastasis. One of the most popular breast cancer cell lines in medical research labs is the MDA-MB-231 epithelial cell line. It was developed from a pleural effusion of a 51-year-old Caucasian woman with metastatic mammary adenocarcinoma1. Based on the IC_{50} results, sesquiterpenes have the lowest IC_{50} value of 0.19 µg/mL in MCF-7 cancer cell line. In addition, from the 19 anticancer studies of sesquitepenes, 17 showed an IC_{50} below 20 µg/mL, indicating strong activity, as compared to monoterpenes with 7 out of 12 anticancer studies below 20 µg/mL. From this, it can be concluded that sesquiterpenes have the best potential anticancer activity compared to other compounds (Figure 4).

3.6. Clinical Trials of Essential Oils for Breast Cancer

Clinical trials for the use of essential oils as a breast cancer treatment on breast cancer patients are currently being conducted. Peppermint essential oil has undergone clinical testing on 100 breast cancer patients receiving outpatient care at the Imam Khomeini Hospital cancer center. Patients were randomized into interventions with control groups. The intervention group received aromatherapy with peppermint essential oil, whereas the control group received saline solution. From this study, it is suggested that the use of anti-nausea medications in combination with aromatherapy using peppermint essential oil can lessen nausea and vomiting during acute phase drug use [136]. The impact of inhaling ginger aromatherapy on patients with breast cancer symptoms, such as nausea, vomiting, and poor health-related quality of life (HRQoL), has also been investigated. In the acute period, essential oil inhalation significantly reduced visual analog scale (VAS) nausea scores when compared to the placebo [137]. The findings of recent clinical trials for the treatment of nausea and vomiting brought on by medication are generally positive. It is postulated that specific medicines using the components of essential oils would be developed to treat cancer in the future.

3.7. Toxicity and Side Effect of Essential Oils

The largest and most diversified collection of naturally occurring substances is comprised of terpenes, also referred to as terpenoids. They are mostly present in plants and make up the bulk of essential oils made from plants. Terpenes have numerous bioactive and pharmacological properties, as well as a variety of medical applications. Terpenes also provide for flexibility in the route of administration and the reduction of side effects in addition to these qualities. Terpenes are natural substances that are unlikely to harm healthy cells or have any negative side effects, which attracts many researchers to explore their potential as a cancer treatment [138]. Castilhos et al. conducted toxicity research of essential oils in 2017, and the findings show that these compounds exhibit relative selectivity to the predator *Chrysoperla externa*; nevertheless, some compounds also showed sublethal effects on reproduction. Carvacrol and thymol, two phenolic monoterpenoids, were less toxic than natural pyrethrins (the toxicity standard) in these bioassays, but they were more acutely lethal than other terpenoids screened, with an LD_{50} of 20,000 g/g. R-(+)-limonene was found to have sublethal impacts on fecundity and fertility, whereas oregano oil merely had a fecundity-related effect [139].

On laboratory animals, usually rats, the potential toxicity of various essential oils and their constituents was studied. The median lethal dose (LDso) test was used to assess acute toxicity in rats, and the results showed that most essential oils have an LDso of 1–20 g/kg, indicating low toxicity. Some essential oils, such as lemon oil, have an LDso of greater than 5 g/kg in humans. Therefore, the deadly dose for an adult weighing 70 kg would be 350 g, which is difficult to achieve under normal circumstances. The EOs from Boldo leaf, Chenopodium, Mentha pulegium (pennyroyal), Satureja hortensis (savory), and Thuja are a few notable outliers; they showed an LDsp between 0.1 and 1 g/kg in rats, signaling a high toxicity and recommending the need for necessary care while using them. Since some of the resultant compounds, such as the oxidation products of limonene, are potentially skin sensitizers, essential oils are subject to oxidative deterioration. Alipanah et al. (2021) formulated *Citrus sinensis* and *Citrus limon* essential oils using chitosan nanoparticles to improve the anti-breast cancer. This idea is also a solution to protect the instability of limonen and keep them from being easily oxidized [140]. Additionally, in order to boost activity against breast cancer cells, Valizadeh et al. (2021) used chitosan to create na-noparticles of Syzygium aromaticum essential oils with the predominant amount of egeunol (MDA-MB-468) [109]. Therefore, proper essential oil storage is required to maintain their potency and lower the likelihood of negative responses. Essential oils should be kept in tightly sealed containers in a refrigerator or in a cool, dark location (brown bottles). Although the Flavor and Extract Manufacturers Association (FEMA) awarded most essential oils the GRAS (generally recognized as safe) certification, it should be noted that these oils were only tested as flavors with very low concentrations in the tested goods. Particular hazardous consequences, either local or systemic, for a concentrated essential oil, could manifest under certain conditions [141,142].

4. Author Perspective

Essential oils are plant-made active metabolites that serve as the plants' internal defense mechanisms. The active metabolites found in essential oils are potent antioxidants that aid in blocking ROS [143,144], and hence have medicinal efficacy to treat a variety of ailments. Due to their antioxidant properties, various research reviews claim that the mechanism of essential oils which inhibit ROS can be used in anti-cancer therapy. The molecular mechanism of anti-cancer essential oils as has been studied extensively due to the high prevalence of cancer patients, particularly breast cancer. From the studies, the terpene group is effective in inducing apoptosis and inhibiting the growth of breast cancer cells, such as the MCF-7 and MDA-MB231 cell lines. Despite extensive preclinical and clinical research using cancer cells and patients, the advantages of essential oils as a treatment for chemotherapy side effects, such as nausea and vomiting, have recently been revealed. In the fight against breast cancer, the outcomes of essential oils against cancer or cancer's side effects are highly substantial. It is envisaged that the physicochemical qualities of essential oils, such as stability in varied environments, can be improved in the future. It is also hoped that the active substance of essential oils will be incorporated into drug development for breast cancer.

5. Conclusions

As potential anti-breast cancer medicines, essential oils have demonstrated promising bioactivity and pharmacological characteristics in vitro and in vivo. Terpenoids, primarily monoterpenes, sesquiterpenes, diterpenes, and triterpenes, make up the majority of the essential oils' chemical composition. According to this review, essential oils from 18 out of 28 sesquiterpenes with IC_{50} values ≤ 20 µg/mL were reported in 17 out of the 19 anti-breast cancer investigations. This demonstrated that sesquiterpenes are the most effective anti-breast cancer agent, with an IC_{50} on the MCF-7 627 cancer cells line of 0.19 µg/mL. We believe that essential oils might eventually be employed as a primary or adjuvant therapy for the treatment of breast cancer, since they are highly effective and rarely exhibit side effects.

Author Contributions: M.A.M. and M.M. conceptualized the study. M.A.M., M.M., I.G. and A.Z. analyzed the data. M.A.M., N.K.K.I. and M.M. wrote the paper. M.M. collected the funding. M.M. and N.K.K.I. edited the paper. All authors have read and agreed to the published version of the manuscript.

Funding: Academic Leadership Grant with contract no. 2203/UN6.3.1/PT.00/2022.

Institutional Review Board Statement: Not applicable.

Informed Consent Statement: Not applicable.

Data Availability Statement: The study did not report any data.

Acknowledgments: The authors gratefully acknowledge Rector Universitas Padjadjaran, Indonesia for the Academic Leadership Grant with contract no. 2203/UN6.3.1/PT.00/2022.

Conflicts of Interest: The authors declare no conflict of interest.

References

1. Hortobagyi, G.N.; de la Garza Salazar, J.; Pritchard, K.; Amadori, D.; Haidinger, R.; Hudis, C.A.; Khaled, H.; Liu, M.-C.; Martin, M.; Namer, M. The global breast cancer burden: Variations in epidemiology and survival. *Clin. Breast Cancer* **2005**, *6*, 391–401. [CrossRef] [PubMed]
2. Zendehdel, M.; Niakan, B.; Keshtkar, A.; Rafiei, E.; Salamat, F. Subtypes of benign breast disease as a risk factor for breast cancer: A systematic review and meta-analysis protocol. *Iran. J. Med. Sci.* **2018**, *43*, 1. [PubMed]
3. Ferlay, J.; Soerjomataram, I.; Ervik, M.; Dikshit, R.; Eser, S.; Mathers, C.; Rebelo, M.; Parkin, D.; Forman, D.; Bray, F. *Cancer Incidence and Mortality Worldwide*; IARC: Lyon, France, 2012.
4. Fasching, P.A.; Ekici, A.B.; Adamietz, B.R.; Wachter, D.L.; Hein, A.; Bayer, C.M.; Häberle, L.; Loehberg, C.R.; Jud, S.M.; Heusinger, K.; et al. Breast Cancer Risk—Genes, Environment and Clinics. *Geburtshilfe Frauenheilkd* **2011**, *71*, 1056–1066. [CrossRef] [PubMed]
5. Youlden, D.R.; Cramb, S.M.; Yip, C.H.; Baade, P.D. Incidence and mortality of female breast cancer in the Asia-Pacific region. *Cancer Biol. Med.* **2014**, *11*, 101. [PubMed]
6. Gnant, M.; Harbeck, N.; Thomssen, C. St. Gallen/Vienna 2017: A Brief Summary of the Consensus Discussion about Escalation and De-Escalation of Primary Breast Cancer Treatment. *Breast Care* **2017**, *12*, 102–107. [CrossRef]
7. Han, S.-J.; Guo, Q.-Q.; Wang, T.; Wang, Y.-X.; Zhang, Y.-X.; Liu, F.; Luo, Y.-X.; Zhang, J.; Wang, Y.-L.; Yan, Y.-X. Prognostic significance of interactions between ER alpha and ER beta and lymph node status in breast cancer cases. *Asian Pac. J. Cancer Prev.* **2013**, *14*, 6081–6084. [CrossRef]
8. Siegel, R.L.; Miller, K.D.; Fuchs, H.E.; Jemal, A. Cancer statistics, 2022. *CA Cancer J. Clin.* **2022**, *72*, 7–33. [CrossRef]
9. Gautama, W. Breast Cancer in Indonesia in 2022: 30 Years of Marching in Place. *Indones. J. Cancer* **2022**, *16*, 2. [CrossRef]
10. Sledge, G.W.; Mamounas, E.P.; Hortobagyi, G.N.; Burstein, H.J.; Goodwin, P.J.; Wolff, A.C. Past, present, and future challenges in breast cancer treatment. *J. Clin. Oncol.* **2014**, *32*, 1979–1986. [CrossRef]
11. Jatoi, I.; Sung, H.; Jemal, A. The Emergence of the Racial Disparity in U.S. Breast-Cancer Mortality. *N. Engl. J. Med.* **2022**, *386*, 2349–2352. [CrossRef]
12. Tran, N.; Pham, B.; Le, L. Bioactive compounds in anti-diabetic plants: From herbal medicine to modern drug discovery. *Biology* **2020**, *9*, 252. [CrossRef]
13. Privitera, G.; Luca, T.; Castorina, S.; Passanisi, R.; Ruberto, G.; Napoli, E. Anticancer activity of *Salvia officinalis* essential oil and its principal constituents against hormone-dependent tumour cells. *Asian Pac. J. Trop. Biomed.* **2019**, *9*, 24–28. [CrossRef]
14. Spyridopoulou, K.; Fitsiou, E.; Bouloukosta, E.; Tiptiri-Kourpeti, A.; Vamvakias, M.; Oreopoulou, A.; Papavassilopoulou, E.; Pappa, A.; Chlichlia, K. Extraction, Chemical Composition, and Anticancer Potential of *Origanum onites* L. Essential Oil. *Molecules* **2019**, *24*, 2612. [CrossRef]
15. Rajivgandhi, G.; Saravanan, K.; Ramachandran, G.; Li, J.-L.; Yin, L.; Quero, F.; Alharbi, N.S.; Kadaikunnan, S.; Khaled, J.M.; Manoharan, N. Enhanced anti-cancer activity of chitosan loaded *Morinda citrifolia* essential oil against A549 human lung cancer cells. *Int. J. Biol. Macromol.* **2020**, *164*, 4010–4021. [CrossRef]
16. Asif, M.; Yehya, A.H.; Dahham, S.S.; Mohamed, S.K.; Shafaei, A.; Ezzat, M.O.; Majid, A.S.A.; Oon, C.E.; Majid, A.M.S.A. Establishment of in vitro and in vivo anti-colon cancer efficacy of essential oils containing oleo-gum resin extract of *Mesua ferrea*. *Biomed. Pharmacother.* **2019**, *109*, 1620–1629. [CrossRef]
17. Rattanamaneerusmee, A.; Thirapanmethee, K.; Nakamura, Y.; Chomnawang, M. Differentiation-inducing effect in human colon cancer cells of essential oils. *Pharm. Sci. Asia* **2018**, *45*, 154–160. [CrossRef]
18. Rezaie-Tavirani, M.; Fayazfar, S.; Heydari-Keshel, S.; Rezaee, M.B.; Zamanian-Azodi, M.; Rezaei-Tavirani, M.; Khodarahmi, R. Effect of essential oil of Rosa Damascena on human colon cancer cell line SW742. *Gastroenterol. Hepatol. Bed. Bench.* **2013**, *6*, 25–31.
19. Zare, E.; Jamali, T.; Ardestani, S.K.; Kavoosi, G. Synergistic effect of Zataria Multiflora essential oil on doxorubicin-induced growth inhibition of PC3 cancer cells and apoptosis. *Complement. Ther. Clin. Pract.* **2021**, *42*, 101286. [CrossRef]
20. Russo, A.; Cardile, V.; Graziano, A.C.E.; Avola, R.; Bruno, M.; Rigano, D. Involvement of Bax and Bcl-2 in Induction of Apoptosis by Essential Oils of Three Lebanese Salvia Species in Human Prostate Cancer Cells. *Int. J. Mol. Sci.* **2018**, *19*, 292. [CrossRef]
21. Khanavi, M.; Enayati, A.; Shams Ardekani, M.; Akbarzadeh, T.; Karimpour Razkenari, E.; Eftekhari, M. Cytotoxic activity of *Juniperus excelsa* M. Bieb. Leaves Essent. Oil Breast Cancer Cell Lines. *Res. J. Pharm.* **2019**, *6*, 1–7.
22. Nguyen, T.K.; Le Nguyen, T.N.; Nguyen, K.; Nguyen, H.V.; Tran, L.T.; Ngo, T.X.; Pham, P.T.; Tran, M.H. Machine learning-based screening of MCF-7 human breast cancer cells and molecular docking analysis of essential oils from *Ocimum basilicum* against breast cancer. *J. Mol. Struct.* **2022**, *1268*, 133627. [CrossRef]
23. Ruttanapattanakul, J.; Wikan, N.; Chinda, K.; Jearanaikulvanich, T.; Krisanuruks, N.; Muangcha, M.; Okonogi, S.; Potikanond, S.; Nimlamool, W. Essential Oil from *Zingiber ottensii* Induces Human Cervical Cancer Cell Apoptosis and Inhibits MAPK and PI3K/AKT Signaling Cascades. *Plants* **2021**, *10*, 1419. [CrossRef] [PubMed]
24. Santos, P.A.; Avanço, G.B.; Nerilo, S.B.; Marcelino, R.I.; Janeiro, V.; Valadares, M.C.; Machinski, M. Assessment of Cytotoxic Activity of Rosemary (*Rosmarinus officinalis* L.), Turmeric (*Curcuma longa* L.), and Ginger (*Zingiber officinale* R.) Essential Oils in Cervical Cancer Cells (HeLa). *Sci. World J.* **2016**, *2016*, 9273078. [CrossRef] [PubMed]
25. Hartwell, L.H.; Kastan, M.B. Cell cycle control and cancer. *Science* **1994**, *266*, 1821–1828. [CrossRef] [PubMed]
26. Evan, G.I.; Vousden, K.H. Proliferation, cell cycle and apoptosis in cancer. *Nature* **2001**, *411*, 342–348. [CrossRef]
27. Dumars, C.; Ngyuen, J.-M.; Gaultier, A.; Lanel, R.; Corradini, N.; Gouin, F.; Heymann, D.; Heymann, M.-F. Dysregulation of macrophage polarization is associated with the metastatic process in osteosarcoma. *Oncotarget* **2016**, *7*, 78343. [CrossRef]

28. Maffini, M.V.; Soto, A.M.; Calabro, J.M.; Ucci, A.A.; Sonnenschein, C. The stroma as a crucial target in rat mammary gland carcinogenesis. *J. Cell Sci.* **2004**, *117*, 1495–1502. [CrossRef]
29. Sgroi, D.C. Preinvasive breast cancer. *Annu. Rev. Pathol.* **2010**, *5*, 193. [CrossRef]
30. Harris, L.N.; Ismaila, N.; McShane, L.M.; Andre, F.; Collyar, D.E.; Gonzalez-Angulo, A.M.; Hammond, E.H.; Kuderer, N.M.; Liu, M.C.; Mennel, R.G. Use of biomarkers to guide decisions on adjuvant systemic therapy for women with early-stage invasive breast cancer: American Society of Clinical Oncology Clinical Practice Guideline. *J. Clin. Oncol.* **2016**, *34*, 1134. [CrossRef]
31. Ziperstein, M.J.; Guzman, A.; Kaufman, L.J. Evaluating breast cancer cell morphology as a predictor of invasive capacity. *Biophys. J.* **2016**, *110*, 621a. [CrossRef]
32. West, A.-K.V.; Wullkopf, L.; Christensen, A.; Leijnse, N.; Tarp, J.M.; Mathiesen, J.; Erler, J.T.; Oddershede, L.B. Division induced dynamics in non-Invasive and invasive breast cancer. *Biophys. J.* **2017**, *112*, 123a. [CrossRef]
33. Hanif, M.A.; Nisar, S.; Khan, G.S.; Mushtaq, Z.; Zubair, M. Essential Oils. In *Essential Oil Research: Trends in Biosynthesis, Analytics, Industrial Applications and Biotechnological Production*; Malik, S., Ed.; Springer International Publishing: Cham, Switzerland, 2019; pp. 3–17. [CrossRef]
34. Shasby, G. Erratum. Inhibition of motility by NEO100 through the calpain-1/RhoA pathway. *J. Neurosurg.* **2019**, *133*, 1262. [CrossRef]
35. Antonioli, G.; Fontanella, G.; Echeverrigaray, S.; Delamare, A.P.L.; Pauletti, G.F.; Barcellos, T. Poly (lactic acid) nanocapsules containing lemongrass essential oil for postharvest decay control: In vitro and in vivo evaluation against phytopathogenic fungi. *Food Chem.* **2020**, *326*, 126997. [CrossRef]
36. Yuan, C.; Wang, Y.; Liu, Y.; Cui, B. Physicochemical characterization and antibacterial activity assessment of lavender essential oil encapsulated in hydroxypropyl-beta-cyclodextrin. *Ind. Crops Prod.* **2019**, *130*, 104–110. [CrossRef]
37. Lapkina, E.; Zaharova, T.; Tirranen, L. Component composition of essential oil of Artemisia salsoloides willd and its antimicrobial properties. *Chem. Plant Raw Mater.* **2017**, *3*, 157–162. [CrossRef]
38. Jamali, T.; Kavoosi, G.; Ardestani, S.K. In-vitro and in-vivo anti-breast cancer activity of OEO (Oliveria decumbens vent essential oil) through promoting the apoptosis and immunomodulatory effects. *J. Ethnopharmacol.* **2020**, *248*, 112313. [CrossRef]
39. Karakaya, S.; Koca, M.; Yılmaz, S.V.; Yıldırım, K.; Pınar, N.M.; Demirci, B.; Brestic, M.; Sytar, O. Molecular docking studies of coumarins isolated from extracts and essential oils of zosima absinthifolia link as potential inhibitors for Alzheimer's disease. *Molecules* **2019**, *24*, 722. [CrossRef]
40. Oboh, G.; Olasehinde, T.A.; Ademosun, A.O. Essential oil from lemon peels inhibit key enzymes linked to neurodegenerative conditions and pro-oxidant induced lipid peroxidation. *J. Oleo Sci.* **2014**, *63*, 373–381. [CrossRef]
41. Cassel, E.; Vargas, R.M.F.; Martinez, N.; Lorenzo, D.; Dellacassa, E. Steam distillation modeling for essential oil extraction process. *Ind. Crops Prod.* **2009**, *29*, 171–176. [CrossRef]
42. Božović, M.; Navarra, A.; Garzoli, S.; Pepi, F.; Ragno, R. Esential oils extraction: A 24-hour steam distillation systematic methodology. *Nat. Prod. Res.* **2017**, *31*, 2387–2396. [CrossRef]
43. Masango, P. Cleaner production of essential oils by steam distillation. *J. Clean. Prod.* **2005**, *13*, 833–839. [CrossRef]
44. Muchtaridi, M.; Diantini, A.; Subarnas, A. Analysis of Indonesian Spice Essential Oil Compounds That Inhibit Locomotor Activity in Mice. *Pharmaceuticals* **2011**, *4*, 590–602. [CrossRef]
45. Flodin, C.; Helidoniotis, F.; Whitfield, F.B. Seasonal variation in bromophenol content andbromoperoxidase activity in Ulva lactuca. *Phytochemistry* **1999**, *51*, 135–138. [CrossRef]
46. Jiao, G.; Yu, G.; Wang, W.; Zhao, X.; Zhang, J.; Ewart, S.H. Properties of polysaccharides in several seaweeds from Atlantic Canada and their potential anti-influenza viral activities. *J. Ocean Univ. China* **2012**, *11*, 205–212. [CrossRef]
47. Ferhat, M.A.; Meklati, B.Y.; Chemat, F. Comparison of different isolation methods of essential oil from Citrus fruits: Cold pressing, hydrodistillation and microwave 'dry' distillation. *Flavour Fragr. J.* **2007**, *22*, 494–504. [CrossRef]
48. Aladić, K.; Jokić, S.; Moslavac, T.; Tomas, S.; Vidović, S.; Vladić, J.; Šubarić, D. Cold pressing and supercritical CO_2 extraction of hemp (*Cannabis sativa*) seed oil. *Chem. Biochem. Eng. Q.* **2014**, *28*, 481–490. [CrossRef]
49. Ferhat, M.-A.; Boukhatem, M.N.; Hazzit, M.; Meklati, B.Y.; Chemat, F. Cold pressing, hydrodistillation and microwave dry distillation of citrus essential oil from Algeria: A comparative study. *Electron. J. Biol. S* **2016**, *1*, 30–41.
50. Lu-Martínez, A.A.; Báez-González, J.G.; Castillo-Hernández, S.; Amaya-Guerra, C.; Rodríguez-Rodríguez, J.; García-Márquez, E. Studied of *Prunus serotine* oil extracted by cold pressing and antioxidant effect of *P. longiflora* essential oil. *J. Food Sci. Technol.* **2021**, *58*, 1420–1429. [CrossRef]
51. Mehraban, M.S.A.; Shirzad, M.; Ahmadian-Attari, M.M.; Shakeri, R.; Kashani, L.M.T.; Tabarrai, M.; Shirbeigi, L. Effect of rose oil on gastroesophageal reflux disease in comparison with omeprazole: A double-blind controlled trial. *Complement. Ther. Clin. Pract.* **2021**, *43*, 101361. [CrossRef]
52. Avila, R.; Santos, S.; Araujo, D.; Vidal, V.; Macêdo, J. Semantic Links Using SKOS Predicates. *Procedia Comput. Sci.* **2017**, *112*, 467–473. [CrossRef]
53. Moradi, S.; Fazlali, A.; Hamedi, H. Microwave-Assisted Hydro-Distillation of Essential Oil from Rosemary: Comparison with Traditional Distillation. *Avicenna J. Med. Biotechnol.* **2018**, *10*, 22–28.
54. Gavahian, M.; Farahnaky, A. Ohmic-assisted hydrodistillation technology: A review. *Trends Food Sci. Technol.* **2018**, *72*, 153–161. [CrossRef]

55. Dawidowicz, A.L.; Olszowy, M. Does antioxidant properties of the main component of essential oil reflect its antioxidant properties? The comparison of antioxidant properties of essential oils and their main components. *Nat. Prod. Res.* **2014**, *28*, 1952–1963. [CrossRef]

56. Cox-Georgian, D.; Ramadoss, N.; Dona, C.; Basu, C. Therapeutic and medicinal uses of terpenes. In *Medicinal Plants 2019*; Springer: Cham, Switzerland, 2019; pp. 333–359.

57. Blank, P.N.; Shinsky, S.A.; Christianson, D.W. Structure of sesquisabinene synthase 1, a terpenoid cyclase that generates a strained [3.1. 0] bridged-bicyclic product. *ACS Chem. Biol.* **2019**, *14*, 1011–1019. [CrossRef]

58. Zielińska-Błajet, M.; Feder-Kubis, J. Monoterpenes and their derivatives—Recent development in biological and medical applications. *Int. J. Mol. Sci.* **2020**, *21*, 7078. [CrossRef]

59. Rostro-Alanis, M.d.J.; Báez-González, J.; Torres-Alvarez, C.; Parra-Saldívar, R.; Rodriguez-Rodriguez, J.; Castillo, S. Chemical composition and biological activities of oregano essential oil and its fractions obtained by vacuum distillation. *Molecules* **2019**, *24*, 1904. [CrossRef]

60. Saleh, A.M.; Al-Qudah, M.A.; Nasr, A.; Rizvi, S.A.; Borai, A.; Daghistani, M. Comprehensive analysis of the chemical composition and in vitro cytotoxic mechanisms of *Pallines spinosa* flower and leaf essential oils against breast cancer cells. *Cell. Physiol. Biochem.* **2017**, *42*, 2043–2065. [CrossRef]

61. Wu, M.; Li, T.; Chen, L.; Peng, S.; Liao, W.; Bai, R.; Zhao, X.; Yang, H.; Wu, C.; Zeng, H. Essential oils from Inula japonica and Angelicae dahuricae enhance sensitivity of MCF-7/ADR breast cancer cells to doxorubicin via multiple mechanisms. *J. Ethnopharmacol.* **2016**, *180*, 18–27. [CrossRef]

62. Matejić, J.; Šarac, Z.; Ranđelović, V. Pharmacological activity of sesquiterpene lactones. *Biotechnol. Biotechnol. Equip.* **2010**, *24*, 95–100. [CrossRef]

63. Yu, Y.P.; Landsittel, D.; Jing, L.; Nelson, J.; Ren, B.; Liu, L.; McDonald, C.; Thomas, R.; Dhir, R.; Finkelstein, S.; et al. Gene expression alterations in prostate cancer predicting tumor aggression and preceding development of malignancy. *J. Clin. Oncol.* **2004**, *22*, 2790–2799. [CrossRef]

64. Yousuf Dar, M.; Shah, W.A.; Mubashir, S.; Rather, M.A. Chromatographic analysis, anti-proliferative and radical scavenging activity of Pinus wallichina essential oil growing in high altitude areas of Kashmir, India. *Phytomedicine* **2012**, *19*, 1228–1233. [CrossRef] [PubMed]

65. Kim, H.S.; Lee, E.H.; Ko, S.R.; Choi, K.J.; Park, J.H.; Im, D.S. Effects of ginsenosides Rg3 and Rh2 on the proliferation of prostate cancer cells. *Arch. Pharm. Res.* **2004**, *27*, 429–435. [CrossRef] [PubMed]

66. Do N Fontes, J.E.; Ferraz, R.P.; Britto, A.C.; Carvalho, A.A.; Moraes, M.O.; Pessoa, C.; Costa, E.V.; Bezerra, D.P. Antitumor effect of the essential oil from leaves of *Guatteria pogonopus* (Annonaceae). *Chem. Biodivers.* **2013**, *10*, 722–729. [CrossRef] [PubMed]

67. Hussain, A.I.; Anwar, F.; Nigam, P.S.; Ashraf, M.; Gilani, A.H. Seasonal variation in content, chemical composition and antimicrobial and cytotoxic activities of essential oils from four Mentha species. *J. Sci. Food Agric.* **2010**, *90*, 1827–1836. [CrossRef] [PubMed]

68. Jena, L.; McErlean, E.; McCarthy, H. Delivery across the blood-brain barrier: Nanomedicine for glioblastoma multiforme. *Drug Deliv. Transl. Res.* **2020**, *10*, 304–318. [CrossRef]

69. Buckle, J. Use of aromatherapy as a complementary treatment for chronic pain. *Altern. Health Med.* **1999**, *5*, 42–51.

70. Quassinti, L.; Lupidi, G.; Maggi, F.; Sagratini, G.; Papa, F.; Vittori, S.; Bianco, A.; Bramucci, M. Antioxidant and antiproliferative activity of *Hypericum hircinum* L. subsp. majus (Aiton) N. Robson essential oil. *Nat. Prod. Res.* **2013**, *27*, 862–868. [CrossRef]

71. Bayala, B.; Bassole, I.H.; Gnoula, C.; Nebie, R.; Yonli, A.; Morel, L.; Figueredo, G.; Nikiema, J.B.; Lobaccaro, J.M.; Simpore, J. Chemical composition, antioxidant, anti-inflammatory and anti-proliferative activities of essential oils of plants from Burkina Faso. *PLoS ONE* **2014**, *9*, e92122. [CrossRef]

72. Tangrea, J.; Helzlsouer, K.; Pietinen, P.; Taylor, P.; Hollis, B.; Virtamo, J.; Albanes, D. Serum levels of vitamin D metabolites and the subsequent risk of colon and rectal cancer in Finnish men. *Cancer Causes Control* **1997**, *8*, 615–625. [CrossRef]

73. Carnesecchi, S.; Schneider, Y.; Ceraline, J.; Duranton, B.; Gosse, F.; Seiler, N.; Raul, F. Geraniol, a component of plant essential oils, inhibits growth and polyamine biosynthesis in human colon cancer cells. *J. Pharm. Exp.* **2001**, *298*, 197–200.

74. Rasoanaivo, P.; Fortuné Randriana, R.; Maggi, F.; Nicoletti, M.; Quassinti, L.; Bramucci, M.; Lupidi, G.; Petrelli, D.; Vitali, L.A.; Papa, F.; et al. Chemical composition and biological activities of the essential oil of *Athanasia brownii* Hochr. (Asteraceae) endemic to Madagascar. *Chem. Biodivers.* **2013**, *10*, 1876–1886. [CrossRef]

75. Murata, S.; Shiragami, R.; Kosugi, C.; Tezuka, T.; Yamazaki, M.; Hirano, A.; Yoshimura, Y.; Suzuki, M.; Shuto, K.; Ohkohchi, N.; et al. Antitumor effect of 1, 8-cineole against colon cancer. *Oncol. Rep.* **2013**, *30*, 2647–2652. [CrossRef]

76. Akrout, A.; Gonzalez, L.A.; El Jani, H.; Madrid, P.C. Antioxidant and antitumor activities of Artemisia campestris and Thymelaea hirsuta from southern Tunisia. *Food Chem. Toxicol.* **2011**, *49*, 342–347. [CrossRef]

77. El-Najjar, N.; Chatila, M.; Moukadem, H.; Vuorela, H.; Ocker, M.; Gandesiri, M.; Schneider-Stock, R.; Gali-Muhtasib, H. Reactive oxygen species mediate thymoquinone-induced apoptosis and activate ERK and JNK signaling. *Apoptosis* **2010**, *15*, 183–195. [CrossRef]

78. Zhao, M.; Bu, Y.; Feng, J.; Zhang, H.; Chen, Y.; Yang, G.; Liu, Z.; Yuan, H.; Yuan, Y.; Liu, L.; et al. SPIN1 triggers abnormal lipid metabolism and enhances tumor growth in liver cancer. *Cancer Lett.* **2020**, *470*, 54–63. [CrossRef]

79. Wu, S.; Wei, F.X.; Li, H.Z.; Liu, X.G.; Zhang, J.H.; Liu, J.X. Chemical composition of essential oil from *Thymus citriodorus* and its toxic effect on liver cancer cells. *Zhong Yao Cai* **2013**, *36*, 756–759.

80. Paik, S.Y.; Koh, K.H.; Beak, S.M.; Paek, S.H.; Kim, J.A. The essential oils from *Zanthoxylum schinifolium* pericarp induce apoptosis of HepG2 human hepatoma cells through increased production of reactive oxygen species. *Biol. Pharm. Bull.* **2005**, *28*, 802–807. [CrossRef]

81. Su, Y.C.; Hsu, K.P.; Wang, E.I.; Ho, C.L. Composition and in vitro anticancer activities of the leaf essential oil of *Neolitsea variabillima* from Taiwan. *Nat. Prod. Commun.* **2013**, *8*, 531–532. [CrossRef]

82. Lortet-Tieulent, J.; Ferlay, J.; Bray, F.; Jemal, A. International Patterns and Trends in Endometrial Cancer Incidence, 1978–2013. *J. Natl. Cancer Inst.* **2018**, *110*, 354–361. [CrossRef]

83. Bou, D.D.; Lago, J.H.; Figueiredo, C.R.; Matsuo, A.L.; Guadagnin, R.C.; Soares, M.G.; Sartorelli, P. Chemical composition and cytotoxicity evaluation of essential oil from leaves of Casearia sylvestris, its main compound α-zingiberene and derivatives. *Molecules* **2013**, *18*, 9477–9487. [CrossRef]

84. El-Readi, M.Z.; Eid, H.H.; Ashour, M.L.; Eid, S.Y.; Labib, R.M.; Sporer, F.; Wink, M. Variations of the chemical composition and bioactivity of essential oils from leaves and stems of *Liquidambar styraciflua* (Altingiaceae). *J. Pharm. Pharm.* **2013**, *65*, 1653–1663. [CrossRef] [PubMed]

85. Sun, X.Y.; Zheng, Y.P.; Lin, D.H.; Zhang, H.; Zhao, F.; Yuan, C.S. Potential anti-cancer activities of Furanodiene, a Sesquiterpene from *Curcuma wenyujin*. *Am. J. Chin. Med.* **2009**, *37*, 589–596. [CrossRef] [PubMed]

86. Ferraz, R.P.; Cardoso, G.M.; Da Silva, T.B.; Fontes, J.E.; Prata, A.P.; Carvalho, A.A.; Moraes, M.O.; Pessoa, C.; Costa, E.V.; Bezerra, D.P. Antitumour properties of the leaf essential oil of *Xylopia frutescens* Aubl. (Annonaceae). *Food Chem.* **2013**, *141*, 196–200. [CrossRef] [PubMed]

87. Manjamalai, A.; Kumar, M.J.; Grace, V.M. Essential oil of *Tridax procumbens* L. induces apoptosis and suppresses angiogenesis and lung metastasis of the B16F-10 cell line in C57BL/6 mice. *Asian Pac. J. Cancer Prev.* **2012**, *13*, 5887–5895. [CrossRef] [PubMed]

88. Seal, S.; Chatterjee, P.; Bhattacharya, S.; Pal, D.; Dasgupta, S.; Kundu, R.; Mukherjee, S.; Bhattacharya, S.; Bhuyan, M.; Bhattacharyya, P.R.; et al. Vapor of volatile oils from *Litsea cubeba* seed induces apoptosis and causes cell cycle arrest in lung cancer cells. *PLoS ONE* **2012**, *7*, e47014. [CrossRef]

89. Keawsa-ard, S.; Liawruangrath, B.; Liawruangrath, S.; Teerawutgulrag, A.; Pyne, S.G. Chemical constituents and antioxidant and biological activities of the essential oil from leaves of Solanum spirale. *Nat. Prod. Commun.* **2012**, *7*, 955–958. [CrossRef]

90. Jacob, E.A. Complete Blood Cell Count and Peripheral Blood Film, Its Significant in Laboratory Medicine: A Review Study. *Am. J. Lab. Med.* **2016**, *1*, 34–57.

91. Rashid, S.; Rather, M.A.; Shah, W.A.; Bhat, B.A. Chemical composition, antimicrobial, cytotoxic and antioxidant activities of the essential oil of *Artemisia indica* Willd. *Food Chem.* **2013**, *138*, 693–700. [CrossRef]

92. Saab, A.M.; Guerrini, A.; Sacchetti, G.; Maietti, S.; Zeino, M.; Arend, J.; Gambari, R.; Bernardi, F.; Efferth, T. Phytochemical analysis and cytotoxicity towards multidrug-resistant leukemia cells of essential oils derived from Lebanese medicinal plants. *Planta Med.* **2012**, *78*, 1927–1931. [CrossRef]

93. Salehi, F.; Behboudi, H.; Kavoosi, G.; Ardestani, S.K. Incorporation of Zataria multiflora essential oil into chitosan biopolymer nanoparticles: A nanoemulsion based delivery system to improve the in-vitro efficacy, stability and anticancer activity of ZEO against breast cancer cells. *Int. J. Biol. Macromol.* **2020**, *143*, 382–392. [CrossRef]

94. Azadi, M.; Jamali, T.; Kianmehr, Z.; Kavoosi, G.; Ardestani, S.K. In-vitro (2D and 3D cultures) and in-vivo cytotoxic properties of *Zataria multiflora* essential oil (ZEO) emulsion in breast and cervical cancer cells along with the investigation of immunomodulatory potential. *J. Ethnopharmacol.* **2020**, *257*, 112865. [CrossRef]

95. Rojas-Armas, J.P.; Arroyo-Acevedo, J.L.; Palomino-Pacheco, M.; Herrera-Calderón, O.; Ortiz-Sánchez, J.M.; Rojas-Armas, A.; Calva, J.; Castro-Luna, A.; Hilario-Vargas, J. The essential oil of *Cymbopogon citratus* stapt and carvacrol: An approach of the antitumor effect on 7, 12-dimethylbenz-[α]-anthracene (DMBA)-induced breast cancer in female rats. *Molecules* **2020**, *25*, 3284. [CrossRef]

96. Namshir, J.; Shatar, A.; Khandaa, O.; Tserennadmid, R.; Shiretorova, V.G.; Nguyen, M.C. Antimicrobial, antioxidant and cytotoxic activity on human breast cancer cells of essential oil from *Pinus sylvestris*. var mongolica needle. *Mong. J. Chem.* **2020**, *21*, 19–26. [CrossRef]

97. Xing, X.; Ma, J.-H.; Fu, Y.; Zhao, H.; Ye, X.-X.; Han, Z.; Jia, F.-J.; Li, X. Essential oil extracted from *Erythrina corallodendron* L. leaves inhibits the proliferation, migration, and invasion of breast cancer cells. *Medicine* **2019**, *98*, e17009. [CrossRef]

98. Zito, P.; Labbozzetta, M.; Notarbartolo, M.; Sajeva, M.; Poma, P. Essential oil of *Cyphostemma juttae* (Vitaceae): Chemical composition and antitumor mechanism in triple negative breast cancer cells. *PLoS ONE* **2019**, *14*, e0214594. [CrossRef]

99. El-Abid, H.; Amaral, C.; Cunha, S.C.; Augusto, T.V.; Fernandes, J.O.; Correia-da-Silva, G.; Teixeira, N.; Moumni, M. Chemical composition and anti-cancer properties of *Juniperus oxycedrus* L. essential oils on estrogen receptor-positive breast cancer cells. *J. Funct. Foods* **2019**, *59*, 261–271. [CrossRef]

100. Bagheri, S.M.; Asl, A.A.; Shams, A.; Mirghanizadeh-Bafghi, S.A.; Hafizibarjin, Z. Evaluation of Cytotoxicity Effects of Oleo-Gum-Resin and Its Essential Oil of Ferula assa-foetida and Ferulic Acid on 4T1 Breast Cancer Cells. *Indian J. Med. Paediatr. Oncol.* **2017**, *38*, 116–120.

101. Estanislao Gómez, C.; Aquino Carreño, A.; Pérez Ishiwara, D.; San Martín Martínez, E.; Morales López, J.; Pérez Hernández, N.; García, G. *Decatropis bicolor* (Zucc.) Radlk essential oil induces apoptosis of the MDA-MB-231 breast cancer cell line. *BMC Complement. Altern. Med.* **2016**, *16*, 266. [CrossRef]

102. Periasamy, V.S.; Athinarayanan, J.; Alshatwi, A.A. Anticancer activity of an ultrasonic nanoemulsion formulation of *Nigella sativa* L. essential oil on human breast cancer cells. *Ultrason. Sonochem.* **2016**, *31*, 449–455. [CrossRef]

103. Yeo, S.K.; Ali, A.Y.; Hayward, O.A.; Turnham, D.; Jackson, T.; Bowen, I.D.; Clarkson, R. β-Bisabolene, a sesquiterpene from the essential oil extract of opoponax (*Commiphora guidottii*), exhibits cytotoxicity in breast cancer cell lines. *Phytother. Res.* **2016**, *30*, 418–425. [CrossRef]

104. De Mel, Y.; Perera, S.; Ratnaweera, P.B.; Jayasinghe, C.D. Novel insights of toxicological evaluation of herbal medicine: Human based toxicological assays. *Asian J. Pharm. Pharmacol.* **2017**, *3*, 41–49.

105. Lee, J.-H.; Lee, K.; Lee, D.H.; Shin, S.Y.; Yong, Y.; Lee, Y.H. Anti-invasive effect of β-myrcene, a component of the essential oil from *Pinus koraiensis* cones, in metastatic MDA-MB-231 human breast cancer cells. *J. Korean Soc. Appl. Biol. Chem.* **2015**, *58*, 563–569. [CrossRef]

106. Wang, H.; Chen, X.; Li, T.; Xu, J.; Ma, Y. A myrsinol diterpene isolated from a traditional herbal medicine, LANGDU reverses multidrug resistance in breast cancer cells. *J. Ethnopharmacol.* **2016**, *194*, 1–5. [CrossRef]

107. Suhail, M.M.; Wu, W.; Cao, A.; Mondalek, F.G.; Fung, K.-M.; Shih, P.-T.; Fang, Y.-T.; Woolley, C.; Young, G.; Lin, H.-K. Boswellia sacra essential oil induces tumor cell-specific apoptosis and suppresses tumor aggressiveness in cultured human breast cancer cells. *BMC Complement. Altern. Med.* **2011**, *11*, 129. [CrossRef]

108. Kumar, P.S.; Febriyanti, R.M.; Sofyan, F.F.; Luftimas, D.E.; Abdulah, R. Anticancer potential of *Syzygium aromaticum* L. in MCF-7 human breast cancer cell lines. *Pharmacogn. Res.* **2014**, *6*, 350. [CrossRef] [PubMed]

109. Valizadeh, A.; Khaleghi, A.A.; Alipanah, H.; Zarenezhad, E.; Osanloo, M. Anticarcinogenic Effect of Chitosan Nanoparticles Containing *Syzygium aromaticum* Essential Oil or Eugenol toward Breast and Skin Cancer Cell Lines. *BioNanoScience* **2021**, *11*, 678–686. [CrossRef]

110. Abu-Dahab, R.; Kasabri, V.; Afifi, F.U. Evaluation of the Volatile Oil Composition and Antiproliferative Activity of *Laurus nobilis* L. (Lauraceae) on Breast Cancer Cell Line Models. *Rec. Nat. Prod.* **2014**, *8*, 136–147.

111. Yu, J.-Q.; Lei, J.-C.; Zhang, X.-Q.; Yu, H.-D.; Tian, D.-Z.; Liao, Z.-X.; Zou, G.-l. Anticancer, antioxidant and antimicrobial activities of the essential oil of *Lycopus lucidus* Turcz. var. hirtus Regel. *Food Chem.* **2011**, *126*, 1593–1598. [CrossRef] [PubMed]

112. Ashmawy, A.M.; Ayoub, I.M.; Eldahshan, O.A. Chemical composition, cytotoxicity and molecular profiling of *Cordia africana* Lam. on human breast cancer cell line. *Nat. Prod. Res.* **2021**, *35*, 4133–4138. [CrossRef]

113. Ashmawy, A.; Mostafa, N.; Eldahshan, O. GC/MS analysis and molecular profiling of lemon volatile oil against breast cancer. *J. Essent. Oil Bear. Plants* **2019**, *22*, 903–916. [CrossRef]

114. Nagappan, T.; Ramasamy, P.; Wahid, M.E.A.; Segaran, T.C.; Vairappan, C.S. Biological activity of carbazole alkaloids and essential oil of *Murraya koenigii* against antibiotic resistant microbes and cancer cell lines. *Molecules* **2011**, *16*, 9651–9664. [CrossRef]

115. Afoulous, S.; Ferhout, H.; Raoelison, E.G.; Valentin, A.; Moukarzel, B.; Couderc, F.; Bouajila, J. Chemical composition and anticancer, antiinflammatory, antioxidant and antimalarial activities of leaves essential oil of *Cedrelopsis grevei*. *Food Chem. Toxicol.* **2013**, *56*, 352–362. [CrossRef]

116. Hamzeloo-Moghadam, M.; Aghaei, M.; Fallahian, F.; Jafari, S.M.; Dolati, M.; Abdolmohammadi, M.H.; Hajiahmadi, S.; Esmaeili, S. Britannin, a sesquiterpene lactone, inhibits proliferation and induces apoptosis through the mitochondrial signaling pathway in human breast cancer cells. *Tumor Biol.* **2015**, *36*, 1191–1198. [CrossRef]

117. Karimian, H.; Fadaeinasab, M.; Moghadamtousi, S.Z.; Hajrezaei, M.; Zahedifard, M.; Razavi, M.; Safi, S.Z.; Mohan, S.; Khalifa, S.A.; El-Seedi, H.R. The chemopreventive effect of *Tanacetum polycephalum* against LA7-induced breast cancer in rats and the apoptotic effect of a cytotoxic sesquiterpene lactone in MCF7 cells: A bioassay-guided approach. *Cell. Physiol. Biochem.* **2015**, *36*, 988–1003. [CrossRef]

118. Fallahian, F.; Aghaei, M.; Abdolmohammadi, M.H.; Hamzeloo-Moghadam, M. Molecular mechanism of apoptosis induction by Gaillardin, a sesquiterpene lactone, in breast cancer cell lines. *Cell Biol. Toxicol.* **2015**, *31*, 295–305. [CrossRef]

119. Nakagawa-Goto, K.; Chen, J.-Y.; Cheng, Y.-T.; Lee, W.-L.; Takeya, M.; Saito, Y.; Lee, K.-H.; Shyur, L.-F. Novel sesquiterpene lactone analogues as potent anti-breast cancer agents. *Mol. Oncol.* **2016**, *10*, 921–937. [CrossRef]

120. Yang, B.; Zhao, Y.; Lou, C.; Zhao, H. Eupalinolide O, a novel sesquiterpene lactone from *Eupatorium lindleyanum* DC., induces cell cycle arrest and apoptosis in human MDA-MB-468 breast cancer cells. *Oncol. Rep.* **2016**, *36*, 2807–2813. [CrossRef]

121. Pragna Lakshmi, T.; Vajravijayan, S.; Moumita, M.; Sakthivel, N.; Gunasekaran, K.; Krishna, R. A novel guaiane sesquiterpene derivative, guai-2-en-10α-ol, from *Ulva fasciata* Delile inhibits EGFR/PI3K/Akt signaling and induces cytotoxicity in triple-negative breast cancer cells. *Mol. Cell. Biochem.* **2018**, *438*, 123–139. [CrossRef]

122. Salehi, F.; Jamali, T.; Kavoosi, G.; Ardestani, S.K.; Vahdati, S.N. Stabilization of Zataria essential oil with pectin-based nanoemulsion for enhanced cytotoxicity in monolayer and spheroid drug-resistant breast cancer cell cultures and deciphering its binding mode with gDNA. *Int. J. Biol. Macromol.* **2020**, *164*, 3645–3655. [CrossRef]

123. Abedinpour, N.; Ghanbariasad, A.; Taghinezhad, A.; Osanloo, M. Preparation of nanoemulsions of mentha piperita essential oil and investigation of their cytotoxic effect on human breast cancer lines. *BioNanoScience* **2021**, *11*, 428–436. [CrossRef]

124. El Gaafary, M.; Hafner, S.; Lang, S.J.; Jin, L.; Sabry, O.M.; Vogel, C.V.; Vanderwal, C.D.; Syrovets, T.; Simmet, T. A novel polyhalogenated monoterpene induces cell cycle arrest and apoptosis in breast cancer cells. *Mar. Drugs* **2019**, *17*, 437. [CrossRef]

125. Duran, G.G.; Duran, N.; Ay, E.; Kaya, D.A.; Kaya, M.G.A. Synergistic activities of the essential oils Hypericum perforatum with methotrexate on human breast cancer cell line MCF-7. In Proceedings of the International Conference on Advanced Materials and Systems (ICAMS), Qingdao, China, 26–27 March 2016; pp. 245–250.

126. Fu, J.; Yu, J.; Chen, J.; Xu, H.; Luo, Y.; Lu, H. In vitro inhibitory properties of sesquiterpenes from *Chloranthus serratus* on cell motility via down-regulation of LIMK1 activation in human breast cancer. *Phytomedicine* **2018**, *49*, 23–31. [CrossRef]
127. Aghaei, M.; Ghanadian, M.; Sajjadi, S.E.; Saghafian, R. Pimpinelol, a novel atypical sesquiterpene lactone from Pimpinella haussknechtii fruits with evaluation of endoplasmic reticulum stress in breast cancer cells. *Fitoterapia* **2018**, *129*, 198–202. [CrossRef]
128. Liu, Y.R.; Cai, Q.Y.; Gao, Y.G.; Luan, X.; Guan, Y.Y.; Lu, Q.; Sun, P.; Zhao, M.; Fang, C. Alantolactone, a sesquiterpene lactone, inhibits breast cancer growth by antiangiogenic activity via blocking VEGFR2 signaling. *Phytother. Res.* **2018**, *32*, 643–650. [CrossRef]
129. Chudzik, M.; Korzonek-Szlacheta, I.; Król, W. Triterpenes as potentially cytotoxic compounds. *Molecules* **2015**, *20*, 1610–1625. [CrossRef]
130. Bishayee, A.; Ahmed, S.; Brankov, N.; Perloff, M. Triterpenoids as potential agents for the chemoprevention and therapy of breast cancer. *Front. Biosci.* **2011**, *16*, 980–996. [CrossRef]
131. Xue, Z.; Li, J.; Cheng, A.; Yu, W.; Zhang, Z.; Kou, X.; Zhou, F. Structure identification of triterpene from the mushroom Pleurotus eryngii with inhibitory effects against breast cancer. *Plant Foods Hum. Nutr.* **2015**, *70*, 291–296. [CrossRef]
132. Subash-Babu, P.; Alshammari, G.M.; Ignacimuthu, S.; Alshatwi, A.A. Epoxy clerodane diterpene inhibits MCF-7 human breast cancer cell growth by regulating the expression of the functional apoptotic genes Cdkn2A, Rb1, mdm2 and p53. *Biomed. Pharmacother.* **2017**, *87*, 388–396. [CrossRef]
133. Shadi, S.; Saeidi, H.; Ghanadian, M.; Rahimnejad, M.R.; Aghaei, M.; Ayatollahi, S.M.; Iqbal Choudhary, M. New macrocyclic diterpenes from *Euphorbia connata* Boiss. with cytotoxic activities on human breast cancer cell lines. *Nat. Prod. Res.* **2015**, *29*, 607–614. [CrossRef] [PubMed]
134. Emami, S.A.; Asili, J.; HosseinNia, S.; Yazdian-Robati, R.; Sahranavard, M.; Tayarani-Najaran, Z. Growth inhibition and apoptosis induction of essential oils and extracts of *Nepeta cataria* L. on human prostatic and breast cancer cell lines. *Asian Pac. J. Cancer Prev.* **2016**, *17*, 125–130. [CrossRef] [PubMed]
135. Mouchira, A.C. Chemical composition and anticancer activity of *Achillea fragrantissima* (Forssk.) Sch. Bip.(Asteraceae) essential oil from Egypt. *J. Pharmacogn. Phytother.* **2017**, *9*, 1–5. [CrossRef]
136. Eghbali, M.; Varaei, S.; Hosseini, M.; Yekaninejad, S.; Shahi, F. The Effect of Aromatherapy with Peppermint Essential Oil on Nausea and Vomiting in the Acute Phase of Chemotherapy in Patients with Breast Cancer. *J. Babol Univ. Med. Sci.* **2018**, *20*, 66–71. [CrossRef]
137. Lua, P.L.; Salihah, N.; Mazlan, N. Effects of inhaled ginger aromatherapy on chemotherapy-induced nausea and vomiting and health-related quality of life in women with breast cancer. *Complement. Med.* **2015**, *23*, 396–404. [CrossRef]
138. Singh, S.; Chaurasia, P.K.; Bharati, S.L. A mini-review on the safety profile of essential oils. *MOJ Biol. Med.* **2022**, *7*, 33–36.
139. Castilhos, R.V.; Grützmacher, A.D.; Coats, J.R. Acute Toxicity and Sublethal Effects of Terpenoids and Essential Oils on the Predator *Chrysoperla externa* (Neuroptera: Chrysopidae). *Neotrop. Entomol.* **2018**, *47*, 311–317. [CrossRef]
140. Alipanah, H.; Farjam, M.; Zarenezhad, E.; Roozitalab, G.; Osanloo, M. Chitosan nanoparticles containing limonene and limonene-rich essential oils: Potential phytotherapy agents for the treatment of melanoma and breast cancers. *BMC Complement. Med.* **2021**, *21*, 186. [CrossRef]
141. Lis-Balchin, M. *Aromatherapy Science: A Guide for Healthcare Professionals*; Pharmaceutical Press: London, UK, 2005; Volume 1.
142. Tisserand, R.; Young, R. 3—Toxicity. In *Essential Oil Safety*, 2nd ed.; Tisserand, R., Young, R., Eds.; Churchill Livingstone: St. Louis, MI, USA, 2014; pp. 23–38.
143. Ferreira, P.; Cardoso, T.; Ferreira, F.; Fernandes-Ferreira, M.; Piper, P.; Sousa, M.J. *Mentha piperita* essential oil induces apoptosis in yeast associated with both cytosolic and mitochondrial ROS-mediated damage. *FEMS Yeast Res.* **2014**, *14*, 1006–1014. [CrossRef]
144. Jo, J.-R.; Park, J.S.; Park, Y.-K.; Chae, Y.Z.; Lee, G.-H.; Park, G.-Y.; Jang, B.-C. Pinus densiflora leaf essential oil induces apoptosis via ROS generation and activation of caspases in YD-8 human oral cancer cells. *Int. J. Oncol.* **2012**, *40*, 1238–1245. [CrossRef]

MDPI

St. Alban-Anlage 66

4052 Basel

Switzerland

Tel. +41 61 683 77 34

Fax +41 61 302 89 18

www.mdpi.com

Applied Sciences Editorial Office

E-mail: applsci@mdpi.com

www.mdpi.com/journal/applsci